AAPS Introductions in the Pharmaceutical Sciences

Volume 7

Series Editor

Claudio Salomon, National University of Rosario, Rosario, Argentina

The *AAPS Introductions in the Pharmaceutical Sciences* book series is designed to support pharmaceutical scientists at the point of knowledge transition. Springer and the American Association of Pharmaceutical Scientists (AAPS) have partnered again to produce a second series that juxtaposes the *AAPS Advances in the Pharmaceutical Sciences* series. Whether shifting between positions, business models, research project objectives, or at a crossroad in professional development, scientists need to retool to meet the needs of the new scientific challenges ahead of them. These educational pivot points require the learner to develop new vocabulary in order to effectively communicate across disciplines, appreciate historical evolution within the knowledge area with the aim of appreciating the current limitations and potential for growth, learn new skills and evaluation metrics so that project planning and subsequent evolution are evidence-based, as well as to simply "dust the rust off" content learned in previous educational or employment settings, or utilized during former scientific explorations. The *Introductions* book series will meet these needs and serve as a quick and easy-to-digest resource for contemporary science.

Avi Domb • Boaz Mizrahi • Shady Farah
Editors

Biomaterials
and Biopolymers

Editors
Avi Domb
School of Pharmacy-Faculty of Medicine
The Hebrew University of Jerusalem
and Centre for Cannabis Research
and the Institute of Drug Research
The Alex Grass Centre for Drug
Design and Synthesis
Jerusalem, Israel

Boaz Mizrahi
Laboratory for Bio-materials
Faculty of Biotechnology and Food
Engineering
Technion-Israel Institute of Technology
Haifa, Israel

Shady Farah
The Laboratory for Advanced
Functional/Medicinal
Polymers & Smart Drug Delivery
Technologies, The Wolfson Faculty of
Chemical Engineering
Technion-Israel Institute of Technology
Haifa, Israel

ISSN 2522-834X ISSN 2522-8358 (electronic)
AAPS Introductions in the Pharmaceutical Sciences
ISBN 978-3-031-36137-1 ISBN 978-3-031-36135-7 (eBook)
https://doi.org/10.1007/978-3-031-36135-7

© American Association of Pharmaceutical Scientists 2023
Jointly published with American Association of Pharmaceutical Scientists
This work is subject to copyright. All rights are reserved by the Publisher, whether the whole or part of the material is concerned, specifically the rights of translation, reprinting, reuse of illustrations, recitation, broadcasting, reproduction on microfilms or in any other physical way, and transmission or information storage and retrieval, electronic adaptation, computer software, or by similar or dissimilar methodology now known or hereafter developed.
The use of general descriptive names, registered names, trademarks, service marks, etc. in this publication does not imply, even in the absence of a specific statement, that such names are exempt from the relevant protective laws and regulations and therefore free for general use.
The publishers, the authors, and the editors are safe to assume that the advice and information in this book are believed to be true and accurate at the date of publication. Neither the publishers nor the authors or the editors give a warranty, express or implied, with respect to the material contained herein or for any errors or omissions that may have been made. The publishers remain neutral with regard to jurisdictional claims in published maps and institutional affiliations.

Editorial Contact: Charlotte Nunes

This Springer imprint is published by the registered company Springer Nature Switzerland AG
The registered company address is: Gewerbestrasse 11, 6330 Cham, Switzerland

Paper in this product is recyclable.

Preface

Biomaterials and biopolymers have revolutionized modern medicine and biology, already helping hundreds of millions of patients worldwide on an annual basis. They can be natural, semi-synthetic, or fully synthetic entities to be used in medical applications either to enhance, support, or replace damaged tissue, or serve as a biological function allowing patients to regain mobility and improve their quality of life. Biomaterial unique properties make them suitable for a range of medical devices, including implants, drug delivery systems, prosthetics, wound dressings, and tissue engineering. A key feature is biocompatibility, thus allowing them to interact with living tissue, to imply a function without causing adverse reactions. Several recent studies indicate that for successful modern biomaterials engineering, several aspects should be taken into account including: synthesis and properties, biocompatibility and host response, the manufacturing process, and a profound understanding of the medical and biotechnological applications specifically focusing on the target site and the physiological microenvironment. Worldwide ongoing research efforts are focused on developing superior biomaterials and biopolymers with improved biocompatibility, degradation rates, drug loading, and mechanical strength, leading to the creation of innovative treatments and devices in medicine, thus boosting both novel treatments and technologies, as well as medical implants developments that can improve patient outcomes and quality of life. With ongoing research efforts and development, the potential for biomaterials and biopolymers to transform healthcare is vast, offering hope for a "healthier" future. In this book, leading biomaterials and biopolymers formulations reported to exhibit unique features and bio properties are covered.

This book intends to serve as textbook on biomaterials and biopolymers for faculty and students, and it thus contains a broad introduction and basic terms, followed by major developments over the years with some emphasis on recent developments and future prospects. As a check on your understanding, each chapter ends with 10 multiple-choice quiz questions. These questions are organized so that you can work them as you proceed through the text. We suggest answering the question yourself, and only then to check the answers.

This volume contains 12 chapters covering the dominant biomaterials with biomedical applications to address unmet medical needs and challenging diseases. The first section provides an overview of biomaterials, classes, uses, and basic properties. The second section is dedicated to the concept of drug delivery systems along with the role of biomaterials in different drug delivery applications. The third section focuses on biomaterial and implant applications, including applications in cosmetic surgeries, medical devices, and equipment. This chapter also introduces the concept of living biomaterials, and polymers that can be covalently bound to a cargo as either bioactive or inert carriers. The last part discusses the biocompatibility of biomaterials and a detailed discussion on biopolymer sterilization methods for both natural and synthetic biodegradable biopolymers.

Chapter "Introduction to Biomaterials" provides an overview of biomaterials and highlights the significance of biomaterials in modern medicine and biology. The chapter also describes the classes of biomaterials and, for each class, the unique properties that make them attractive for various applications in different fields, including medical implants and devices, pharmaceutics, tissue engineering, food packaging, cosmetics, and environmental applications.

Chapter "Biodegradable Polymers" discusses biodegradable polymers, their synthesis, biodegradability, and biocompatibility, along with their advantages and disadvantages for various biomedical applications, including drug delivery and tissue engineering. The chapter focuses on degradable biopolymers that can provide a safe and effective way of preparing devices/implantable materials for various biomedical applications.

Chapter "Natural and Semi-natural Polymers" focuses on natural and semi-natural polymers isolated from living organisms such as plants, animals, and microorganisms, and it presents the most relevant biopolymers used in biomedicine, their classifications, and some of their applications with emphasis on polysaccharides and proteins.

Chapter "Fundamentals and Biomedical Applications of Smart Hydrogels" provides an overview of hydrogels as three-dimensional elastic networks containing a large amount of water formed from crosslinked hydrophilic polymer chains, which possess tunable tissue-like physicochemical properties. This chapter thoroughly introduces the definition, classification, formation, properties, typical representative, and biomedical applications of natural and synthetic hydrogels. Then, it focuses on several typical hydrogels, such as self-healing hydrogels, injectable hydrogels, and stimuli-responsive hydrogels, including shape memory hydrogels and hydrogel actuators and their major attractive applications in biomedical fields such as contact lenses, hygiene products, drug delivery, tissue engineering, and wound healing.

Chapter "Engineering Biomaterials for Nucleic Acid-Based Therapies" focuses on biomaterials for the delivery of nucleic acids. Engineering biomaterials for nucleic acid-based therapies involve the design and optimization of materials and formulations that can protect nucleic acids from degradation, deliver them to desired target cells, facilitate cellular uptake, and promote endosomal escape. This chapter is focused on the engineering of such materials from a chemical, formulation, and manufacturing perspective.

Chapter "Mechanics of Biomaterials for Regenerative Medicine" defines the diverse types of biomaterials and describes their mechanical characteristics. Conventional methods for measurement of the mechanical properties of biomaterials and the mechanical behavior of tissues and biomaterials for regenerative medicine are described, as well as functional biomechanical tests for different applications. The mechanical properties of biomaterials play a critical role in designing and developing medical products and selecting suitable materials for various applications. These properties are discussed.

Chapter "Biomaterials for Controlled Drug Delivery Applications" overviews the concept of drug delivery systems along with the role of biomaterials in different drug delivery applications. Initially, the chapter starts by introducing the fundamentals of drug delivery systems, including the classification of drugs-based biopharmaceutics, why there is a need for controlled drug delivery, different routes of drug administration, the pharmacokinetics of drug delivery systems, and the different release kinetics of drugs. These discussions provide a brief understanding of a particular type of drug and disease model, and what type of biomaterials should be designed. In the second part of the chapter, we focus on the design considerations for controlled drug delivery systems, the role of biomaterials for controlled drug delivery applications, and different biomaterials for drug delivery applications.

Chapter "Biomaterials Application: Implants" provides an overview of biomaterial and implant applications in cosmetic surgeries, medical devices, and equipment. Structural components derived from biomaterials can successfully mimic the function of tissue and integrate with a biological system. These topics are discussed. This chapter focuses on use in skeletal, skin, cardiac, neuronal, and ocular implants, and it discusses the recent progress in biomaterials and implants that can truly mimic the function of specific tissues to improve healing outcomes.

Chapter "Living Biomaterials" introduces the concept of living biomaterials as a combination of live organisms such as bacteria or cells with traditional biomaterials, explores various systems, and focuses on the interactions and complexity of these multi-component systems. The integration of functional microorganisms into polymeric matrices imposes stringent requirements on material composition and engineering that are extensively discussed. Then the chapter presents some applications of living biomaterials with an emphasis on the medical field and presents two case studies that describe in detail specific systems in an attempt to illustrate this growing field.

Chapter "Therapeutic Polymer Conjugates and Their Characterization" presents polymers that can be covalently bound to cargo as either bioactive or inert carriers of such therapeutics to impart protective characteristics, such as biodistribution, first-pass clearance reduction, specific targeting, and immune evasion. The different conjugation chemistry strategies, how they should be considered for targeted applications, and how one might reconcile them with different desired downstream characterization as well as in vivo uses are also discussed. In addition, selectively cleavable linkers that may be utilized to engineer cargo release at specific locations under unique physiological conditions, such as low pH, oxygen, or biomolecule-laden environments, are presented.

Chapter "Biocompatibility of Polymers" discusses the biocompatibility of biomaterials and how it remains a great challenge for manufacturers during their development. This chapter outlines the material-host interactions and describes the factors that need to be considered when evaluating biocompatibility of a material. Various methods for biocompatibility assessments are also discussed in detail.

In Chapter "Sterilization Techniques of Biomaterials (Implants and Medical Devices)", a detailed discussion on biopolymer sterilization methods for both natural and synthetic biodegradable biopolymers is presented. The most commonly used sterilization methods that have been applied on biopolymers, including dry-heat sterilization, steam-autoclaving, irradiation (gamma, ultraviolet, X-rays and electron beam), chemical treatment (ethylene oxide), gas plasma, and supercritical fluid sterilization, are comprehensively reviewed. The sterilization techniques with their advantages and disadvantages are discussed with examples.

Jerusalem, Israel	Avi Domb
Haifa, Israel	Boaz Mizrahi
Haifa, Israel	Shady Farah

Contents

Introduction to Biomaterials 1
Pulikanti Guruprasad Reddy, Ravi Saklani, Manas Kumar Mandal,
and Abraham J. Domb

Biodegradable Polymers 33
Mudigunda V. Sushma, Aditya kadam, Dhiraj Kumar, and Isha Mutreja

Natural and Semi-natural Polymers 55
Katia P. Seremeta and Alejandro Sosnik

Fundamentals and Biomedical Applications of Smart Hydrogels 71
Qi Wu, Eid Nassar-Marjiya, Mofeed Elias, and Shady Farah

Engineering Biomaterials for Nucleic Acid-Based Therapies 95
Parveen Kumar, Umberto Capasso Palmiero, and Piotr S. Kowalski

Mechanics of Biomaterials for Regenerative Medicine 119
Yevgeniy Kreinin, Iris Bonshtein, and Netanel Korin

Biomaterials for Controlled Drug Delivery Applications 135
Krishanu Ghosal, Merna Shaheen-Mualim, Edwar Odeh,
Nagham Moallem Safuri, and Shady Farah

Biomaterials Application: Implants 159
Aditya Ruikar, Chase Bonin, Gauri S. Kumbar,
Yeshavanth Kumar Banasavadi-Siddegowda, and Sangamesh G. Kumbar

Living Biomaterials 183
Caroline Hali, Adi Gross, and Boaz Mizrahi

Therapeutic Polymer Conjugates and Their Characterization 197
Victor M. Quiroz, Joshua Devier, and Joshua C. Doloff

Biocompatibility of Polymers................................... 235
Ruba Ibrahim, Abraham Nyska, and Yuval Ramot

**Sterilization Techniques of Biomaterials
(Implants and Medical Devices)** 255
Chau Chun Beh

Index.. 271

About the Editors

Avi Domb earned B.Sc. degrees in Chemistry (Bar Ilan, 1978), Pharmacy (Hebrew University, 1984) and Law (Hebrew University 2007). He earned Diplomas in Polymer and Textile Chemistry (Hebrew University, 1980) and Executive Business Management (Hebrew University, 1997) and a Ph.D. in Organic Chemistry (Hebrew University, 1985). He had a Postdoctoral Fellowship at Syntex Research (1984–1985), and continued at the Department of Surgery, The Children's Hospital, Harvard Medical School (1985–1986) and the Department of Chemical Engineering, MIT (1986–1987).

In 1987 he joined the Biological Institute in Nes-Ziona, Israel, and in 1988 he served as head of Drug Delivery Laboratories at Nova Pharmaceutical Corporation, Baltimore, USA. Since 1991, he has been a faculty member at the School of Pharmacy, Faculty of Medicine, The Hebrew University. Between 2012 and 2017, he took leave to head the Division of Identification and Forensic Sciences (DIFS), Israel police, Rank: Brigadier General. During 2014–2016, he was President of the Azrieli College of Engineering Jerusalem, and from 2018, he has been the head of the School of Pharmacy, Faculty of Medicine, The Hebrew University. In 2021 he joined the Ministry of Science and Technology as Chief Scientist.

Boaz Mizrahi earned a B.Pharm. (Pharmacy) degree from the Hebrew University (1998) and a Diploma in Pharmacy from Hadassah Medical Hospital. He then earned a Ph.D. in Medicinal Chemistry (Hebrew University, 2008). Prof. Mizrahi had a Postdoctoral Fellowship at the Department of Chemical Engineering/Koch Institute at MIT (2012) and at Children's Hospital, Harvard Medical School (2012). He was appointed as professor at the Faculty of Biotechnology and Food Engineering of the Technion, Israel Institute of Technology in October 2013. Today Prof. Mizrahi is the head of the Interdisciplinary Program of Biotechnology of the Technion. Prof. Mizrahi combines his greatest passions – medicine and chemistry – to promote the development of innovative biotechnological and engineered solutions to everyday challenges in medicine.

Shady Farah earned his B.Sc. in Medicinal Chemistry from the School of Pharmacy, Faculty of Medicine, The Hebrew University of Jerusalem (HUJI), Israel (2008), and then completed his M.Sc. (2009, direct track to Ph.D.) and Ph.D. in Medicinal Chemistry at HUJI, Faculty of Medicine, Israel (2015). Prof. Farah was a postdoctoral fellow at Chemical Engineering Department and Koch Institute at MIT and BCH/Harvard Medical School in the group of Prof. Daniel Anderson/Prof. Robert Langer (2014–2019). In 2019, he joined the Technion, where he is heading The Laboratory for Advanced Functional/Medicinal Polymers and Smart Drug Delivery Technologies at the Wolfson Faculty of Chemical Engineering – Technion (www.TheFarahLab.com). His main research interests are in the fields of medicinal polymers, biomaterials, implants and drug delivery: advanced functional and biodegradable polymers, smart materials, drugs crystallizations, controlled and localized drug delivery for chronic diseases such as diabetes and cancer, cells encapsulation and living therapeutics and 3D printing of implants. Prof. Farah won several prestigious awards and recognitions, among them: Named in the 100-list of Global MIT Technology Review's 35 Innovators Under 35 competition (2021). Neubauer Asst. Prof. Award (2019), MAOF Fellowship for Outstanding Young Researchers (2019), and in 2023, Prof. Farah was selected to the Global Young Academy.

Introduction to Biomaterials

Pulikanti Guruprasad Reddy, Ravi Saklani, Manas Kumar Mandal, and Abraham J. Domb

Abstract This book chapter highlights the significance of biomaterials in modern medicine and biology. Biomaterials can be used to replace or repair damaged or missing body parts, allowing patients to regain mobility and improve their quality of life. The unique properties of biomaterials make them suitable for a range of medical devices, including implants, prosthetics, cardiovascular devices, drug delivery systems, wound dressings, and tissue engineering. Biocompatibility is a key characteristic of biomaterials, allowing them to interact with living tissue without causing adverse reactions. Mechanical properties and durability, which cause biomaterials to degrade in the body over time, are other crucial characteristics. Biomaterials are classified into natural and synthetic biomaterials. Each class has its unique advantages, making them suitable for various applications in the medical field. For instance, metals and ceramics are commonly used as cardiovascular, orthopaedic, dental implants and vascular stents due to their strength and biocompatibility, while polymers are used in wound dressings, drug delivery systems, and tissue engineering due to their versatility and customization. The unique properties of natural biomaterials make them attractive for various applications in different fields. Some of these applications include medical implants and devices, pharmaceutics, tissue engineering, food packaging, cosmetics, and environmental applications. Ongoing research efforts are focused on developing biomaterials with improved biocompatibility, mechanical strength, and degradation rates, leading to the creation of innovative treatments and devices in medicine. Overall, biomaterials have the potential to revolutionize healthcare by enabling the development of novel treatments and devices that can improve patient outcomes and quality of life. With ongoing research and development, the potential for biomaterials to transform healthcare is vast, offering hope for the future.

P. G. Reddy · R. Saklani · M. K. Mandal · A. J. Domb (✉)
School of Pharmacy-Faculty of Medicine, The Hebrew University of Jerusalem, and Centre for Cannabis Research and the Institute of Drug Research,
The Alex Grass Centre for Drug Design and Synthesis, Jerusalem, Israel
e-mail: avid@ekmd.huji.ac.il

Keywords Natural & synthetic biomaterials · Biomaterials History · Biocompatibility · Medical implants · Tissue engineering & regeneration · Drug delivery · Diagnosis · Wound healing

1 Introduction

Materials are used to engineer different things all around us. However, engineering is present not only around us, but also within us. The materials used for that purpose are called biomaterials, a special category of materials used for engineering in/on our bodies. The human tissues, organs, or body systems occasionally fail to perform their normal functions, and these disorders are sometimes treated with the medications, i.e., drugs. All disorders, however, cannot be treated with drugs. They require one-of-a-kind biomaterials. In its broadest definition, the term "biomaterial"can refer to all the substances other than food and drugs that are utilized for biomedical purposes, i.e., to treat or diagnose illnesses or to repair, enhance, or replace tissues, organs, or physiological functions. This is from a bandage to an artificial pacemaker or a dental implant. All are biomaterials that have been engineered using a distinct set of materials designed to work well with a human body (Fig. 1).

Humans have been using biomaterials since prehistoric times, and now progress in biomedical research has been steadily accelerating to meet ever-increasing needs in healthcare and medicine practices. The introduction of medical devices made

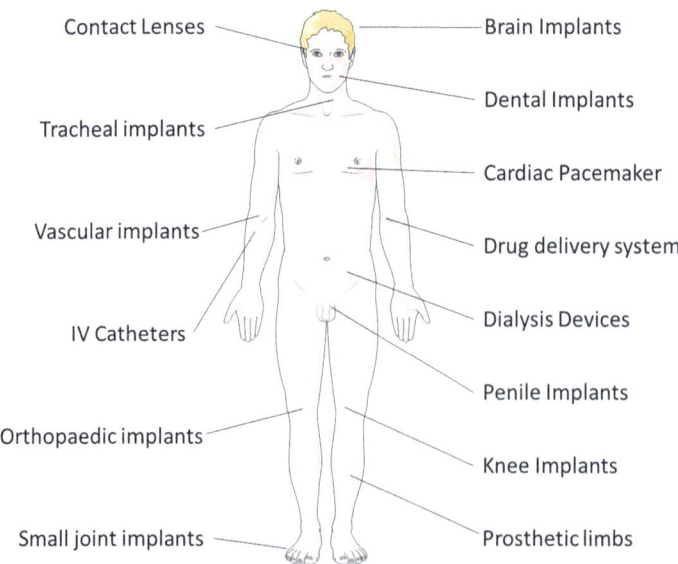

Fig. 1 Examples of various biomaterial systems used in humans

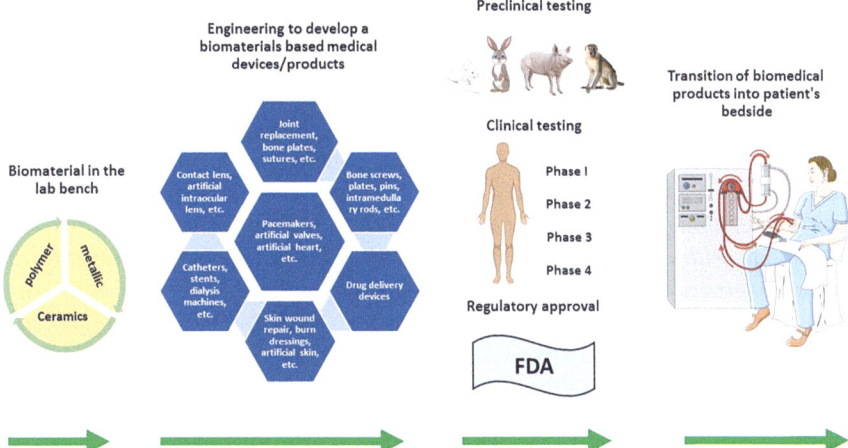

Fig. 2 Schematic illustrations of steps involved in the translation of a biomaterial into a clinical device

from biomaterials improved the quality of human life; as a result, millions of lives have been saved. Advancements in biomaterials research have solved many medical problems in both therapeutics and diagnostics. Therefore, translation of biomaterials into medical devices is a clinically important aspect clearly dependent on several factors such as biomaterials engineering, clinical realities, in vivo testing on humans and animals, industry involvement in developing suitable biomaterial devices and commercialization, etc. Figure 2 depicts a schematic representation of the path from biomedical advances to clinical use. Thousands of biomedical medical devices and diagnostic products have been used to aid in restoration to normal body functions of human tissues or organs after their deterioration. Currently, over 6000 different types of biomedical devices are listed in the Medical Device Product Classification Database, which is governed by the Food and Drug Administration's Centre for Medical Devices and Radiological Health [1].

2 Definition of Biomaterials

Biomaterial research is a highly dynamic and ever-changing discipline, with ever-changing definitions of biomaterial. Biomaterials are seen differently at different times due to changes in their application, and numerous attempts have been made to describe them [2, 3].

An agreement was established among a group of biomaterial scientists at a European Society for Biomaterials meeting in 1987, and a contested definition of biomaterials was derived: "***A non-viable material utilized in a medical device, designed to interact with biological systems***". However, the definition evolved and the reference to non-viability was later deleted.

In 1999 David F. William defined biomaterials as *"a substance intended to interface with biological systems to assess, treat, augment, or replace any tissue, organ, or function of the body"* in his Dictionary of Biomaterials.

Taking into account the dynamic developments in the area of biomaterial research and biomaterials definition, William discussed these changes in his recent leading opinion article, "On the nature of biomaterials", and he redefined the definition of biomaterials as: *"A biomaterial is a substance that has been engineered to take a form which, alone or as part of a complex system, is used to direct, by control of interactions with components of living systems, the course of any therapeutic or diagnostic procedure, in human or veterinary medicine"*.

Biomaterials are defined by the American National Institute of Health as *"any substance or combination of substances, other than drugs, synthetic or natural in origin, that can be used for any period of time, that augments or replaces partially or completely any tissue, organ, or function of the body, in order to maintain or improve the quality of life of the individual"*.

Considering all the definitions in a broad sense, *a biomaterial can be defined as a material engineered to acquire a form that can affect the course of any therapeutic or diagnostic procedure through interactions with the biological systems*. These are all rather narrow definitions, but they will all be elaborated on throughout this book.

3 History of Biomaterials

Biomaterials are being used by humans either deliberately or unknowingly since prehistoric times. The biomedical implants and prostheses were found in human skeletons and skulls during the excavation of sites attributed to different civilizations of antiquity – Egyptian, Roman, Greek, and Etruscan. A spear point was discovered inserted into the hip of a Kennewick man's approximately 9000-year-old remains. Even though the spear tip may have been an accidental implant bearing very little relation to the modern biomaterials, it is one of the earliest examples of an exogenous material that has been well tolerated and lodged in the human body. The use of skin grafting utilizing autogenous forehead skin for the restoration of the nose and skin from the cheek for the treatment of damaged earlobes is documented in one of the first surgical manuals from 600 BC. According to ancient Mayan documents from 600 BC people then used seashells as dental implants. It is also widely documented that the Middle Ages Europeans used catgut for suturing. The Ancient Egyptians used linen thread for wound treatment. It is noteworthy that humans have been studying biomaterials for a long time and have had great success despite not having the necessary knowledge of material science, sterilization, or biocompatibility.

Materials of natural origin and other metals were employed in Greece and Rome from the seventh century BC to the fourth century AD for the treatment of wounds and other health issues. During the sixteenth century in Europe, dental repairs were made using metals like gold and silver, and bone fractures were repaired by using

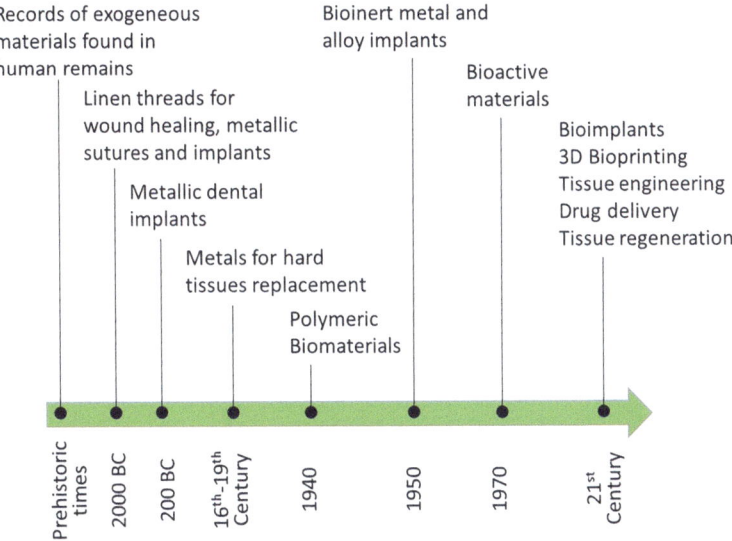

Fig. 3 History of biomaterials evolution from prehistoric times to twenty-first century

the iron threads. The industrial revolution in the nineteenth century saw the development of X-rays and anaesthesia, and surgeries were performed under sterile conditions. The use of metal in internal body repairs was also initiated. After World War II, science and technology combined to develop prostheses, medical devices, or implants, such as heart valves, bone plates, hip joints, and cardiac pacemakers, to repair or replace damaged body parts or tissues. Following that, the progress on biomaterials was nucleated from several scientific communities, and scientific journals were established to collect the progress on biomaterials. Because of the numerous complications caused by implantable biomaterials, scientists coined the term "biocompatibility" for the first time in 1968. Ideally, the developed biomaterials should be biocompatible, meaning they should not be toxic or harmful to the biological system. Currently, biomaterials are used extensively in the medical field for drug delivery, tissue engineering, regenerative medicine, 3D bio printing of the organs, and many more applications [4]. A pictorial summary of the history of biomaterials from prehistoric times to twenty-first century is given in Fig. 3.

4 Biomaterial Characteristics

As stated above, there is no agreement on the definition of biomaterials, and it is constantly evolving. As a result, it is critical to understand the characteristics of biomaterials to aid in determining which materials are biomaterials and which are not. However, the requirements of biomaterials vary and depend on the type of

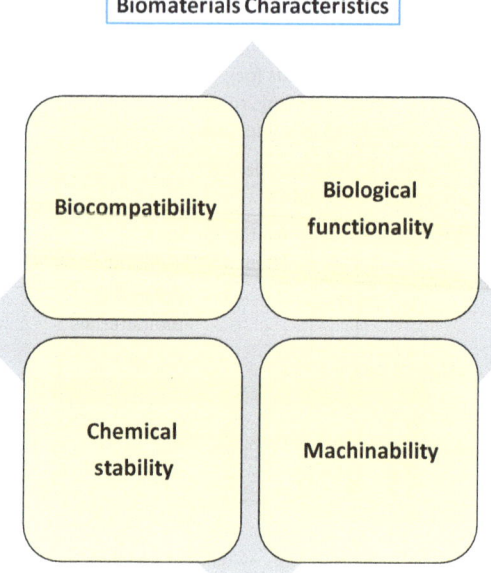

Fig. 4 Characteristics of biomaterials

biomedical application. The objective of a biomaterial is to treat, improve, or replace a body tissue or function, thus all the biomaterials must have the following four characteristics (Fig. 4).

4.1 Biocompatibility

Biomaterials that interact directly with the human body require specific design expertise. Biocompatibility is not only a vital characteristic of biomaterials, but it is also what designates something as a biomaterial. Technically, any substance can be used as a biomaterial, but to succeed the material must be biocompatible with which the biological system it intends to interact.

Biocompatibility assessment is intended to confirm the ability of a material to prevent negative responses and correctly carry out the specific biomedical task. However, assessment of biocompatibility is an intricate process, as there is not any specific definition or exact measurement to declare a material as a biomaterial. Usually, biocompatibility is characterized in context of its use for a specific application. Thus, any material performing well for a specific biomedical application cannot be declared biocompatible for all the biomedical applications.

For example, ceramic materials are widely used in dental implants and show good biocompatibility. They can cause incompatibility, however, when used with the vascular system like blood clotting. So, it is important to define biocompatibility specifically for a system like biocompatibility with cardiovascular system or with soft or hard tissues. Biocompatibility may have to be specifically measured for each specific biomedical application. It is also important to take functional evaluation into consideration while determining biocompatibility. For example, teflon (polytetrafluoroethylene) is biologically inert in the context that it does not trigger any adverse biological reaction with the living tissues. However, when teflon is used for the temporomandibular joint replacement, its fragmentation can trigger serious foreign body responses leading to erosion of the adjacent structures. Thus, its biocompatibility can be argued, and it gives us an insight into the importance of functional evaluation of the entire implant system. Biocompatibility of a material is determined by combination of many aspects like its chemical composition, intended use, duration of use, and physical shape. An example is given in the review by T. G. Moizhess, where carcinogenesis not only depends on the chemical composition of the biomaterial but also on shape of the implant system. The tumorigenic polymer system when implanted caused lower carcinogenicity after perforation and showed complete loss of carcinogenicity on fragmentation [5].

Some basic and important principles of biocompatibility are defined by U.S. Food and Drug Administration (FDA). FDA evaluates the biocompatibility of the complete biomaterial device system and not only the component materials of the system. ISO 10993-1 establishes norms for the biological assessment of the medical devices, substantiating that biological assessments must be "performed on the final medical device or representative samples from the final device or materials processed in the same manner as the final medical device (including sterilization, if needed)".

4.2 Biological Functionality

Over the years, there has been a transition from permissive bio-inert materials to bioactive biomaterials, and biological functionality is an important characteristic of a biomaterial. Biological functionality implies a set of characteristics that a biomaterial must possess or comply with in its intended biological function. For example, an artificial valve has to close and open as per the requirement in its biological functionality. A controlled-release drug delivery system should provide a controlled drug release at a predetermined rate in its biological functionality. A biomaterial is designed with dynamic functionality and anticipated to integrate with the biological complexity and perform the desired functions in the body. So, numerous parameters are tested before a biomaterial is introduced into the market, just like drugs testing.

4.3 Chemical Stability

Chemical stability implies the ability of a biomaterial to maintain its integrity after implantation into its site of application. After surgical implantation, a biomaterial has to bear the attack of the body's physiological environment. Under these conditions, some biomaterials are oxidized and can produce toxic by-products. Consequently, an implant can degrade and not perform its intended function. For example, one of the major limitations of metallic biomaterial is its chemical instability due to prolonged contact with the biological fluids, leading to corrosion (sum of electrochemical phenomena). So, a biomaterial should be chemically inert and stable to perform its function properly, not causing any harm to the body. It should be resistant to biological ageing for the intended duration of use, or if it is supposed to be used for a limited period, it should be biodegradable with no toxic by-products.

4.4 Machinability

Machinability implies that the material should be mouldable into a biomedical device with the potential to be sterilized by any standard sterilization method (autoclave, UV sterilization, alcohol disinfection, ethylene oxide gas disinfection, etc.). The material should not be impaired after sterilization. A biomaterial should have suitable manufacturability, i.e., the ability to manufacture the biomedical device easily for its anticipated use with high reliability. A biomaterial device should be robust to bear the strain, stress, shear, and mixture of these forces anticipated at their intended site of action in the body. Hardness, yield strength, tensile strength, and elasticity are some of the important characteristics of a biomaterial that should be considered and assessed prior to implantation. For example, for the materials to be used in hard tissue applications, the mechanical properties are of utmost importance. Wear and tear are often the main reasons for the failure of the implants. The device should have suitable mechanical properties appropriate for their intended site use. For example, a bone replacement biomaterial should possess good mechanical properties as it is anticipated to face significant mechanical stress at the implantation site. Similarly, a heart valve has to keep opening and closing many times a day and for a prolonged period of time, so it should be resistant to wear and tear on prolonged duration of use.

To summarize a biomaterial should:

- Be biocompatible, i.e., non-toxic, non-inflammatory, non-carcinogenic, etc. Its degradation product should not be harmful to the body and excretable.
- Have desirable chemical and physical and chemical stability.
- Have suitable processability for the anticipated application.
- Have suitable mechanical properties to bear the anticipated stress at the site of application for the intended period of use.
- Carry the anticipated function at its site of implantation.
- Be easy to process and sterilizable by validated sterilization techniques without any impairment.
- Be cost-effective and reliable.

5 Classification of Biomaterials

Biomaterials can be classified on the basis of their source or application. Based on the source and material properties, the biomaterials can be classified into natural and synthetic biomaterials. The natural biomaterials are further classified into protein and polysaccharide-based biomaterials. The synthetic biomaterials are classified into metallic, polymeric, and ceramic biomaterials (Fig. 5). All these classes of biomaterials exhibit different physicochemical, mechanical, and biological properties, which help them to enable their function in or on the human body [5].

5.1 Natural Biomaterials

Natural biomaterials are obtained from living species, from either plants or animals [6]. The use of natural biomaterials in biomedical applications is not new; humans have been using them since ancient times. However, natural biomaterials have some drawbacks. The biomaterials themselves have a certain level of variability due to inherent differences between material sources that may make producing high levels of reproducibility difficult. Furthermore, unless the material is extremely pure, these substances may occasionally cause immunological reactions. Natural biomaterials are divided into two types 1) protein origin biomaterials and 2) polysaccharide origin biomaterials.

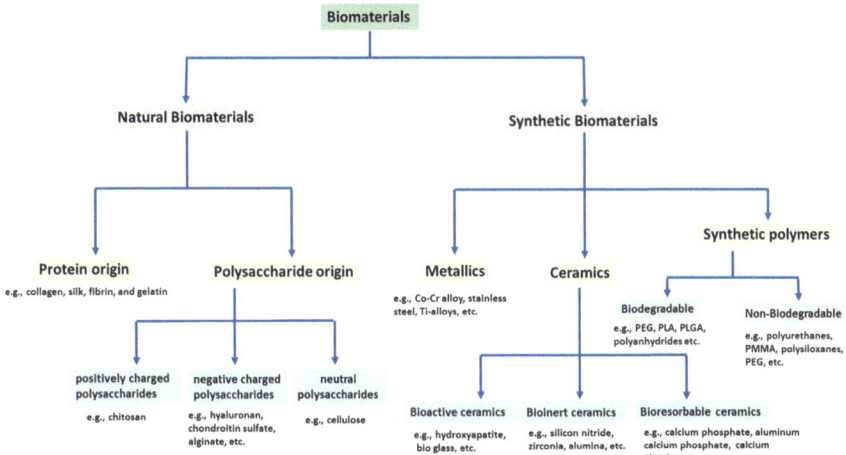

Fig. 5 Classification of biomaterials based on their source and material properties

5.1.1 Protein Origin Biomaterials

Protein origin biomaterials are the primary structural components for many tissues. Proteins are essentially amino acid polymers arranged in a three-dimensional folded form. They are used for the fabrication of sutures, haemostatic agents, tissue scaffolds, and drug delivery systems. Examples of protein origin biomaterials include collagen, silk, fibrin, and gelatin (Fig. 6).

Collagen Collagen is the most common protein in the human body and is essential for the development of skin and other musculoskeletal tissues. Collagen is a main structural component in the connective tissue. In mammals about 20–35% of whole-body protein content contains collagen. Based on the structure, currently 28 different forms of collagen are identified in the human body. All these structures consist of at least one triple helix. Among them, Type-I collagen is most prevalent and accounts for nearly 90% of collagen in the human body.

Collagen has attracted interest as an appropriate matrix material for tissue engineering, since it is the main component of the extracellular matrix and acts as a natural substrate for cell adhesion, proliferation, and differentiation. Collagen is one of the key initiators of the coagulation cascade because of its high thrombogenicity, which has led to its usage as a haemostatic agent. For a variety of surgical indications, several collagen-based haemostats are currently available or in clinical studies.

Collagen undergoes enzymatic degradation in the presence of collagenases and metalloproteinases and yields amino acids. Because of their enzymatic biodegradability, interesting physicochemical, mechanical, and biological properties, collagen has been processed into different forms such as sheets, tubes,

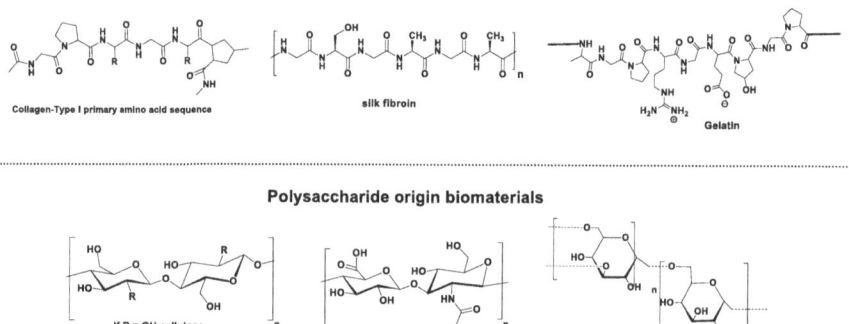

Fig. 6 Chemical structures of protein and polysaccharide origin biomaterials

sponges, foams, nanofibrous matrices, powders, fleeces, injectable viscous solutions, and dispersions for biomedical applications. Several forms of collagen derivatives are used as scaffold materials for cardiovascular, musculoskeletal, and nervous tissue engineering applications.

Due to collagen's high reactivity, a variety of compounds, such as polyepoxy substances, hexamethylene-diisocyanate, carbodiimides, difunctional or multifunctional aldehydes, and succinimidyl ester polyethylene glycol, can be used to create cross-link collagen. Due to highly favourable interactions of collagen matrix with the proteins, collagen can be used in the protein delivery applications. The applicability of collagen also extends into gene delivery owing to its good Injectability and good protective property from the enzymatic degradation. The composite biomaterials of collagen are also used in different orthopaedic applications. For example, one of the FDA-approved composite "collagrafts" comprised of fibrillar collagen, hydroxyapatite, and tricalcium phosphate is used as a biodegradable synthetic bone graft material.

Silk Silk is a natural polymeric protein fibre produced by different insect species. The first record of silk use as a biomaterial dates back to Chinese and Indus civilizations around 2500 BC. Silk possesses excellent biocompatibility, self-assembly, mechanical properties, controllable structure formation, and interesting morphology. Such characteristics make silk-based biopolymers valuable candidates in the biomedical field for wound healing and tissue engineering. The silk fibroin forms three-dimensional structures such as sponges, foams, and scaffolds, which are used in tissue engineering, disease models, and implantable devices. Silk is also used in biomedical textiles as wound dressing material due to its good toughness, tensile strength, and ductility. Silk has good mechanical properties, softness, and antibacterial properties, contributing to its use in hygiene and healthcare products including materials used in hospital wards and operating rooms. Silk materials are employed in operating theatres as patient drapes and surgical equipment such as gowns, caps, masks, cover cloths, etc. Silk fibroin-based biomaterials are also used as skin grafts, bone grafts, as well in the preparation of artificial skin. Silk-based biomaterials are also used in other applications such as regenerating ligaments, tendons, and cardiac tissues. Silk is also used as a suture in cardiovascular, ophthalmic, and neurological surgeries, owing to its good knot strength, ease of processing, low propensity to rip through tissue, and good biocompatibility.

Fibrin Fibrin is a non-globular protein, which is involved in the clotting of blood. Fibrin has been one of the most well-known biomaterials since it was first extensively purified in the 1940s. The application of fibrin as a biomaterial is very diverse in the medical field. For tissue engineering applications, fibrin gels are created by mixing fibrinogen, NaCl, thrombin, and $CaCl_2$. Its key advantages include being highly biocompatible, having the ability to control tissue regeneration, and fast polymerizing reaction. Fibrin-based biomaterials are useful for tissue engineering applications because they can act as a scaffold for tissue healing, and they promote cellular proliferation. Tisseel was the first FDA-approved tissue sealant based on fibrin that is available in the market for clinical applications. Later, other sealant brands such as Beriplast and Biocol entered the market. ARTISSfibrin is another fibrin-based sealant (Baxter International Inc., Deerfield, Illinois, USA) available in the market. It is used to adhere autologous skin grafts in wound healing. Fibrin glue is used in the postoperative process in place of sutures or dressings to enhance healing and lessen scarring.

Gelatin Gelatin is a molecular derivative of type I collagen. Like collagen, the major structural components in the gelatin include glycine, proline, and hydroxyproline. Gelatin is primarily created when the triple helical structure of collagen is irreversibly hydrolysed by methods like heat and enzymatic denaturation, leaving behind randomly coiled domains. Therefore, gelatin has a very similar molecular composition to collagen, but it is less organized than collagen. Gelatin is a fibrous insoluble protein and is a major component in skin, bone, and connective tissues. Gelatin can be extracted from several sources like cattle bones, fish, pig skins, and some insects. Compared to the extracellular matrix proteins such as collagen, fibronectin, and laminin, gelatin has gained more popularity as a biomaterial in the biomedical field due to the following reasons.

(i) Gelatin is readily available at less cost.
(ii) The solubility of gelatin is much higher when compared to extracellular matrix proteins.
(iii) The molecular structure of gelatin is similar to the structure of collagen and contains important binding sites for cell attachment.
(iv) Gelatin is highly biocompatible, biodegradable, and doesn't show any toxic effects, and antigenicity to the human cells.

Gelatin is highly useful in cell and tissue culture applications and also used as a plasma expander, wound dressing, adhesive, and absorbent pad for surgical application. Gelatin is used in pharmaceutical products including hard capsules, soft capsules, and tablets. The gelatin film is used in both hard and soft capsules to mask the taste and odour of the medications inside while also shielding them from the outside environment. Gelatin is employed in tablet applications as a natural binding and disintegrating agent as well as an excellent tablet coating that protects the active ingredients against oxidation and light damage. It is frequently found in body care items such as facial creams, suntan lotions, shampoos, and moisturizing creams.

5.1.2 Polysaccharide Origin Biomaterials

Polysaccharides are long-chain polymeric biomaterials composed of monosaccharide units bound together by glycosidic linkages glycosidic linkages.The polysaccharides are abundant in nature, and they can be extracted from different sources such as algae, plants, microorganism, and animals. Examples of polysaccharide biomaterials obtained from algae include alginate, agarose, galactans, and carrageenan. Cellulose, pectin, guar gum, and starch are examples of plant-derived polysaccharides. Similarly, the polysaccharides extracted from the microorganism include xanthan gum, dextran, gellan gum, pullulan, and bacterial cellulose. Hyaluronan, chondroitin, heparin, chitin, and chitosan are polysaccharide biomaterials extracted from animals. Polysaccharides possess different physiochemical and biological properties that depend on their chemical structure, functional groups, charge, etc. Based on the charge, the polysaccharide-based biomaterials are classified into three sub-classes: 1) positively charged polysaccharides; 2) negatively charged polysaccharides; 3) neutral polysaccharides. Examples of positively charged, negatively charged, and neutral polysaccharide biomaterials include chitosan, hyaluronan, and cellulose respectively. In this section, we describe a few commonly employed polysaccharide biomaterials (Fig. 6) for medical applications:

Cellulose Cellulose $(C_6H_{10}O_5)_n$ is the most prevalent natural polysaccharide with adjustable characteristics. It is used in diverse biomedical applications such as drug delivery and tissue engineering. It is composed of several d-glucose units connected through β-(1–4) glycosidic bonds. The polysaccharide chains in the cellulose are aligned in parallel to form microfibrils with high tensile strength. Cellulose is the structural component for primary cell wall of plants. Cellulose is usually obtained from plants, fungi, algae, animals, and certain species of bacteria (example: acetobacter xylinum). Cellulose is a tasteless and odourless biodegradable neutral polysaccharide.

Depending on the source of origin, cellulose is available in a variety of compositions and morphologies. The high intramolecular hydrogen bonding of cellulose is responsible for its high solution viscosity, strong crystallization propensity, and capacity to form fibrillar threads. Owing to their good biocompatibility, cellulose-based biomaterials are widely used in drug delivery applications, wound healing, and as tablet binders, coating materials, and viscosity modifiers. Cellulose has a wide range of uses in cartilage tissue engineering. It offers a platform for cell growth and development. The microcrystalline cellulose and powdered cellulose derivatives such as E460i and E460ii are widely used as inactive fillers in tablet manufacturing.

Chitosan Chitosan is a prevalent polysaccharide biomaterial found in nature. It is a liner polysaccharide and is made up of randomly distributed deacetylated unit "β-(1 → 4)-linked D-glucosamine" and acetylated unit "N-acetyl-D-glucosamine". The free $-NH_2$ and $-OH$ groups in the structure of chitosan enable it to react with a wide variety of chemical entities to form chitosan-based biomaterials. The distinctive properties of chitosan such as biocompatibility, its affinity towards biomolecules, and suitable chemical structure for chemical modifications allow its utility as a prominent material for biomedical applications. Several factors such as crystallinity, molecular weight of the polymer, and degree of deacetylation influence the physiochemical properties of the chitosan. The applicability of chitosan-based biomaterials has been found in tissue engineering, regenerative medicine, drug delivery, and wound dressings.

The chitosan salts with organic acids such as succinic or lactic acid are used as haemostatic agents. Due to its positive charge, protonated chitosan attracts platelets, causing them to quickly clump together and form thrombuses. Chitosan derivatives are used as artificial kidney semi-permeable membranes. The properties of chitosan derivatives such as optical clarity, mechanical stability, sufficient optical correction, gas permeability, wettability, and immunological compatibility make them useful in the development of ocular bandage lenses.

Hyaluronic Acid Hyaluronic acid is an example of anionic polysaccharide. It is widely distributed in connective, epithelial, and neural tissues. Hyaluronic acid is a glycosaminoglycan comprised of alternating N-acetyl glucosamine and D-glucuronic acid units connected through β (1–4) and β (1–3) linkages respectively. Hyaluronic acid is a naturally occurring substance found primarily in extracellular matrix, synovial fluid in joints, the umbilical cord, skin, the vitreous humour of the eye, and articular cartilage. Due to the presence of a carboxyl group in its structure, hyaluronic acid is classed as a naturally occurring negatively charged polysaccharide. Hyaluronic acid can absorb a significant amount of water and swell up to 1000 times its original volume. Many researchers use this property to build hyaluronic acid-based drug delivery systems. Additionally, hyaluronic acid is a strong candidate for chemical functionalization due to the availability of free carboxyl and hydroxyl groups. Owing to its properties such as hydrophilicity and biocompatibility, hyaluronic acid has a wide range of uses in biomedical field, particularly in the design of drug delivery systems and tissue engineering. The hyaluronic acid is an FDA-approved biomaterials that frequently is used to treat knee osteoarthritis through intra-articular injection. It is also used in some of the eye formulations to create artificial tears to treat dry eye disease.

Dextran Dextran is a hydrophilic homopolysaccharide, which is made up of α-1,6-linked D-glucopyranose units with a low percentage of α-1,2-, α-1,3-, or α-1,4-linked side chains. Dextran is a bacterially derived polymer. It is derived from sucrose by the catalytic action with dextransucrase or enzymatic hydrolysis of maltodextrin by dextrinase. Dextran is a water-soluble polysaccharide and is usually stable in mild acidic and basic environments. Due to the high abundance of hydroxylic groups in dextran, it can be derivatized using chemical and physical crosslinking to produce various scaffolds like spheres, tubules, and hydrogels. In the body, dextran is slowly broken down by human enzymes and microbial enzymes present in the gastrointestinal system. The characteristics of dextran such as low toxicity, biocompatibility, biodegradability, and suitability for chemical modifications allow its use in the biomedical field. Dextran is clinically used as an antithrombotic agent, plasma volume expander, viscosity-reducing agent, etc. It is also used to reduce the inflammatory response. Dextran and its derivatives are also used in drug delivery and tissue engineering applications.

Alginate Alginate is a naturally occurring, anionic, and hydrophilic polysaccharide, which is usually obtained from brown algae (examples: *Laminaria hyperborea*,

Laminaria digitata, *Laminaria japonica*, etc.). It is a block copolymer of (1–4)-linked β-D-mannuronic acid (M-block) and α-L-guluronic acid (G-Block) monomers. The physicochemical properties of the alginate vary with the length of the block, and the M/G ratio as well as the arrangement of repeating units in its structure. The carboxylic and cis diol groups found in alginate monomers enable the chemical functionalization of this polysaccharide with various moieties such as amino acids, alkyl groups, etc. Because of their biocompatibility, mechanical flexibility, crosslinking activity, and gelling qualities, alginate-based biomaterials are employed in various biomedical applications.

Alginates can form ionic gels in the presence of divalent metal cations like Ca^{2+}, where the high amount of G-clock in the alginate reacts with the Ca^{2+} and forms an egg-box-like conformational arrangement. Alginates can form hydrogels with a variety of cross-linking agents. The hydrogels of alginate are structurally similar to extracellular matrices of the living tissues. Therefore, the applications of alginate-based hydrogels are extensively found in wound healing, drug delivery, and cell transplantation in tissue engineering. In drug delivery, the controlled release of drugs and other macromolecular proteins from the hydrogel matrix of alginate depends on the cross-linker types used to prepare alginate hydrogel. Moreover, alginate gels can be administered orally or injected into the body in a minimally invasive manner, allowing for a wide range of pharmaceutical applications. Due to their high biocompatibility and low toxicity, the alginate derivatives are also used in the food industry and dentistry.

Pullulan Pullulan is an exopolysaccharide obtained from yeast-like fungus Aureobasidium pullulans. Chemically, pullulan is comprised of maltotriose repeating units connected by α-(1–6) glycosidic bonds. It is a linear unbranched polysaccharide with a molecular formula of $(C_6H_{10}O_5)_n$. Pullulans are available with molecular weights in the range of 4.5×10^4 to 6×10^5 Da. The hydroxyl groups present in the structure of pullulans provide an opportunity for chemical modification with a variety of amino acids, alkyl chains, metal nanoparticles, etc. The characteristics of pullulans such as biodegradability, no toxicity, and nonmutagenicity allow their utility in drug delivery and tissue engineering applications.

Chondroitin Sulphate Chondroitin sulphate (CS) is an anionic heteropolysaccharide, comprised of repeating disaccharide units of β-1,3-linked d-glucuronic acid and β-1,3-linked N-acetyl galactosamine having sulphate groups at various carbon positions. CS is classified into five subgroups based on the position of the sulphate group: chondroitin-4-sulphate; chondroitin-2,4 sulphate/dermatan sulphate; chondroitin-6-sulphate; chondroitin-2,6-sulphate; and chondroitin-4,6-sulphate. The molecular weight of CS varies in the range of 20–25 kDa. The unique characteristics of CS such as biocompatibility, biodegradability, mucoadhesion, and hydrophilicity allow its wide utility in the biomedical field. CS possesses anti-inflammatory, antioxidant, antithrombotic, anticoagulating, and immunomodulatory properties. Hence, CS is used in the biomedical field for managing cardiovascular, cancer, wound healing, and joints-related pathologies. CS is used as a natural supplement for the treatment of osteoarthritis. CS is an FDA-approved biomaterial to use as a skin substitute. CS is also used in the drug delivery applications.

5.2 Synthetic Biomaterials

This class of biomaterials are synthesized in the laboratory for biomedical purposes [7]. The synthetic biomaterials are further subdivided into ceramic, metallic, and polymeric biomaterials. The characteristics, applications, and disadvantageous of these classes of synthetic biomaterials are summarized in Table 1.

5.2.1 Ceramic Biomaterials

Ceramic biomaterials and their composites have received significant importance in the biomedical field due to their properties such as resistance to corrosion, high mechanical strength, high stiffness, hardness, high compression strength, wear resistance, durability, and low density. They can be effective with compressive force, but poor with tension force. Ceramic biomaterials are electric and thermal insulators. They are widely used bioactive inert materials in the human body. Ceramic materials are widely used in orthopaedics and dentistry. Ceramic biomaterials are categorized into three types: 1) ceramic bioactive materials, 2) ceramic bio-inert materials, and 3) ceramic bioresorbable materials.

Bioactive ceramic materials interact with the surrounding bone and soft tissue after being inserted them into the human body. Their implantation into the living bone causes a time-dependent kinetic modification of the surface. Some of the common examples of ceramic bioactive materials include synthetic hydroxyapatite [$Ca_{10}(PO_4)_6(OH)_2$], glass ceramic, and bioglass.

On the other hand, examples of bio-inert materials include stainless steel, titanium, alumina, partly stabilized zirconia, and ultra-high-molecular-weight polyethylene. These materials have limited interaction with the surrounding tissue once implanted in the human body. In general, a fibrous capsule may grow around bio-inert implants. Therefore, the bio-functionality of bio-inert materials is dependent on tissue integration through the implant. Bio-inert materials that are especially compatible with body parts include silicon nitride (Si_3N_4), zirconia (ZrO_2), alumina (Al_2O_3), and pyrolytic carbon. They are widely used in hip prostheses, bone scaffolds, spinal fusion implants, hip joint replacement, and dental implants.

Bioresorbable refers to a substance that, when introduced into the human body, begins to disintegrate and is gradually replaced by a new tissue. This class of ceramics is comprised mainly of calcium. Some of the examples of bioresorbable ceramic materials include calcium phosphate, aluminium calcium phosphate,

Table 1 Classification of synthetic biomaterials; their characteristics, applications and limitations

Biomaterial	Examples	Characteristics	Applications	Limitations
Metallic biomaterials	Ti-alloys, stainless steel, co-Cr alloys, ta alloys, mg alloys, Ni-Ti alloys, etc.	Desirable mechanical properties such as stiffness, resistance, tensile strength, compressive stress etc.	Metallic biomaterials used in orthopaedics (used as artificial joints, plates, and screws), orthodontics (used as braces, and dental implants), cardiovascular implants (used in artificial hearts, stents, artificial heart valves, pacemaker leads), neurosurgical devices, vascular stents, and load-bearing implants such as hip and knee replacements.	High modulus, cytotoxicity, easy corrosion, and metal ion sensitivity.
Ceramic biomaterials	**Ceramic bioactive materials** (ex: Hydroxyapatite, glass ceramic, and bio glass) **Bio-inert ceramic materials** (ex: Stainless steel, titanium, alumina, zirconia, and ultra-high-molecular-weight polyethylene) **Bioresorbable ceramic materials** (ex: Calcium phosphate, aluminium calcium phosphate, tricalcium phosphate, calcium aluminates, zinc sulphate calcium phosphate, zinc calcium phosphorous oxide, etc.)	Resistance to corrosion, high mechanical strength, high stiffness, hardness, high compression strength, wear resistant, durability, and low density	Commonly used in dentistry, orthopaedics, and cardiovascular implants.	Hard, not flexible, brittle, etc.

(continued)

Table 1 (continued)

Biomaterial	Examples	Characteristics	Applications	Limitations
Polymeric biomaterials	**Synthetic degradable polymers** (ex: PLA, PGA, PLGA, PCL, PVA, polyanhydrides, poly-(alkyl cyanoacrylates), etc.)	Easy synthesis, tuneable mechanical properties, tuneable degradation kinetics, producing non-toxic by-products after polymer degradation, good biocompatibility, etc.	Drug delivery, orthopaedics, dentistry, plastic surgery, cardiovascular applications, regenerative medicine, 3D printing technology, etc.	Poor mechanical strength
	Synthetic non-degradable polymers (ex: PMMA, polyurethanes, polysiloxanes, poly(ethylene), poly(propylene), poly(styrene), poly(ethylene glycol), etc.)	Robust mechanical properties, easily mouldable, readily available, etc.	Used in cosmetic surgeries, used in dentistry for the fabrication of dental prosthetics, artificial teeth, shape memory devices, cardiovascular implants, wound healing, bone regeneration, etc.	Responsible for generating immunogenic reactions, non-degradable.

tricalcium phosphate, calcium aluminates, zinc sulphate calcium phosphate, zinc calcium phosphorous oxide, and ferric calcium phosphorous oxide. These compounds undergo hydrolytic breakdown in the body at their implantation site. The degraded products produced from the process are absorbed by the body and eliminated through the standard metabolic process. They are highly useful as dental restorative products as well as in orthopaedic applications. They are used as artificial bones, teeth, knees, hips, tendons, and ligaments.

In order to apply ceramic biomaterials to medical applications they must possess a certain set of characteristics including non-allergic, bio-functional, bio-compatible, non-carcinogenic, non-toxic, and non-inflammatory. The applicability of ceramic materials has also been well explored in the dental field because of their high inertness to the biological fluid like saliva, aesthetically favourable appearance, and excellent compressive strength. Their utility is also explored in drug delivery applications. The black pyrolytic carbon ceramic materials are easy to make and have good biocompatibility. They are widely used in cardiovascular implants. Owing to their good strength, they are also used in composite implant materials supporting components for tensile loading applications, e.g., loading of artificial ligaments and tendons.

5.2.2 Metallic Biomaterials

Metallic biomaterials have been given significant importance in the biomedical field because of their excellent thermal, electrical, and mechanical properties. They have been widely used in the preparation of artificial heart valves, pacemaker leads, and vascular stents. They are also used in load-bearing implants such as hip and knee replacements. Metallic biomaterials are also used as electrodes due to their high conductivity. However, there are certain drawbacks to the metallic biomaterials such as high modulus, cytotoxicity, easy corrosion, and metal ion sensitivity that reduce their utility as implant materials.

The physicochemical and mechanical properties of metallic biomaterials can be fine-tuned by incorporating multi-metal functionality. This can be achieved by surface modification, such as surface structuring or coating with bioactive ceramic materials and polymer thin films. The resultant chemically modified biomaterials obtained from this method are called "bio-metallic alloys", which are highly inert to the biological system; hence, they possess high corrosion resistance, long-term stability, and reliable mechanical strength. In addition, they also possess excellent tensile strength, fracture toughness, and fatigue stress. These characteristics allow them to be used in orthopaedics (used as artificial joints, plates, and screws), orthodontics (used as braces, and dental implants), cardiovascular implants (used in artificial hearts, and stents), neurosurgical devices, etc. The first metal alloy implant, "Sherman Vanadium Steel", was developed to manufacture bone fracture plates and screws. Some of the other commonly used metallic biomaterials include pure Ti, Ti-6Al-4 V, Co-Cr alloys, stainless steel, noble metal alloys, and shape memory alloys.

The metallic biomaterials are majorly classified into three major groups: Ti-alloys, stainless steel, and Co-Cr alloys (Fig. 7).

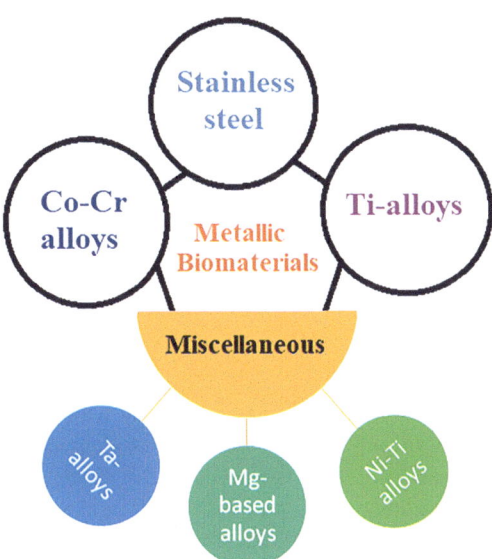

Fig. 7 Classification of metallic biomaterials

Stainless Steel Stainless steel is resistant to rust and corrosion and is an alloy of iron. It contains chromium along with carbon, other non-metals, and metals to produce desirable characteristics. The corrosion resistance of stainless steel is attributable to chromium, which forms a passive layer that protects the material and self-heal in the presence of oxygen and water. Reducing the Ni-content in the stainless steel offers improved protection from corrosion. Due to characteristics such as good toughness, biocompatibility, and easily producible with low cost, stainless steel is widely used in biomedical applications. Different types of stainless steel are produced by industries, among them austenitic stainless steel is most widely used for the orthopaedic applications in the forms of screws, plates, and hip-nails apparatus.

Co-Cr Alloys Cobalt-based alloys were initially utilized as medical implants in the 1930s. They are one of the hardest and biocompatible alloys. They possess high wear and corrosion resistance properties and are used as orthopaedic implants. Cobalt-based alloys were the first alloys to be used in dental implants. Their applications also extend to joint and fracture fixing. There are a variety of biocompatible Co-based composites available, including CoCrMo, CoCr, and Ni-free CoCrW alloys. Cobalt-chromium alloys are generally classified into two types: (i) Co-Cr-Mo alloy and (ii) Co-Cr-Ni-Mo alloy. Bearing good corrosion protection, high loading capacity, stronger fatigue and elastic-modulus, Co-Cr-Mo alloys have been used for many years in dentistry, and recently in orthopaedic treatment, especially for joint reconstructions. Co-Cr-Ni-Mo's are used in the high loading prosthesis for joint rebuilding of knee and hip problems in biomedical treatment. Co-based alloys are extremely resistant to corrosion.

Titanium Alloys Ti and its alloys have received interest in the biomedical field especially in the orthopaedic and dental applications because of their good mechanical properties, corrosion resistance, and good biocompatibility. Commercially, pure titanium and the alloy of Ti-6Al-4 V are frequently used metallic biomaterials applied for dental and orthopaedic applications. However, the long-term use of Ti-6Al-4 V implants causes an allergic reaction in human tissue due to leaching of aluminium and vanadium. As a result, new alloys free from cytotoxic components are now being developed.

Recently magnesium (Mg), iron (Fe), tantalum (Ta), and zinc (Zn) alloys have received significant interest as biodegradable metallic alloys. Among them, Mg is essential for human metabolism, and it possesses enough tensile strength, and resistance to the fracture. It is a lightweight element; therefore, Mg-based alloys are used to support the load-bearing applications, particularly applied in stents and small fracture repairs. Mg is well known to provide stimulatory effects on the generation of new bone tissues. The biodegradability of the Mg-based metallic scaffolds and implants is fine-tuned by changing the ratio and type of the alloying metals. However, the performance of biodegradable metallic implants is significantly undermined by characteristics such as the risk of infection and inflammation. This phenomenon significantly leads to the loss of tissues or cells in the proximity of the implant. Another challenge associated with metallic biomaterials is their biodegradation that leads to premature loss of the mechanical strength of the implant system.

5.2.3 Polymeric Biomaterials

Polymers are long-chain macromolecules comprised of covalently bonded repeating monomer units. Polymers constitute the largest class of diverse biomaterials. Depending on their origin, they are classified as synthetic or natural polymeric biomaterials. Owing to their unique properties such as biocompatibility, controlled biodegradability, non-toxicity, and remodelling, polymeric biomaterials have attracted considerable interest in the biomedical field. In this section, we provide a brief summary and applications of the synthetic-derived polymeric biomaterials. The natural polymeric biomaterials are already summarized in the Sect. 2.1. Natural biomaterials.

Based on degradation behaviour, synthetic polymers are classified into two classes: 1) synthetic degradable polymeric biomaterials and 2) synthetic non-degradable polymeric biomaterials.

Synthetic Degradable Polymeric Biomaterials

This class of biodegradable polymers has gained significant interest in the medical field, because of their interesting physicochemical properties and biodegradability [8]. The first example of a synthetic biodegradable polymer produced is poly(glycolic acid) (PGA) in 1954. Some other examples for synthetic biomaterials include polylactic acid (PLA), polyethylene glycol (PEG), poly (lactic-co-glycolic acid) (PLGA), polycaprolactone (PCL), and polydioxanone (PDS). These synthetic biodegradable polymeric materials are alternatives to the natural biodegradable polymers. They undergo controlled degradation into small monomer species under in vivo conditions due to hydrolysis or enzymatic actions occurring in the body. The degraded products of polymeric biomaterials are non-toxic, which do not show any inflammatory reactions in the body. The degraded products are easily metabolizable and are excretable. For example, PLGA undergoes hydrolytic cleavage into biocompatible monomers glycolic acid and lactic acid. The biodegradation of the polymeric structures is affected by its molecular weight, hydrophilicity, hydrophobicity, etc. Examples of some of the most used synthetic polymer for biomedical applications are summarized below (Fig. 8):

Polylactic Acid (PLA) PLA is a thermoplastic polymer obtained from the condensation reaction of lactic acid or ring-opening polymerization of lactide in the presence of an organometal catalyst like stannous octate. It is one of the largely consumed polymeric bioplastics. PLA occurs in the form of L-PLA, D-PLA, and in the mix-

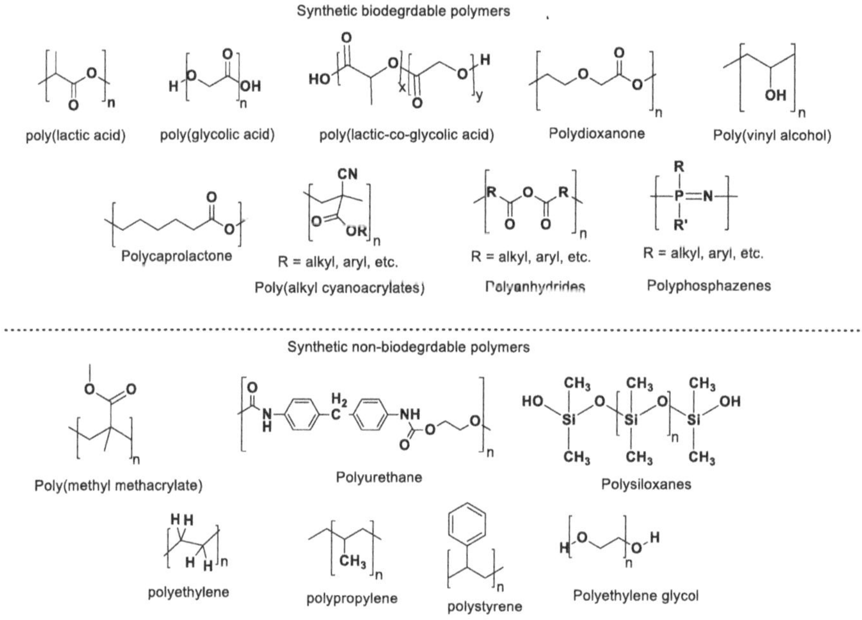

Fig. 8 Chemical structures of synthetic biodegradable polymeric biomaterials and non-biodegradable polymeric biomaterials

ture of D,L-PLA. PLA undergoes hydrolytic degradation through a desertification process and leads to the formation of lactic acid monomer as a by-product. Lactic acid is biocompatible and can be easily eliminated from the body through a natural biochemical pathway. Therefore, PLA and PLA-based derivatives are often used in the biomedical implants in the form of screws, anchors, pins, plates, rods, and as a mesh. The average biodegradation time of PLA varies from 6 months to 2 years based on its structure and molecular weight. Due to good control over the biodegradation rate of PLA and its derivatives, they are widely used for the fabrication of different drug delivery systems. The composite materials of PLA are also extended for different biomedical applications. For instance, the composite of PLA/tricalcium phosphate is used as a scaffold for bone engineering. PLA derivatives are also used in the 3D printing technology to print artificial biological structures.

Polyglycolic Acid (PGA) Like PLA, PGA is also a thermoplastic polymer obtained from the condensation of glycolic acid or ring opening polymerization from glycolide in the presence of an organometal catalyst such as stannous octate. PGA is a linear polymeric biomaterial used as implants in vascular and orthopaedic surgeries. Both PLA and PGA are FDA-approved synthetic polymers used to produce resorbable stitches. The degraded products of PGA are water soluble and are easily eliminated from the body completely between 2 to 3 months. Medically, PGA is used to produce implantable medical devices such as anastomosis rings, pins, rods, plates, and screws. Its application also extends to tissue engineering and drug delivery.

Owing to its properties such as high tensile strength, easy handling, and excellent knotting ability, PGA is used in subcutaneous sutures, intracutaneous closures, and abdominal and thoracic surgeries. Copolymers of PGA such as poly(lactic-*co*-glycolic acid), poly(glycolide-*co*-caprolactone), and poly(glycolide-*co*-trimethylene carbonate) are widely used in the preparation of surgical sutures.

Poly(Lactide-*Co*-Glycolide) (PLGA) PLGA is a copolymer of the lactide and glycolide and is obtained from the ring-opening polymerization of these two monomers in the presence of stannous octate catalyst. PLGA is an FDA-approved polymer used in a variety of biomedical devices such as grafts, sutures, implants, prosthetic devices, and surgical sealant films. The physicochemical properties such as crystallinity, glass transition temperature, solubility, and degradation properties of PLGA can be varied by changing the molar ratio of the lactide and glycolide content in the polymer. For example, increasing glycolic acid content in the copolymer leads to faster degradation of the PLGA and is vice versa when it contains high content of lactide in the structure. PLGA undergoes hydrolytic degradation in the presence of water and breaks to non-toxic by-products, lactic acid, and glycolic acid. These products are easily eliminated from the body. The glass transition temperature of PLGA varies from 40 to 60 °C. Because of the controlled biodegradability, biocompatibility, and suitable glass transition temperature, PLGA is widely used in various drug delivery applications. PLGA is also used as scaffolds for bone, skin, cartilage, and nerve regeneration applications.

Polydioxanone (PDS) PDS is one the ether-ester-linked synthetic biodegradable polymers obtained from the ring opening polymerization of p-dioxanone in the presence of an organometallic catalyst like zirconium acetylacetone or zinc L-lactate. The glass transition temperature of PDS varies in the range of $-10°C$ to $0°C$. Due to the low glass transition temperature, PDS is not thermally stable as are other biodegradable synthetic polymers such as PLA and PLGA. Therefore, polymer processing for medical applications is usually kept under as much as low possible temperatures to avoid depolymerization of PDS. The characteristics of PDS such as high biocompatibility, and mechanical flexibility, enable its utility in the biomedical field for the preparation of surgical sutures. The applicability of PDS also extends to several other biomedical fields including tissue engineering, orthopaedics, drug delivery, maxillofacial surgery, plastic surgery, and cardiovascular applications.

Polycaprolactone (PCL) PCL is an aliphatic biodegradable polyester obtained from the ring-opening polymerization of ε-caprolactone in the presence of a metal-based catalyst like Stannous (II) octoate. It is a hydrophobic, semi-crystalline, and thermoplastic polymer. It undergoes slow ester hydrolysis under physiological conditions over a variable period of time that depends on the PCL molecular weight, degree of crystallinity, and degradation conditions. Therefore, PCL is widely used in controlled drug delivery applications. It is used for a variety of medical applications including sutures, wound dressing, cardiovascular tissue engineering, nerve regeneration, and bone tissue engineering. Few PCL-based drug delivery devices and sutures are FDA approved. However, PCL has some limitations such as slow degradation rate, poor mechanical properties, and low cell adhesion properties that shorten its utility in the biomedical field, particularly in tissue engineering. Preparation of composite biomaterials with PCL leads to remarkably improved mechanical properties, controllable degradation rates, and enhanced bioactivity that are suitable for a variety of biomedical applications.

Polyvinyl Alcohol (PVA) PVA is 1,3-diol linkage synthetic biodegradable polymer, which is usually obtained from the hydrolysis of polyvinyl acetate. PVA is a crystalline material soluble in the water. PVA forms strong and ultrapure hydrogels without any cross-linking agents. It also exhibits good biocompatibility with low protein adhesion tendency. Owing to these properties, PVA is widely used for the fabrication of vascular stents, cartilages, contact lenses, drug delivery carriers, etc.

Recently, some other classes of synthetic biodegradable polymers such as polyanhydrides, polyphosphazenes, block copolymers with PEG, and poly-(alkylcyanoacrylates) have received importance for a variety of biomedical applications. Among them, research on polyanhydride biodegradable polymers is currently a popular topic of research due to their controlled surface erosion, and shorter average half-lives properties. Polyanhydrides shows controlled drug release property for a variety of drugs, hence they are extensively used as a drug delivery carrier. Poly(sebacic acid-co-1,3- bis(p-carboxyphenoxy)propane) is an example of a polyanhydride used for the fabrication of FDA-approved Gliadel wafers used for the localized brain delivery of 1,3- bis(2-chloroethyl)-N-nitrosourea against brain cancer. Polyanhydride copolymers of erucic acid dimer and sebacic acid are used for the fabrication of septacin implants for the controlled delivery of gentamicin sulphate in the treatment of osteomyelitis.

Synthetic Non-degradable Polymeric Biomaterials

This class of polymers does not undergo biodegradation; however, because of their prominent physicochemical properties, these polymers are widely used in the biomedical field. Some of the common non-degradable polymeric biomaterials used for the medical purpose are summarized as follows (Fig. 8):

Poly-Methyl-Methacrylate (PMMA) PMMA is obtained by the free radical polymerisation of methyl methacrylate monomers. PMMA is lightweight, transparent, and it possess good mechanical properties. PMMA is used in orthopaedic applications. PMMAs possess good biocompatibility, hence they are widely used for the preparation of intraocular lenses. PMMA is also used in the cosmetic surgeries and in dentistry for the fabrication of dental prosthetics, artificial teeth, and orthodontic appliances.

Polyurethanes (PUs) PUs are alternating polymers obtained from the reaction between diisocyanates (hard segments) and polyols (soft segments) in the presence of light or catalyst (example: dibutyltin dilaurate, DABCO, etc.). In 1937, Otto Bayer and his co-workers developed the first PUs. By changing the ratio of hard and soft segments during the synthesis of polymer, variety of PU derivatives are designed for biomedical applications such as rubber, fibres, films, paints, coatings, elastomers, foams, gels, etc. For many decades, PU derivatives have been used in the biomedical field due to their prominent physicochemical characteristics such as high tensile strength, good durability, fatigue resistance, and excellent biocompatibility. PU derivatives are used as scaffolds in tissue engineering, shape memory devices, nontoxic implants, various cardiovascular implants, wound healing, and bone regeneration.

Polysiloxanes Polysiloxanes are prepared by the hydrolysis of alkyl silicon or polysilicon halides. Polysiloxanes are comprised of a Si-O backbone and with functional groups such as methyl are typically attached to the Si atom. Changing the −Si − O− chain lengths, side groups, and different cross-linking agents produces a variety of polysiloxane-based derivatives with interesting physicochemical characteristics. Poly(dimethylsiloxane) (PDMS) is one of the common examples of this class of polymers, which is frequently used as an implantable device for a long time in the biomedical field. Polysiloxanes are used in the fabrication of elastomers, gels, lubricants, foams, and adhesives. Polysiloxanes are hydrophobic and possess good biocompatibility, electrical insulation, and bio-durability characteristics. The applications of polysiloxanes and its derivatives include finger and toes joints, heart valve prostheses, blood oxygenation membranes, breast implants, artificial ventricles, wound dressings, plastic surgery, penile prostheses, intraocular lenses, vitreous humour, etc. They are used in orthopaedics as hand and foot joint implants material. They are also extensively used in cosmetic implants for aesthetic and reconstructive plastic surgeries.

Poly(Ethylene), Poly(Propylene), and Poly(Styrene) Poly(ethylene) (PE), poly(propylene) (PP), and poly(styrene) (PS) are common synthetic non-biodegradable polymers used for medical applications. They are thermoplastic polymers. PE, PP and PS are obtained by the radical or metal-mediated polymerization of their corresponding monomers, i.e., ethylene, propylene, and styrene respectively. High-density PE possesses strong intermolecular forces and tensile strength due to a low degree of branching; hence it is widely used in the fabrication of highly durable hip and knee prostheses. Moreover, high-density PE is also used to construct implants that have been used for facial and cranial reconstruction. The copolymers of PE, i.e., poly(ethylene-*co*-vinyl acetate) is an FDA-approved biomaterial that has been widely used in the fabrication of drug delivery systems. Ocusert and progestasert are some examples of poly(ethylene-co-vinyl acetate)-based drug delivery systems. The applications PP are well known and found in syringe bodies. Due to their hard and brittle nature, the PS is widely used for the fabrication of tissue culture flasks and dishes. The copolymerization of PS with butadiene allows the fabrication of catheters and medical devices for perfusion and dialysis with improved elasticity.

Poly(Ethylene Glycol) Poly(ethylene glycol) or poly(ethylene oxide) is an ether-linked hydrophilic polymer. PEG is obtained by the polymerization of ethylene oxide in the presence of an acidic or basic catalyst. The high molecular weight PEG can form hydrogels. PEG possesses high hydrophilicity, bio inertness, and outstanding biocompatibility, which allow its utility as a suitable candidate for biomedical applications. PEG-based copolymers are widely used for the delivery of a variety of drugs. They are also used as tissue engineering scaffolds, medical devices, and implants. A variety of PEG-based block copolymers are used for the fabrication of injectable drug delivery carriers. One example is pluronics or poloxamers, which is a triblock copolymer, composed of two hydrophilic PEG blocks and one hydrophobic poly (propylene oxide). Poloxamers are extensively studied as a non-biodegradable carrier for the delivery of a wide variety of drugs. PEG is also used as an inactive ingredient in the pharmaceutical industry as a plasticizer, surfactant, ointment, suppository base, capsule lubricant, and so on. Other applications of PEG include bio-sensing, imaging, bone, and tissue engineering.

6 Applications

Today biomaterials are used in every sphere of biomedical science whether it be diagnostic application, or tissue replacement to tissue regeneration. They have expanded their applications especially over the last few decades, from medical equipments to therapeutic medications and emerging regenerative technologies, and it is continuing to expand more [9]. Five major applications of biomaterials are discussed here to clarify the important ideas (Fig. 9).

6.1 Tissue Engineering

Tissue engineering is one of the most prominent applications of biomaterials, i.e., to repair or replace (partially or completely) a tissue/organ in the body to maintain and restore or improve a body function. Tissue engineering is defined by Langer and Vacanti as "a field that applies the principles of biology and engineering to the development of functional substitutes for damaged tissue". Examples of tissue engineering applications are as follows:

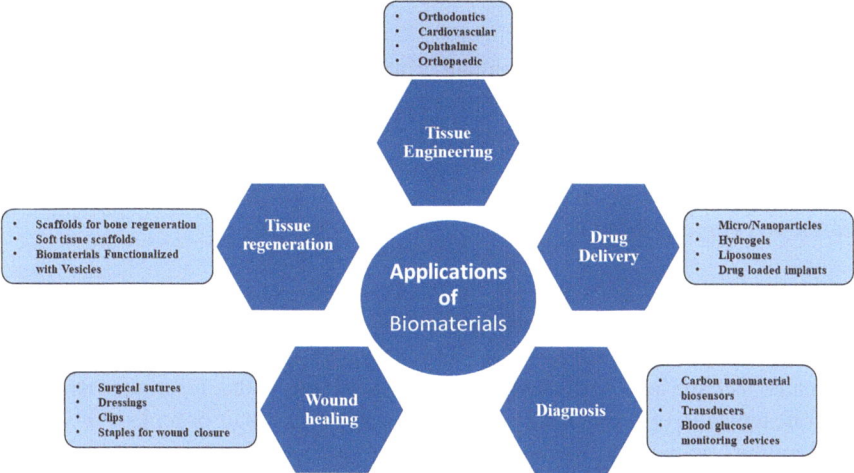

Fig. 9 Applications of biomaterials

6.1.1 Applications in Orthodontics

Biomaterials are used in dentistry for a long time as dental implants. The metallic materials are most widely used biomaterials for dental prosthesis in which the metallic biomaterial is integrated into the dental structure by the process called osseointegration. Zirconia, resin composites, titanium alloys, etc. are other widely used biomaterials for dental implants. Calcium hydroxide is another widely used biomaterial with application in root canal treatment.

6.1.2 Cardiovascular System Applications

Use of biomaterials in cardiovascular system is one of the most widely used applications of biomaterials. Biomaterial implants are used for a long time to treat blocked arteries, failure of cardiac valves, etc. Pathological changes in heart valves obstruct their proper opening and closing function that is usually treated by replacing the malfunctioned valve with an artificial one. A prosthetic heart valve can be either mechanical or biological. Some examples of biomaterials used for the mechanical prosthetic heart valves are silicone, stainless steel, titanium, and pyrolytic carbon. Biological prosthetic heart valves can be comprised of both biological (e.g., heart valve from pig and other human donor) and synthetic components like dacron and polytetrafluoroethylene (PTFE).

Similarly, pathological changes in the architecture of an artery can cause obstruction with the flow of blood that is treated by the use of a stent. Corrosion-resistant metals like nitinol, chromium-cobalt alloy, and stainless steel are widely used materials for the grafting of an artificial stent. Cardiopulmonary bypass systems, pacemakers, vascular grafts, and entire artificial hearts are other important examples of biomaterials applications in a cardiovascular system.

6.1.3 Ophthalmic Applications

Biomaterials are widely used for ophthalmic applications. One of the most important uses of biomaterials in ophthalmic is cataract surgery, i.e., replacement of an opacified eye lens with a prosthetic intraocular lens. Silicone and acrylic are commonly used biomaterials for the fabrication of the intraocular lenses. Other examples of biomaterials application in ophthalmic include artificial tears, contact lenses, and vitreous substitutions.

6.1.4 Orthopaedic Applications

Biomaterials have long been used in orthopaedics. One of the most important uses of biomaterials in orthopaedics is for joint prosthesis to repair or replace the joints of knee, hip, elbow, and shoulder. Zirconium oxide and aluminium oxide are the ceramic biomaterials widely used as bonding and bearing material for joint replacement. Polyacetal and polytetrafluorethylene are important biodegradable and bioresorbable materials used for bone plate applications. Polymeric biomaterials like polylactic acid, polydioxanone, and polyglycolic acid are widely used for the fabrication of screws, plates, and pins for bone fixation. Calcium salts such as phosphate and sulphates are another example of widely used biomaterials as bone substitutes to fill up bone defects.

6.2 Applications in Drug Delivery Systems

Today biomaterials are playing important roles in the medicines for the targeted and controlled drug delivery. Many drug moieties have very poor aqueous solubility or poor bioavailability. Rendering them unsuitable for a dosage form. In such cases, biomaterials like polymers and lipids are used to fabricate a drug delivery system that can circumvent these challenges. Examples of biomaterial applications in medicines include micro and nanoformulations for the controlled drug delivery in the context of releasing drug at a desired rate at their intended site of action. Liposomes, nanoparticles, drug-coated vascular stents, and wafer implants are important examples of biomaterial application in drug delivery.

6.3 Diagnosis

Diagnosis of a disease plays a crucial role in the final outcome of disease treatment. Biomaterials are also used for the diagnosis or biosensing, i.e., to detect and report the presence of specific biomolecules in the body. Biomaterials can be used to recognize and label a target biomolecule. Carbon nanomaterials are widely used as biosensor materials due to their good biocompatibility and optical and electronic properties. Transducers as sensors for brain activity and blood glucose monitoring devices are two important examples of biomaterial applications as a biosensor.

6.4 Wound Healing

Dressings, clips, sutures, bandages, and staples for wound closure are a commonly used application of biomaterials. Polyester copolymers, nylon, silk, bovine tissues, and teflon are examples of biomaterials used for wound healing.

6.5 Tissue Regeneration

Initially, biomaterials were conceived to repair or substitute an impaired biological function or tissue. Over recent years, however, biomaterial science has made great leaps, and it has moved from tissue engineering to tissue regeneration. Polymers and ceramics are widely used as scaffold material for tissue regeneration applications. For example, tri-calcium phosphate and hydroxyapatite are used as scaffold material for bone regeneration owing to their excellent biomechanical and biochemical compatibility with bone tissues. Polymers and copolymers of PGA, PLA, PCL, and PEG are studied as scaffold material for soft tissue regeneration. Recently, biomaterials are being used in tissue regeneration as biomaterials functionalized with extracellular vesicles. Biomaterials are being used as scaffold material seeded with living cells to regenerate or restore a missing biological tissue. Biomaterials are being used and studied as a platform to carry and deliver the mesenchymal stem cell exosomes to promote wound healing and skin regeneration.

7 Conclusion

Biomaterials are an incredibly powerful tool that is changing our lives daily. Biomaterials have expanded their applications especially over the last few decades, from tissue engineering to diagnostics, medical equipment to the therapeutic medications, and emerging regenerative technologies. The wide meaning has changed over time based on research and usage, and it may continue to expand with exciting advancements in the correspondingly emerging branches of biotechnology and medical science. Biomaterials are amongst the most multidisciplinary fields of all the sciences touching almost every sphere of the biomedical science. This chapter presents a broad picture of biomaterials and their applications that will be further elaborated throughout this book. This chapter guides the readers on how to proceed with the further chapters in this book and comprehend this intricate field of science.

Quiz/Multiple Answer Questions

1. **Which of the Following Is Not a Biomaterial:**

    (a) Dental Implants (b) Antibiotics (c) Prosthetic Limbs (d) Heart Pacemaker

    Answer: (b) antibiotics.

    Explanation: Antibiotics are drugs, which do not belong to the category of biomaterials.

2. **Biomaterials can be:**

    (a) Natural (b) Synthetic material (c) Both a and b (d) None of the above.

    Answer: (c) Both a and b.

3. **What is the most important characteristic of a biomaterial:**

    (a) Biocompatibility (b) Biological functionality (c) Chemical stability (d) All.

    Answer: (d) All.

4. **Which of the following is not the application of biomaterials:**

    (a) Drug delivery (b) Tissue replacement (c) Nutrition (d) Orthodontics.

    Answer: (c) Nutrition.

5. **Among the following which is not a natural biomaterial:**

    (a) Polylactic acid (b) Collagen (c) Silk (d) Gelatin.

    Answer: (a) Polylactic acid.

    Explanation: Collagen, silk and gelatine and natural biomaterials are obtained from the biological source. Perhaps, the polylactic acid is a synthetic polymer which is obtained from the polymerization of lactic acid monomer (D,L-Lactide).

6. **Among the following which biomaterial belongs to the class of synthetic non-degradable polymers:**

    (a) poly(methyl methacrylate) (b) polyurethanes (c) both a and b (d) poly(lactide-*co*-glycolide)

    Answer: (c) both a and b.

    Explanation: poly(lactide-*co*-glycolide) consisting ester linkage in the polymer chain, which can undergo hydrolysis in aqueous environment. While the poly(methyl methacrylate) and polyurethanes are non-biodegradable and do not have any ester linkage to undergo hydrolysis.

Introduction to Biomaterials

7. **Among the following statements which one is incorrect for polyanhydrides biomaterials:**

   (a) polyanhydrides are synthetic polymeric biomaterials.
   (b) polyanhydrides belong to the class of natural synthetic polymers.
   (c) polyanhydrides are biodegradable.
   (d) Both A and C.

   Answer: (d) Both A and C.

   Explanation: Polyanhydrides are synthetic polymer biomaterials, which are biodegradable.

8. Biomaterial are not used as.

   (a) Drugs (b) Drug delivery carrier (c) Medical implants (d) Diagnostics.

   Answer: (a) Drugs.

9. Among the following which is not belong to the class of metallic biomaterials:

   Stainless steel (b) Co-Cr alloys (c) Hydroxyapatite (d) Ti and its alloys.

   Answer: (c) Hydroxyapatite.

   Explanation: Hydroxyapatite is an example for ceramic biomaterials. While the stainless steel, Co-Cr alloys and Ti and its alloys belong to the class of metallic biomaterials.

10. Which of the following is a natural biomaterial?

    A) Silicone B) Stainless steel C) Collagen D) Polyethylene.

    Answer: C) Collagen.

    Explanation: Collagen is a naturally occurring protein that is found in connective tissue and is commonly used as a biomaterial in various medical applications.

11. **Which of the following is not a biomaterial?**

    a) Titanium b) Silicone c) Glass d) Polyester.

    Answer: c) Glass.

    Explanation: Glass is not considered a biomaterial because it is not naturally occurring in the body and does not interact with biological systems in the same way as other biomaterials.

12. **What is the primary purpose of biomaterials?**

    (a) To replace damaged or diseased tissue.
    (b) To stimulate tissue regeneration.
    (c) To enhance tissue function.
    (d) All of the above.

Answer: d) All of the above.

Explanation: Biomaterials can be used for a variety of purposes, including replacing damaged or diseased tissue, stimulating tissue regeneration, and enhancing tissue function.

13. **Which of the following is a disadvantage of using metallic biomaterials?**
    (a) Poor biocompatibility.
    (b) Limited mechanical properties.
    (c) High cost.
    (d) Susceptibility to corrosion.

Answer: d) Susceptibility to corrosion.

Explanation: Metallic biomaterials can corrode over time, leading to device failure or release of toxic ions into the surrounding tissue.

14. **What is the primary advantage of using biodegradable biomaterials?**
    (a) They are more biocompatible than non-biodegradable materials.
    (b) They reduce the risk of long-term complications.
    (c) They eliminate the need for removal surgery.
    (d) They are more cost-effective than non-biodegradable materials.

Answer: (b) They reduce the risk of long-term complications.

Explanation: Biodegradable biomaterials can be broken down and absorbed by the body over time, reducing the risk of long-term complications or the need for removal surgery.

References

1. Stanciu L, Diaz-Amaya S (2022) Introductory biomaterials an overview of key concepts
2. Williams DF (1999) The Williams Dictionary of Biomaterials. Liverpool University Press. https://doi.org/10.5949/upo9781846314438
3. Williams DF (2009) On the nature of biomaterials. Biomaterials 30:5897–5909
4. Ratner BD et al (2020) Introduction to biomaterials science. Biomaterials science. Elsevier. https://doi.org/10.1016/b978-0-12-816137-1.00001-5
5. Moizhess TG (2008) Carcinogenesis induced by foreign bodies. Biochemist 73:763–775
6. Joyce K, Fabra GT, Bozkurt Y, Pandit A (2021) Bioactive potential of natural biomaterials: identification, retention and assessment of biological properties. Signal Transduct Target Ther 61(6):1–28
7. Langer R, Tirrell DA (2004) Designing materials for biology and medicine. Nat 428:487–492
8. Domb AJ, Kost J, Wiseman D (1998) Handbook of biodegradable polymers. CRC press
9. Todros S, Todesco M, Bagno A (2021) Biomaterials and their biomedical applications: from replacement to regeneration. PRO 9

Biodegradable Polymers

Mudigunda V. Sushma, Aditya kadam, Dhiraj Kumar, and Isha Mutreja

Abstract Biodegradable polymers are a rapidly growing field driven by increasing concerns about plastic waste and its environmental impact. Polymers prepared from inexpensive and renewable raw materials might be the perfect alternative to plastics, and the properties like biodegradability and biocompatibility make them suitable for various biomedical applications. The biodegradability of the polymers is controllable by altering the monomer concentration and adding hydrolytically degradable groups in the polymeric backbone. These biopolymers can provide a safe and effective way of preparing devices/implantable materials for various biomedical applications. This chapter discusses biodegradable polymers, their synthesis, biodegradability, biocompatibility, along with their advantages and disadvantages for various biomedical applications, including drug delivery and tissue engineering.

M. V. Sushma · A. kadam · I. Mutreja (✉)
MDRCBB, Minnesota Dental Research Center for Biomaterials and Biomechanics, University of Minnesota, Minneapolis, MN, USA
e-mail: imutreja@umn.edu

D. Kumar
Division of Pediatrics Dentistry, School of Dentistry, University of Minnesota, Minneapolis, MN, USA

Graphical Abstract

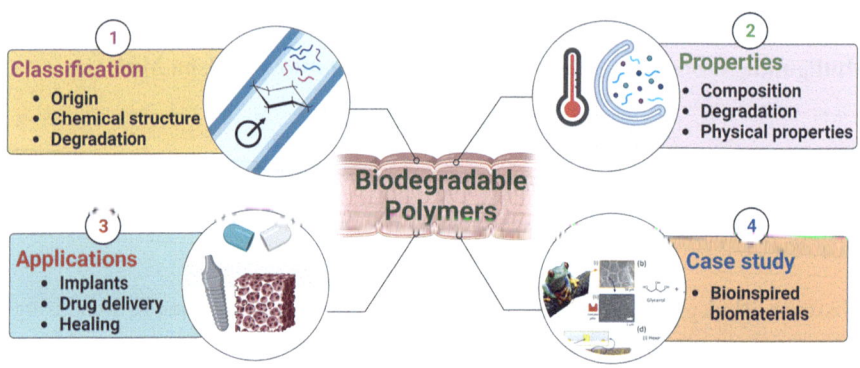

Keywords Classification · Natural versus synthetic polymers · Degradation mechanisms · Biomedical applications

1 Overview

Polymer usage has soared dramatically over other materials due to various potential applications and the ease with which novel compositions with radically different properties can be designed. Modern alchemists have disassembled and repurposed hydrocarbons to create hundreds of compounds in the plastics family [1]. Synthetic chemicals, particularly petroleum-based products, are non-biodegradable and pose significant ecological risks, resulting in severe environmental contamination from waste buildup caused by manufacturing and incineration. For a period, the plastics industry seemed to be a boon and more beneficial to society, but its over usage resulted in devastating consequences on the natural environment and created a massive imbalance in the ecosystem. It is essential to explore alternatives like synthetic polymers to create a superior path in the form of natural polymers.

Recent decades have seen a rise in the need for environment-friendly products that encourage the development of biodegradable properties in past times. Biodegradable formulations are products or materials made from natural ingredients that readily degrade into non-toxic compounds when exposed to water, air, and microbes. These exposures will result in the breakdown of these materials into smaller components. They are typically made from renewable resources and are an alternative to conventional, non-degradable plastics. Biodegradable formulations can be incorporated into biodegradable plastics, agricultural products, and personal care products [2]. Biopolymers constructed from lipids, polysaccharides, DNA, and proteins are low cost and can be utilized from renewable raw materials. They also have a promising alternative to non-biodegradable plastic petroleum products [3].

Biopolymers market sales growth is very modest yet expanding. Due to their excellent biocompatibility and biodegradability properties, many of these materials are preferred over synthetic polymers in the medical, agricultural, engineering, and textile industries. Polysaccharides, like other biopolymers obtained from natural sources, have a wide range of possible applications due to their lack of toxicity and biodegradability. Among polysaccharides, cellulose is the most common polymer. Bio polysaccharides may be derived from a variety of sources, including plants (such as starch and pectin), animals (such as chitin/chitosan), and even microorganisms (e.g., bacterial cellulose). Hence, using microorganisms to manufacture biobased polymers has become more common. Silk is a biodegradable and biocompatible natural protein-based fiber, rendering it suitable as a biodegradable polymer. Silk fibers are intriguing for use in various applications, such as nanomedicine, and drug delivery systems, due to their distinctive features, which include high strength, flexibility, and biodegradability. Silk was developed as a suture material, which fueled the development of bio-based polymers [4]. DNA, the genetic material that carries the instructions for the development and function of all living organisms, has been explored as a potential biomaterial for various applications. One of the primary reasons DNA is being studied as a biomaterial is its ability to self-assemble into complex structures. Researchers have created various three-dimensional structures by manipulating the base-pairing interactions between DNA molecules, including nanostructures and scaffolds. These DNA-based structures have potential applications in drug delivery, tissue engineering, and biosensors. For example, DNA scaffolds can be used to support the growth of cells, while DNA nanoparticles can be designed to deliver drugs to specific target cells in the body [5].

The following are key characteristics of biodegradable polymers [6].

1.1 Composition

Biodegradable polymers are made from renewable resources such as corn, sugarcane, and potato. The most common biodegradable polymers include polylactic acid (PLA), polyhydroxyalkanoates (PHA), and starch-based polymers.

1.2 Degradability

The degradation rate can vary depending on the polymer composition and the environment in which it is exposed. Usually, the biodegradable property is calculated by the time required by the natural polymer to degrade completely into a relatively smaller compound. Microorganisms degrade natural polymers like cellulose, starch, and chitin rapidly. Whereas chemically derived biodegradable polymers like PLA and PHAs need specific environmental conditions, such as temperature, humidity, and oxygen, to affect polymer degradation. pH, moisture, and high-oxygen conditions may enhance biodegradable polymer degradation. The size of the polymer chain also plays a role in its degradability. Microorganisms can more easily break down smaller polymer chains.

1.3 Physical Properties

The physical properties of biodegradable polymers may differ depending on the source and the processing conditions. The glass transition temperature (Tg) of the polymer affects its mechanical properties and the stability of the payload (drug or other biomolecules) in the delivery system. Polymers with a high Tg tend to be brittle and have low flexibility, whereas polymers with a low Tg are more flexible and have a higher drug-loading capacity. The degree of crystallinity of the polymer affects its mechanical properties and the drug release rate. Highly crystalline polymers tend to have slower drug release rates compared to amorphous or semi-crystalline polymers. The surface area of the polymer affects its degradation rate and drug release rate. Polymers with a higher surface area tend to have faster degradation and drug release rates.

1.4 Applications

Biodegradable polymers are used in various applications. So, the design and properties of biopolymers should be varied based on their application and usage. They can be used as a substitute for conventional plastics in products such as bags, food containers, and disposable cutlery. Depending on the application, the properties of the biopolymer like its elasticity, durability, reliability, and sheer and tear stress level should be optimized.

2 Synthesis of Biodegradable Polymers

These polymers can be synthesized using polymerization techniques such as ring-opening polymerization, ring-opening copolymerization, and step-growth polymerization. Polycondensation is a reaction between two monomers to form a polymer chain, releasing a small molecule such as water as a byproduct. An example of a biodegradable polymer synthesized by polycondensation is polylactic acid (PLA). Copolymerization is the process of polymerizing two or more different monomers to form a copolymer, which can have unique properties compared to individual polymers. An example of a biodegradable copolymer is poly(butylene succinate-co-adipate) (PBSA). In addition, blending is a simple method of producing biodegradable polymers. It involves mixing two or more polymers to build a new material with improved properties. In the grafting technique, a biodegradable polymer is grafted onto another polymer to improve its properties. Other crosslinking methods are used to chemically bond the polymer chains to produce a three-dimensional network. This enhances the material's strength and stability. Different chemical and physical methods are used for modifying polymeric materials to improve their properties and performance. For example, adding hydrolytically degradable groups, crosslinking, and modifying the surface can enhance the biodegradability of the polymer[7].

Biodegradable polymers can be classified based on their origin, chemical structure, and degradation mechanisms.

2.1 Based on Origin

(a) Natural biodegradable polymers: These occur naturally in plants and animals. Examples include cellulose, chitin, and collagen.
(b) Synthetic biodegradable polymers: These are synthesized in the laboratory using petrochemicals or renewable resources. Examples include polylactic acid (PLA), polyglycolic acid (PGA), and polyhydroxyalkanoates (PHAs).

2.2 Based on Chemical Structure

(a) Aliphatic polyesters: These are polymers with ester bonds in their backbone structure. Examples include PLA, PGA, and poly(ε-caprolactone) (PCL).
(b) Polyurethanes: These are polymers formed by the reaction of isocyanates with polyols.
(c) Polyamides: These are polymers that have amide bonds in their backbone structure. Examples include nylon-2 and nylon-6.

2.3 Based on Degradation Mechanisms

(a) Hydrolytic degradation: These are polymers that degrade in the presence of water. Examples include PLA, PGA, and PCL.
(b) Enzymatic degradation: These are polymers that are degraded by enzymes. Examples include PHAs and cellulose.
(c) Photodegradable polymers: These are polymers that are degraded by exposure to light. Examples include poly(ε-caprolactone-co-ethylene carbonate) (PCL-co-EC) (Fig. 1).

Fig. 1 Broad classification of biodegradable polymer

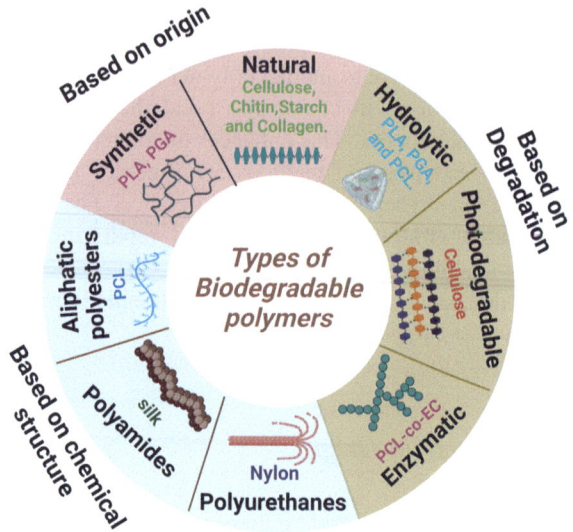

3 Biodegradation Mechanism

Biodegradable polymers undergo degradation by different mechanisms. These biodegradable polymers may degrade by bulk or surface pathways, which are influenced by temperature, humidity, pH, and microbes.

3.1 Bulk Degradation

Bulk degradation of biodegradable polymers refers to the chemical or enzymatic processes that occur over the whole volume of the substance. Biodegradable polymers may degrade by hydrolysis, oxidation, or enzymatic degradation, resulting in the fragmentation of polymer chains into oligomers and monomers.

Hydrolysis is the most prevalent method for the bulk destruction of biodegradable polymers. The ester bonds in polymer chains are broken by water molecules, creating hydroxyl groups and carboxylic acids. This process may occur under various circumstances, including high temperature, acidic or basic pH, and in the presence of enzymes.

Oxidative degradation of biodegradable polymers entails the creation of carbonyl or carboxyl groups by the interaction of polymer chains with oxygen. This process may begin by exposure to ultraviolet light or ambient oxygen and can lead to the loss of mechanical characteristics and deterioration of the material.

Enzymatic degradation: Enzymes, such as lipases or proteases, accelerate the breakdown of biodegradable polymer chains, resulting in their enzymatic destruction. This process may occur in natural conditions, such as soil or water, where microorganisms can break down biodegradable polymers into smaller pieces that can then be digested.

Bulk degradation of biodegradable polymers is essential for controlling the material degradation rate and biocompatibility. By understanding the bulk degradation mechanism, researchers can design biodegradable polymers with tailored properties and degradation rates, making them suitable for a wide range of applications, such as drug delivery, tissue engineering, and environmental remediation.

3.2 Surface Degradation

Surface degradation of biodegradable polymers refers to the breakdown process primarily occurring on the material surface due to environmental exposure or physical pressures. Biodegradable polymers degrade by various surface degradation modes, including corrosion, erosion, abrasion, and microbial activity. When biodegradable polymers are exposed to an environment that induces chemical deterioration of the polymer chains, corrosion occurs. This may cause surface pits, fissures, or holes to emerge, leading to mechanical failure of the material. Exposure to acids, bases, or other reactive compounds may cause biodegradable polymers to corrode.

Erosion of biodegradable polymers happens due to mechanical forces acting on the material, such as fluid flow or impingement, causing the substance to be removed from the surface. Surface flaws or roughness may emerge as a consequence of erosion, affecting the mechanical qualities and performance of the material.

Abrasion of biodegradable polymers happens due to mechanical wear and tear, such as friction or rubbing, which may result in the loss of the surface material. Abrasion may cause surface scratches or grooves to emerge, altering the material's aesthetic and functional capabilities.

Microbial degradation of biodegradable polymers happens when microorganisms invade the material's surface and feed on the polymer chains. This may lead to surface flaws or changes in the material's surface chemistry and morphology.

When developing biomedical, environmental, or industrial materials, surface degradation of biodegradable polymers is also significant to address. Understanding the surface degradation processes allows researchers to design techniques to improve the material's resistance to deterioration and prolong its lifetime, making it more appropriate for various applications[8] (Fig. 2).

3.3 Characterization Techniques for Biodegradation

Many approaches, including chemical, physical, and biological methods, can be used to analyze the degradation and characterization of biodegradable polymers. The choice of technique is determined based on the type of polymer,

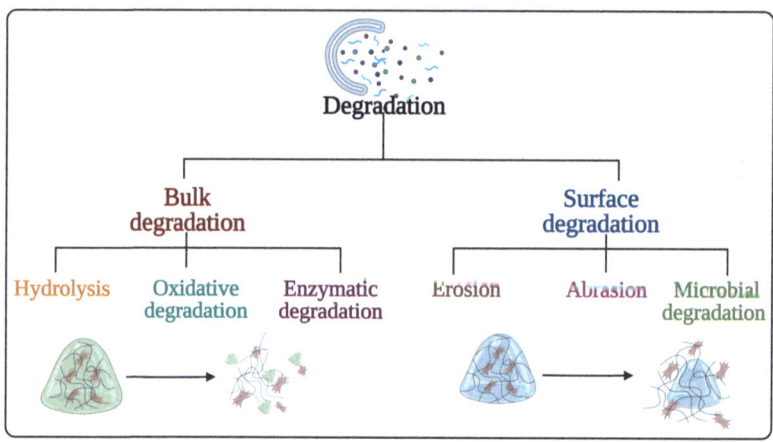

Fig. 2 Mechanism for degradation of biodegradable polymers

degradation process, and environmental factors. Spectroscopic techniques, such as Fourier transform infrared spectroscopy (FTIR) or nuclear magnetic resonance (NMR) spectroscopy, which can detect changes in the functional groups and chemical composition of the polymer chains, are chemical methods for monitoring the degradation of biodegradable polymers. Chromatography methods, such as gel permeation chromatography (GPC), may also assess time-dependent changes in the molecular weight and polydispersity of polymer chains. Mechanical testing, such as tensile testing or dynamic mechanical analysis (DMA), may identify changes in the mechanical characteristics of the material, such as tensile strength or elasticity, to monitor the degradation of biodegradable polymers. Thermal analysis methods, such as differential scanning calorimetry (DSC) or thermogravimetric analysis (TGA), may also be used to assess variations in the thermal characteristics of a material, such as its glass transition temperature or thermal stability. Monitoring the breakdown of biodegradable polymers biologically involves measuring the activity of microorganisms or enzymes in degrading the substance. For instance, respirometry may be used to assess the rate of microbial oxygen consumption as a metric of the polymer's biodegradation rate. The enzymatic breakdown of polymer chains may also be measured using lipase or protease assays. In addition to assessing the physicochemical qualities and performance of biodegradable polymers, characterization techniques may also be used to monitor their breakdown. Microscopy methods such as scanning electron microscopy (SEM) and atomic force microscopy (AFM) may be used to observe the surface morphology and topography of the material. Surface analysis methods, such as X-ray photoelectron spectroscopy (XPS) or contact angle measurements, may also be employed to analyze the material's surface chemistry and wettability [9].

4 Examples

4.1 Synthetic Polymers

PLGA (poly (lactic-co-glycolic acid) is a biodegradable and biocompatible polymer with extensive biomedical applications in tissue engineering and drug delivery. PLGA nanoparticles have been used to deliver various drugs, including small molecules, peptides, proteins, and nucleic acids. PLGA can encapsulate hydrophobic and hydrophilic drugs and be modified to release drugs in a controlled manner. The synthesized particles can be functionalized with imaging agents to visualize specific cells or tissues and monitor diseases. Additionally, in the scaffold configuration, it can be used to support tissue regeneration. They can be seeded with cells and implanted in the body to provide a temporary matrix for cell growth and tissue formation. PLGA nanofibers can be used to create tissue engineering scaffolds or wound dressings. They have a high surface area-to-volume ratio, allowing efficient drug delivery. The hydrolytic attack of the water molecules degrades the ester bond linkage in the PLGA polymer backbone.

Degradable bonds: The by-products of degradation are lactic acid and glycolic acid, which are biocompatible and quickly eliminated from the body through the renal system.

PLGA structure

Polycaprolactone (PCL) is a biodegradable and biocompatible polymer with diverse applications in medicine, agriculture, and the environment. It is a thermoplastic aliphatic polyester that is synthesized from caprolactone monomers. It has been used in the medical field for scaffold preparation, drug delivery systems, and sutures. This polymer is also highly used in agriculture as a biodegradable alternative to synthetic plastics for mulch films, plant pots, and twine. In the environmental field, PCL has been used as a biodegradable alternative for plastic packaging, disposable tableware, and shopping bag products. Due to its biodegradability, biocompatibility, and versatility, PCL has attracted increasing attention as a promising material for broad applications.

Degradable bonds: Hydrolytic degradation of PCL takes place by breaking the ester bond in the structure.

PCL structure

4.2 Natural Polymers

Cellulose is a natural, biodegradable polymer that has gained increasing interest in drug delivery and nanotechnology applications. Cellulose is abundant in nature and can be derived from various sources, including plants, bacteria, and algae. One of the most promising applications of cellulose in drug delivery is due to its unique physicochemical properties, including its high crystallinity and strong intermolecular forces that contribute to its strength and rigidity, while its hydrophilic nature and capacity to form hydrogen bonds with water molecules make it an efficient moisture absorber. In addition, the chemical stability and biodegradability of cellulose have led to its use in a wide range of biomedical applications. Cellulose-based drug carriers have several advantages, including high biocompatibility, low toxicity, and the ability to be easily modified with various functional groups to improve drug loading, stability, and release. Cellulose-based drug carriers can be synthesized in multiple forms, including nanofibers, nanoparticles, and hydrogels, and can be used to deliver a wide range of drugs, including small molecules, proteins, and nucleic acids. Moreover, cellulose-based nanomaterials have shown promising results in cancer therapy. Modified cellulose nanoparticles can target cancer cells, increasing drug concentration in the tumor tissue and reducing off-target effects.

Degradable bonds: Cellulose degradation is done by hydrolysis of β-1,4-linkages in cellulose.

Cellulose structure

Collagen is a fibrous and most abundant protein in the human body. It plays a vital role in maintaining the structural integrity of many tissues, including skin, tendons, cartilage, and bone. It has a unique triple-helical structure composed of three polypeptide chains. The primary sequence of the polypeptide chains determines the specific type of collagen, of which there are over 28 types identified so far. Hydrogen bonds and covalent crosslinks between adjacent chains stabilize the triple-helix structure. Collagen-based nanostructures have recently received much interest because of their potential uses in medication delivery. Collagen nanoparticles can be functionalized with targeting moieties such as antibodies or peptides to target particular cells or organs. It may be used as a scaffold to aid tissue regeneration in various applications, including skin, cartilage, bone, and blood vessels. Because of its gelling and emulsifying capabilities, collagen is a food ingredient. It's used to make sutures, wound dressings, and artificial skin replacements, among other things. It's also employed in orthopedic procedures, including bone transplants and joint replacements. Since collagen-based nanoparticles may be functionalized to attach to specific cell types or tissues, they can be employed for targeted medication delivery.

Degradable Bonds: Collagen can be degraded by heat, acids, and proteases, which disrupt the hydrogen bonds and crosslinks, leading to a loss of structure and function.

Collagen structurelpsum

Gelatin is a protein derived from the hydrolysis of collagen, which breaks down the protein into smaller peptides and amino acids. It has been used in tissue engineering due to its biocompatibility, biodegradability, and ability to form hydrogels. Hydrogels made from gelatin can mimic the extracellular matrix (ECM) of tissues and provide a supportive environment for cell growth and tissue regeneration. The physical and mechanical properties of gelatin hydrogels can be modified by changing the concentration of gelatin, the degree of crosslinking, and adding other components such as polymers or growth factors. Gelatin can be added to other polymers through blending, electrospinning, and crosslinking, among other methods. The properties of a polymer matrix can be enhanced by combining gelatin with polymers such as chitosan, polyethylene glycol, or polyvinyl alcohol. Additionally, gelatin has been chemically modified using different functional groups like methacryloyl or thiol, for example, to alter the material properties and provide additional means of crosslinking. In some cases, gelatin can also be used as a coating material to improve the biocompatibility of implantable devices, such as artificial joints or stents. Using gelatin coatings can reduce the risk of inflammation and rejection by the body's immune system. Due to its biocompatibility, low toxicity, and low cost, gelatin is a commonly used biodegradable polymer in the formulation of microparticles. It is advantageous for sustained drug delivery that the gelatin microparticles can be loaded with drugs and designed to deliver them over a specific time period. In addition, gelatin microparticles can be tailored to target specific sites within the body, thereby enhancing drug efficacy and minimizing adverse effects. Therefore, gelatin microparticles are a suitable option for applications involving drug delivery. Gelatin-based hydrogels can be used to deliver drugs topically to the skin. The hydrogel provides a moist environment that can enhance drug absorption, and the hydrogel's gel texture can help improve patient comfort.

Degradable bonds: Enzymatic degradation is the most common method for degrading gelatin, as it is a protein that can be broken down by enzymes such as proteases. During the process of enzymatic degradation, proteases break down the peptide bonds that hold the amino acid residues in gelatin together. This results in the cleavage of the protein into smaller peptides and eventually into individual amino acids.

Gelatin structure

5 Applications

Biodegradable polymers are a rapidly growing field driven by increasing concerns over non-biodegradable materials' adverse reactions, such as inflammation and tissue rejection, and they may require surgical removal if they fail or cause complications. In recent years, significant advances have been made in developing new biodegradable polymers and optimizing existing ones. The followings are some of the current research areas and future trends in the field of biodegradable polymer [10]. Researchers are actively developing new biodegradable polymers with improved properties such as increased strength, durability, and biodegradability. This is accomplished using new polymerization methods and techniques and incorporating biodegradable additives [2].

5.1 Medical Devices

Biodegradable polymers are an attractive option for medical device applications because they can be designed to degrade over time, eliminating the need for device removal and reducing the risk of long-term complications. Some examples of biodegradable polymers used in medical devices include:

1. Poly(lactic acid): PLA is derived from renewable resources such as corn starch or sugarcane. It has been used to make sutures, screws, pins, and plates that can be absorbed by the body over time.
2. Poly(glycolic acid): PGA is commonly used in medical devices such as sutures and tissue engineering scaffolds. It degrades rapidly in the body, releasing any drugs or growth factors that have been incorporated into the device.
3. Polycaprolactone: PCL has been used in surgical meshes due to its unique combination of mechanical properties and biodegradability. They are commonly used in hernia repair surgeries or designing scaffolds for bone regeneration, providing mechanical support to the damaged tissue and promoting new tissue growth.
4. Polyhydroxyalkanoates (PHAs): PHAs are a family of biodegradable polymers bacteria produce. They have been explored for use in orthopedic implants. PHA-based materials have been shown to support the growth and development of bone cells, making them suitable for bone regeneration applications. PHAs can also be processed into porous scaffolds that can be used to fill bone defects and promote the regeneration of new bone tissue.
5. Polydioxanone: PDO is used in surgical sutures, particularly for closing wounds requiring long-term support. PDO sutures maintain their strength for up to 180 days before breaking down in the body.

5.2 Drug Delivery

Biodegradable polymers can develop drug delivery systems, such as implants or microspheres, that can release drugs over a controlled period. This can improve the efficacy of the drugs and reduce side effects. These materials can be designed to dissolve over time in response to physiological conditions, releasing drugs in a controlled and sustained manner. This can be particularly useful in cases where it is desirable to avoid the accumulation of drugs in the body or where the drugs are required to be delivered directly to the site of an injury or disease. Biodegradable polymers formulate injectable drug delivery systems such as microparticles, nanoparticles, and implantable devices. These systems can provide controlled drug release over days, weeks, or even months, reducing dosing frequency and improving patient compliance. Biodegradable polymers can be functionalized with specific molecular groups that target specific cells or tissues, increasing the local concentration of drugs and reducing systemic side effects. Some examples of biodegradable polymers used in drug delivery include polylactide (PLA), polyglycolide (PGA), and copolymers of PLA and PGA (PLGA). These materials have been extensively studied and are well-established in drug delivery, with numerous FDA-approved products available on the market.

5.3 Tissue Engineering

Biodegradable polymers are used as scaffolds for tissue engineering, providing a supportive structure for tissue repair and regeneration. As the tissue grows, the polymer gradually degrades and is eventually replaced by the regenerated tissue. These polymers serve as scaffolds for tissue growth and repair and provide a supportive structure for cells to attach, grow, and differentiate. The properties of tissue engineering polymers, such as biocompatibility, mechanical strength, and degradation rate, can be tailored to suit the specific needs of different tissue sites. Some commonly used tissue engineering polymers include polylactic acid (PLA), polyglycolic acid (PGA), polycaprolactone (PCL), polyethylene glycol (PEG), polyvinyl alcohol (PVA), and alginate. These polymers can be processed into different forms, such as fibers, films, or porous scaffolds, to provide additional physical, mechanical, and chemical environments for cell differentiation and tissue regeneration (Fig. 3).

5.4 Wound Healing

Biodegradable polymers can be used as wound dressings to promote healing and reduce the risk of infection. The polymers can absorb excess fluids, provide a moist environment for the wound, and gradually degrade over time. There are several

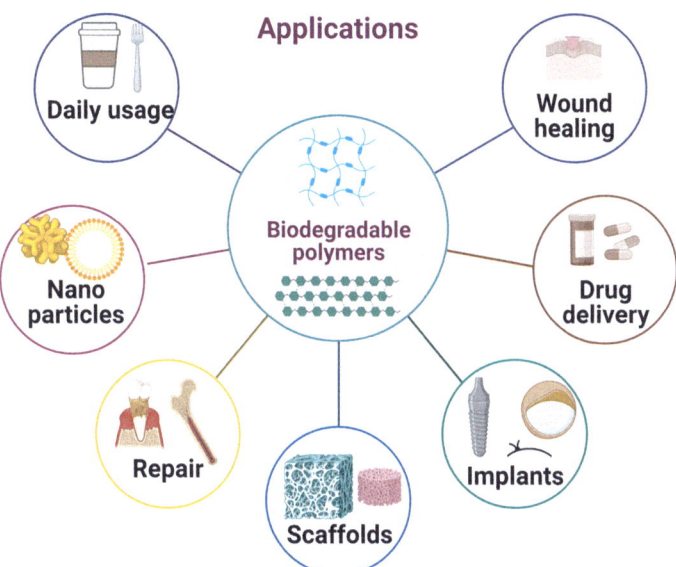

Fig. 3 Different kinds of biopolymer applications

types of biodegradable polymers used in wound healing, including polylactic acid (PLA), polyglycolic acid (PGA), and copolymers (PLGA). These polymers can be formed into various sizes and shapes based on wound severity and can be used as wound dressings and bio-bandages. In addition, these biomaterials promote external wound healing and show internal wound healing properties. The biodegradable nature of these polymers allows them to degrade gradually as the wound heals without leaving any foreign materials in the body for the long term. This reduces the risk of infection and other adverse effects and ensures the wound-healing process is not impeded. Some biodegradable polymers can release growth factors, antimicrobial agents, and other therapeutic agents to promote healing and prevent infections. Overall, biodegradable polymers have shown great potential for wound healing and have been used in various clinical applications with promising results. For example, internal intestinal and stomach wound can be healed using gelatin and other biobased polymers, these biomaterial helps in significant wound healing and avoid the risk of surgery and invasive stitches in the wounded regions.

5.5 Implant Materials

Biodegradable polymers have gained increasing attention in developing medical devices such as heart valves, stents, and nerve guides.

Heart Valves: Biodegradable polymers can be used to develop heart valves to replace damaged or diseased valves. This can eliminate the need for lifelong anticoagulation therapy, which is required for patients with mechanical heart valves. Polyglycolic acid (PGA), poly(glycerol sebacate) (PGS), and polylactic acid (PLA) are commonly used biodegradable polymers in heart valve applications.

Stents: Biodegradable polymers have been used to manufacture stents that slowly degrade over time, reducing the risk of complications such as restenosis and thrombosis. These stents are typically made from polymers such as polylactic acid (PLA), polyglycolic acid (PGA), and their copolymers. These materials have been shown to degrade within 6–12 months, leaving behind a more natural vessel structure.

Nerve guides are used to help damaged nerves regenerate. Biodegradable polymers are used as a scaffold-like structure that supports nerve growth and eventually degrades as the nerve heals. The polymer can be designed to release growth factors or other molecules that promote nerve growth and regeneration. Small tubes bridge gaps in damaged nerves and help regenerate nerve tissue. PGA tubes and poly(caprolactone) (PCL) are commonly used biodegradable polymers in nerve guide applications.

5.6 Biopolymeric DNA Vaccines

These are a type of vaccines that use genetic material to stimulate an immune response against a disease. They typically consist of a small piece of the DNA of the pathogen, such as a virus or bacterium, that causes the disease. The DNA is delivered into the body, usually via injection, and taken up by cells, where it is expressed and processed into viral or bacterial antigens. This stimulates an immune response, including the production of antibodies, that can recognize and neutralize the pathogen if it is encountered in the future. Biopolymeric DNA vaccines have several advantages over traditional vaccines, including the ability to be rapidly manufactured, their stability, and the fact that they do not carry the risk of infecting the recipient with the live pathogen. They have the potential to provide long-lasting protection against diseases. However, some challenges are associated with biopolymeric DNA vaccines, including their relatively low immunogenicity compared to traditional vaccines and the need for repeated dosing or adjuvants to enhance their effectiveness. Overall, biopolymeric DNA vaccines are an exciting new area of research and have the potential to provide a safe and effective way of preventing and treating a range of diseases.

6 Case Study I

6.1 Bioinspired and Biodegradable Polymer-Based Adhesive Films [11]

Main goal: Development of bioadhesives to address the limitations of traditional sealing methods, reducing the need for invasive procedures.
Biodegradable polymer: Poly (glycerol sebacate) (PGS).
Background: Researchers in the medical field are becoming more interested in using bioinspired and biodegradable polymer-based adhesive films because of their impeccable adhesion performance and ability to break down in the body without the need for further surgical procedures. Poly (glycerol sebacate) (PGS) can produce adhesive films of tunable soft microarchitecture because it is bioresorbable and leaves no residue. One research group has manufactured bioinspired oil-coated sticky film using PGS that can maintain adhesion to a wet surface by imitating a frog's toe pad and mucus. Unlike commercial acrylic-based glue, Raman and FTIR spectroscopy showed no liver surface spectra change following the PGS-based film removal. This film is capable of being made with bioinspired oil. Simple models based on the degree of esterification and the interfacial energy difference were utilized to define the parameters under which PGS patterned to construct frog-like adhesive designs. Reproducing the frog-like hexagonal microchannel and concave cup structures covered with aggressive glycerol oil led to a durable residue-free wet adhesion against a variety of non-flat soft organ surfaces, which was achieved by this oil-coated film's wet adhesion (Fig. 4).

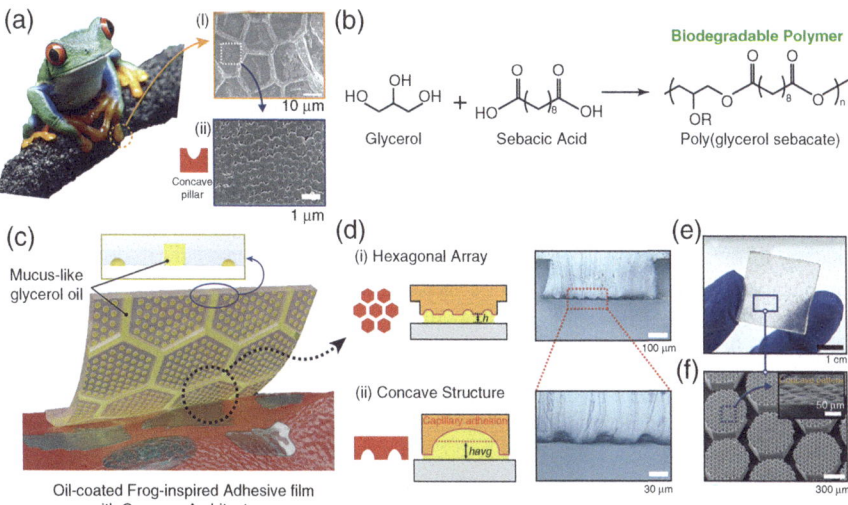

Fig. 4 Schematic illustration. (**d–f**), hierarchical structure of (i) hexagonal array and (ii) hemispherical concave cup enhanced the adhesion and friction forces by increasing the capillary interaction

7 Case Study II

7.1 Bacteria-Responsive Biopolymer-Coated Nanoparticles for Biofilm Penetration and Eradication [12]

Main goal: Development of multi-stimuli-responsive NPs to combat bacterial biofilms.
Biodegradable polymer: Gelatin, chitosan (CS), polyanion, hyaluronic acid (HA).
Background: Biofilm-associated bacterial infections are a major source of morbidity and death among patients globally. Biofilms are complex, three-dimensional bacterial communities encased in a matrix of extracellular polymeric substances (EPS) comprising proteins, polysaccharides, extracellular DNA, and lipids. These bacterial colonies are often seen on surfaces, including necrotic tissue and implants. Biofilms are associated with various illnesses, such as dental caries, urinary tract infections, burn wound infections, and diabetic foot ulcers. To tackle bacterial biofilms, the authors have developed a gelatin NP (GNP) drug delivery system that responds to the acidic environment of the biofilm and the presence of gelatinases and hyaluronidases. Here, the authors used layer-by-layer (LbL) self-assembly to attach a bilayer of the polycation chitosan (CS) and the polyanion hyaluronic acid (HA) to the surface of antibiotic-loaded GNPs. Each layer served a mechanical purpose. GNPs were loaded with the FDA-approved tetracycline antibiotic doxycycline (Doxy) since it is widely employed to treat V. vulnificus infections.

Indeed, they found that HA-CS-Doxy-GNPs displayed pH- and enzyme-responsive drug release characteristics and viability staining indicated severe membrane damage when HA-CS-Doxy-GNPs were applied to premade biofilms, suggesting that these nanoparticles had a high capacity to penetrate and eradicate V. vulnificus biofilms compared with free Doxy. In an ex vivo pig V. vulnificus infection model, these HA-CS-Doxy-GNPs similarly decreased bacterial load, lending credence to their potential for translation. Furthermore, fibroblasts, endothelial cells, and RBCs were all generally biocompatible with these NPs. Since these enzyme triggers and decreased pH are characteristics of many bacterial species, the multi-stimuli-responsive NPs platform may also demonstrate comparable antibiofilm effects against various biofilm-producing bacteria, including additional Gram-negative and Gram-positive bacteria. This adaptable drug delivery system could deliver a combination of drugs, such as those explicitly targeted at biofilms (e.g., antibiofilm peptides) or signaling molecules for infection detection, for efficient, all-encompassing infection treatment and detection (e.g., fluorescent dyes).

8 Future Perspective

Without a doubt, using biodegradable polymers in biomedicine has enormous potential. Drug delivery gene transfection, protein transport, bioimaging and diagnostics, tissue engineering, and biomedical devices are some of the current biomedical applications that use biodegradable polymers. There are a significant number of other biomedical applications as well. In addition to exhibiting their unique properties, which are application specific, they can be easily incorporated with additional required properties. In addition, BPs have several unique benefits, including degradability, compatibility, nontoxicity, and intelligent responsiveness to a range of physiological stimuli, demonstrating significant potential in various applications within biomedicine and agriculture. BPs have a bright and fruitful future ahead of them. Even though much ground has been covered, many challenges remain. Existing BPs still have suboptimal mechanical characteristics, forcing the progress of more complex polymers. Second, despite the reality that many systems made of non-degradable polymers and metals have been used in a wide range of applications, the biomedical use of BPs has not been fully realized due to their high cost of initial production and instability in vivo. This is in contrast to the widespread use of systems made of non-biodegradable polymers and metals. In recent years, many promising and multi-functional biopolymers have been reported, and they show more significant properties than traditional implants and methods. This shows that more research and new polymerization methods must be explored to develop pathbreaking novel biopolymer compounds.

Questions

1. Which option is correct for blending in the context of biodegradable polymers?
    (a) Grafting a biodegradable polymer on another polymer
    (b) Mixing two or more polymers to build a new material with improved properties
    (c) Chemically bonding polymer chains together to create a three-dimensional network
    (d) Polymerizing two or more different monomers to form a copolymer

2. Which of the following is an example biodegradable polymer?
    (a) Polylactic acid (PLA)
    (b) High molecular weight polyethylene
    (c) Polypropylene
    (d) Polyvinylchloride

3. Which monomers are used to synthesize PLGA?

    (a) Poly-L-lysine and glycolic acid
    (b) Lactose and glycine
    (c) Lactic acid and glycolic acid
    (d) Lactic acid and glycine

4. Which of the following methods increases the material's strength and stability?

    (a) Ring-opening polymerization
    (b) Introduction of hydrolytically degradable groups
    (c) Grafting
    (d) Crosslinking

5. Which of the following is the protein-based polymer?

    (a) Polybutylene terephthalate
    (b) Silk
    (c) Polyethylene terephthalate
    (d) Polylactic acid

6. Which polymer is indigestible by humans?

    (a) Gelatin
    (b) Starch
    (c) Cellulose
    (d) Chitosan

7. Implantable medical devices prepared with biodegradable and biocompatible polymers have the following properties except.

    (a) It has minimal risk due to biodegradation and biocompatibility
    (b) No need to remove the implant by surgery
    (c) Produces harmful byproducts
    (d) Does not produce immunogenic responses

8. Why biopolymers are used in drug delivery?

    (a) It releases drugs all at once
    (b) It releases drugs in a controlled manner
    (c) It can release drugs at the targeted site
    (d) Both b and c

9. What are biodegradable polymers?

    (a) The polymers which do not degrade at all
    (b) The polymers do not degrade by enzymes and bacteria
    (c) The polymers degrade by enzymes, and bacteria and produce harmless byproducts
    (d) The polymers prepared from bacteria that do not degrade

10. What is the role of the scaffold in tissue engineering?
    (a) To create non-degradable artificial tissue
    (b) To support the cell attachment, growth, and differentiation
    (c) To prevent body from absorbing tissue
    (d) To make new tissue

Explanations

1. Poly lactic acid is a biodegradable polymer that is derived from resources like sugarcane, and corn starch. it has properties like high strength, stiffness, and good thermal properties.
2. In the context to biodegradable polymers, the term "blending" describes the process of combining two or more distinct polymers to create a composite material with improved properties. Strength, stiffness, thermal conductivity, and biodegradability could all be improved by this process.
3. Lactic and glycolic acid are used as the monomers in a ring-opening polymerization (ROP) reaction to synthesize PLGA. A catalyst, like stannous octoate, and a co-initiator, like benzyl alcohol, are used in the reaction to start the polymerization process. The ratio of lactic acid to glycolic acid controls the properties of PLGA, increasing glycolic acid proportion increases degradation rate, biodegradability, and hydrophilicity.
4. Crosslinking is the process of chemically linking or more polymer chains together to create a three-dimensional network structure. This process can increase the stability and strength of a material by preventing the sliding of chains and increasing the intermolecular forces between the chains.
5. Among the given options, two polymers, polyethylene terephthalate (PET) and polybutylene terephthalate (PBT) are synthetic polymers made up of chemical reactions of synthetic monomers and they are non-biodegradable. Polylactic acid (PLA) is a biodegradable polymer made up of starch, and silk is only a protein-based natural polymer produced by silkworms, is it composed of two proteins sericin and fibroin.
6. Human digestive system can digest simple sugars like, glucose and fructose. The beta 1,4 glycosidic bond present in cellulose cannot be broken down by human digestive enzymes such as amylase.
7. Implantable medical devices prepared with biodegradable polymers would degrade in physiological conditions with time and form harmless byproducts, and there is no need to remove these implants by surgical procedure.
8. Biopolymers are biodegradable and biocompatible, which minimizes chances of toxicity and immune response to the delivery system, additionally, these polymers can be engineered to enhance the drug loading capacity, control release, and targeted delivery.

9. A scaffold is a three-dimensional structure used in tissue engineering those functions as a guide for the growth of new tissue. The scaffold gives cells a framework to stick to and arrange themselves on, directing the growth of new tissue and encouraging regeneration. it mimics the ECM of tissue which supports cell attachment, growth, and differentiation.
10. Biodegradable polymers are polymers that can be degraded naturally by enzymes, microorganisms, and other natural processes into harmless byproducts such as water and carbon dioxide. On the other hand, non-biodegradable polymers are a class of polymers that cannot degrade by these processes and can withstand this environment for many years. Some examples of biodegradable polymers are polylactic acids, polylactic acids, gelatin, and polyvinyl alcohol. In the given example (c) is the only option that defines a biodegradable polymer.

References

1. Bhovi VK, Melinmath SP, Gowda R (2022) Biodegradable polymers and their applications: a review. Mini-Rev Med Chem 22:2081–2101. https://doi.org/10.2174/1389557522666220128152847
2. Alaswad SO, Mahmoud AS, Arunachalam P (2022) Recent advances in biodegradable polymers and their biological applications: a brief review. Polymers (Basel) 14. https://doi.org/10.3390/polym14224924
3. Nanda HS, Yang L, Hu J et al (2022) Editorial: biodegradable polymers for biomedical applications. Front Mater 9:1–3. https://doi.org/10.3389/fmats.2022.944755
4. Shah TV, Vasava DV (2019) A glimpse of biodegradable polymers and their biomedical applications. E-Polymers 19:385–410. https://doi.org/10.1515/epoly-2019-0041
5. Cao D, Xie Y, Song J (2022) DNA hydrogels in the perspective of mechanical properties. Macromol Rapid Commun 43:1–15. https://doi.org/10.1002/marc.202200281
6. Niaounakis M (2013) Introduction to biopolymers. In: Biopolymers reuse, recycling, and disposal. William Andrew Publishing, Oxford
7. Doppalapudi S, Jain A, Khan W, Domb AJ (2014) Biodegradable polymers-an overview. Polym Adv Technol 25:427–435. https://doi.org/10.1002/pat.3305
8. Fink JK (2013) The chemistry of bio-based polymers, 2nd edn. Wiley, Hoboken
9. Sun XS (2013) Overview of plant polymers: resources, demands, and sustainability. Elsevier, Amsterdam
10. Samir A, Ashour FH, Hakim AAA, Bassyouni M (2022) Recent advances in biodegradable polymers for sustainable applications. NPJ Mater Degrad 6
11. Lee YS, Kim DW, Song JH et al (2022) A biodegradable bioinspired oil-coated adhesive film for enhanced wet adhesion. Surf Interf 35:102415
12. Wang Y, Shukla A (2022) Bacteria-responsive biopolymer-coated nanoparticles for biofilm penetration and eradication. Biomater Sci:2831–2843

Natural and Semi-natural Polymers

Katia P. Seremeta and Alejandro Sosnik

Abstract Natural polymers, also known as biopolymers, are polymers produced and isolated from living organisms such as plants, animals and microorganisms. They are biodegradable, avoiding the contamination of the planet, and they often display good biocompatibility and low toxicity with limited adverse effects. One of the main advantages of using biopolymers is the reduction of the dependence on fossil fuels, which contributes to a more sustainable planet. Among the main application fields of biopolymers are the food, the pharmaceutical, and the biomedical industries. However, their use is challenged by their limited mechanical properties (e.g., low tensile strength), which can be overcome by chemically modifying the structure to obtain semi-natural biopolymers with improved thermal and mechanical properties and greater durability. This chapter introduces the most relevant biopolymers used in biomedicine, their classifications, and some of their applications with emphasis on polysaccharides and proteins.

K. P. Seremeta
Instituto de Investigaciones en Procesos Tecnológicos Avanzados – Consejo Nacional de Investigaciones Científicas y Técnicas, Universidad Nacional del Chaco Austral (INIPTA-CONICET-UNCAUS), Presidencia Roque Sáenz Peña, Chaco, Argentina
e-mail: kseremeta@uncaus.edu.ar

A. Sosnik (✉)
Department of Materials Science and Engineering, Technion – Israel Institute of Technology, Haifa, Israel
e-mail: sosnik@technion.ac.il

© American Association of Pharmaceutical Scientists 2023
A. Domb et al. (eds.), *Biomaterials and Biopolymers*, AAPS Introductions in the Pharmaceutical Sciences 7, https://doi.org/10.1007/978-3-031-36135-7_3

Graphical Abstract

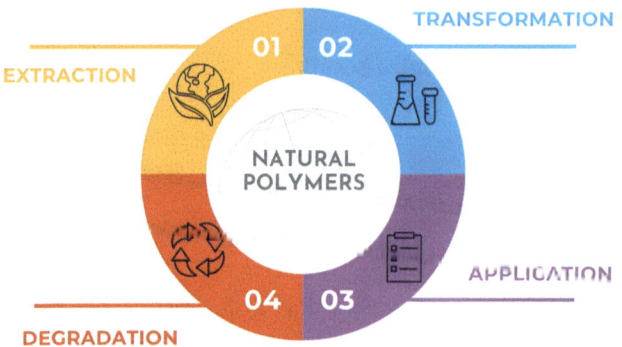

Keywords Natural and semi-natural polymers · Polysaccharides · Protein · Drug delivery · Tissue engineering

1 Introduction

1.1 Definition

Polymers are high molecular weight molecules (macromolecules) formed by repeating units known as monomers with linear or branched/crosslinked chains [1, 2, 5]. The term biopolymer comes from two Greek words, bio and polymer, and means that they are polymers produced and isolated from living organisms such as animals, plants, and microorganisms and therefore are biodegradable (Fig. 1).

1.2 History

For decades, the polymer industry has faced two serious problems: limited access to fossil resources and environmental pollution. The development of biopolymers from renewable sources is advantageous because it reduces the reliance on the traditional fossil-based polymers that are not biodegradable advancing toward a green sustainable life and an eco-friendly environment and reducing the burden of greenhouse gases in the environment. The use of biomass to produce biopolymers

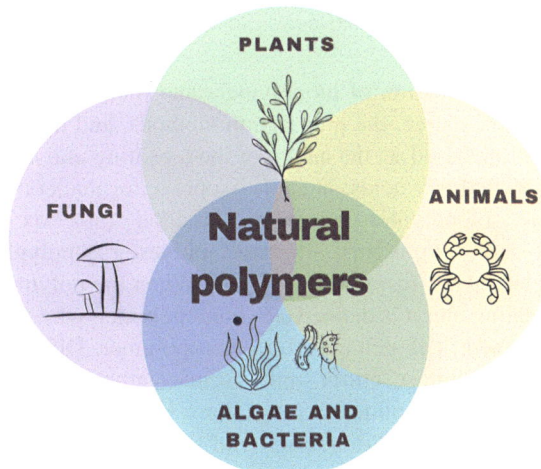

Fig. 1 Main sources of natural polymers

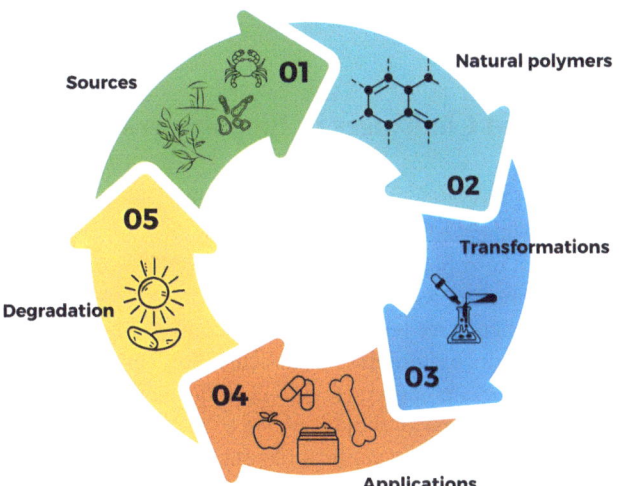

Fig. 2 Cycle of isolation, transformation, use, and biodegradation of natural polymers

constitutes a potential solution to reduce environmental pollution because, as opposed to fossil-based counterparts that are non-biodegradable, they undergo enzymatic degradation under aerobic conditions by microorganisms such as fungi, archaea, and bacteria, resulting in carbon dioxide, water, and mineral salts as final products [5, 12, 13, 15] (Fig. 2).

1.3 Classification

The classification of biopolymers and their products can be done based on the isolation source, the preparation methods, and the potential applications [15]. In addition, based on the nature of the repeating units, namely carbohydrates, amino acids, hydroxy acids, phenols, isoprene, or nucleotides along the polymer backbone, biopolymers can be classified into six families: polysaccharides, polypeptides/proteins, aliphatic polyesters, polyphenols, polyisoprenes, and nucleic acids, respectively [1, 11]. When the polymer is composed of one single repeating unit it is defined as a homopolymer, while when different monomers are combined, it is defined as a heteropolymer. Different types of polysaccharides such as cellulose, chitin, and its semi-synthetic derivative chitosan, starch, pectin, proteins, and peptides, among others, have been recently reviewed by Sivakanthan et al. [13]. In addition, microbial aliphatic polyesters known as polyhydroxyalkanoates have been also extensively investigated. Other synthetic aliphatic polyesters are produced by polycondensation of naturally occurring compounds such as lactic acid, glycolic acid, and caproic acid or ring-opening polymerization reactions of their respective lactones.

2 Advantages and Challenges of Biopolymers

A major advantage of biopolymers is their availability and affordability in nature and as by-products of other industries. In addition, biopolymers show good biocompatibility and are often resorbed or cleared from the body after enzymatic or chemical hydrolysis or metabolism without producing immune reactions or other adverse effects. Thus, biopolymers have been proposed as components of a broad spectrum of biomedical products including pharmaceuticals, implants, and medical devices [6, 9, 15]. At the same time, biopolymers often exhibit short half-life in the biological milieu and poor mechanical properties such as low tensile or compressive strength that are required in certain applications such as ligament or bone prostheses [7]. To overcome these drawbacks, biopolymers can be chemically modified to improve their mechanical properties and slow down their degradation rate. Other disadvantages include variability of the properties according to the source, the isolation method, and geographical and seasonal considerations, and potential contamination with pathogens and traces of immunogenic compounds such as proteins. In the next section, some of the most popular biomedical polysaccharides and proteins are overviewed.

Natural and Semi-natural Polymers

3 Biopolymers and Their Semi-synthetic Derivatives

3.1 Polysaccharides

3.1.1 Cellulose

Cellulose is the most abundant natural polymer formed by a polymer chain of entirely glucopyranose monosaccharides linked by equatorial 1,4-glycosidic bonds. Cellulose is poorly absorbed from the gastrointestinal tract and it cannot be metabolized by humans. In addition, it is not soluble in water and common organic solvents. To overcome these disadvantages, chemically modified derivatives with better properties than crystalline cellulose and multiple applications are obtained mainly by etherification and esterification. For example, cellulose ethers can be used in various areas such as the biomedical, pharmaceutical, food, cosmetic, chemical, textile, paper industries, and biorefineries, among others. It is noteworthy that by several pathways of synthesis cellulose and derivatives can be converted to biofuel such as bioethanol and other chemical compounds.

3.1.2 Chitin and Chitosan

Chitin is the second most abundant natural biopolymer after cellulose consisting of β-(1,4)-linked N-acetyl-D-glucosamine units. This linear polysaccharide is part of the exoskeleton of several marine invertebrates and is produced by mollusks, crustaceans, fungi, insects, and other organisms. Chitosan, the main semi-synthetic derivative of chitin obtained by alkaline deacetylation, is a polycationic polymer formed by different sequence, composition, and molecular chain length of randomly distributed N-acetyl glucosamine and D-glucosamine (Fig. 3). It finds more biomedical applications than chitin due to greater pH-dependent aqueous solubility, as

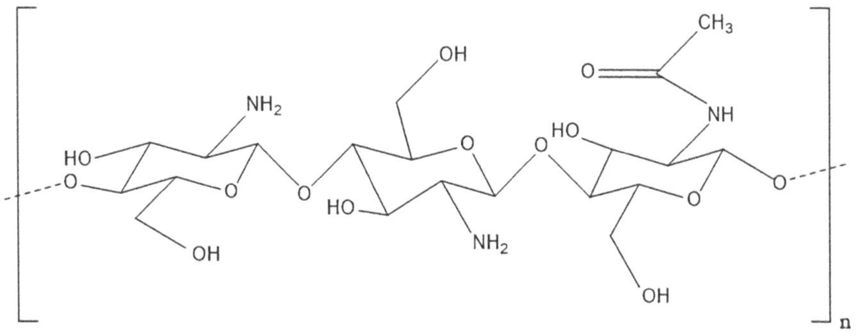

Fig. 3 Chemical structure of chitosan

it is soluble at pH values below 6. In addition, due to its high charge density (one positive charge per glucosamine residue) can electrostatically interact with cell membranes and negatively charged biomolecules. Notably, the amine and hydroxyl groups of chitosan can be modified to functionalize its structure means reactions such as etherification, esterification, carboxymethylation, and crosslinking, among others, allowing a wide range of biomedical applications. This polymer is biodegradable, biocompatible, non-toxic and it is classified as "generally recognized as safe" (GRAS) substance by the US Food and Drug Administration (FDA) and it can be crosslinked with polyanions such as sodium tripolyphosphate to form gels and particles of variable size. Its applications are multiple and range from tissue regeneration, wound healing, and controlled drug delivery systems to cosmetics, food, and textile, among others. It is also among the most popular mucoadhesive (that adheres to mucus) biopolymers.

3.1.3 Alginate

Alginate is a linear polyanionic polysaccharide extracted from the cell walls of brown algae. It is formed by blocks of (1→4)-linked β-D-mannuronate (M) and α-L-guluronate (G) repeating units. This polysaccharide is more water-soluble under slightly basic pH conditions and it can be chemically or enzymatically modified for example by reducing its molecular weight to improve its biological properties and extend its biomedical applications. Alginate also forms ionotropic gels by interaction with divalent and multivalent cations such as calcium(II), so it is used for the development of different types of drug delivery systems. Alginate is biodegradable, biocompatible, non-toxic, low cost and it is also considered as a GRAS compound by the FDA. The most common applications include cosmetic, pharmaceutical, and biomedical devices such as wound dressings and as inert support matrix since it can form porous three-dimensional hydrogels useful in tissue engineering and it has been extensively used in cell immunoisolation in cell therapy.

3.1.4 Agar

Agar is a linear polysaccharide which is extracted from the cell walls of red algae. Agar is composed of two main components: agarose and agaropectin. Agarose, a major component of agar, is a linear polymer containing β-D-galactose and 3,6-anhydro-α-L-galactose linked by glycosidic bonds β(1–4) (neoagarobiose disaccharide) and α(1–3) (agarobiose disaccharide). Agar, like alginate, is poorly soluble in water. This, added to its high viscosity, limits its use. Therefore, derivatives such as oligosaccharides are obtained to improve its biological properties. The applications include the chemical, pharmaceutical, food, and biomedical industries due to its particular characteristic of good rheological properties.

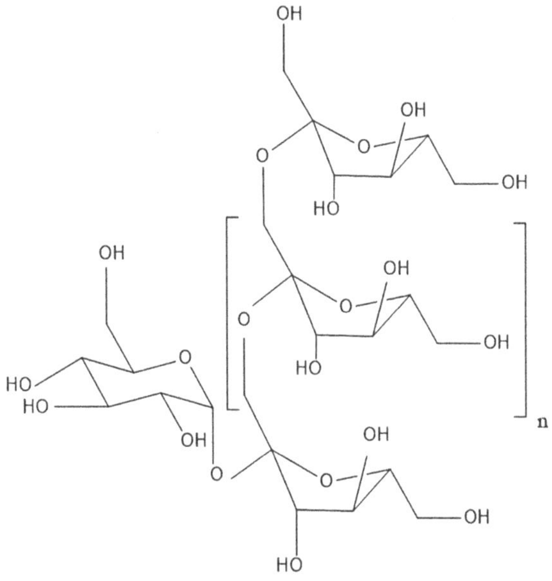

Fig. 4 Chemical structure of inulin

3.1.5 Inulin

Inulin is heterogeneous blend of fructose polymers extracted from many different plant species, some bacteria, and fungi. It is a water-soluble carbohydrate that belongs to the fructans family being a linear fructan composed of fructosyl units [β (2→1) linkage] containing usually one terminal glucose moiety [α (1→2 linkage] per molecule (Fig. 4). The functionality of the inulin depends on its branches and degrees of polymerization. It is not digested in the small intestine but in the large intestine it can be fermented by the microflora. The main application of inulin is in food industry as a sugar and fat replacer because it provides energy with low caloric value.

3.1.6 Pectin

Pectin is a polysaccharide isolated from the cell wall of higher plants, gymnosperms, pteridophytes, bryophytes, and a charophycean green algae, Chara. It is formed by esterified D-galacturonic acid residues linked by α-(1–4) bonds. Its uses are multiple in the cosmetic, food, pharmaceutical, and biomedical fields as stabilizing and gelling agent, in production of films, and as part of drug delivery systems and medical devices, among others.

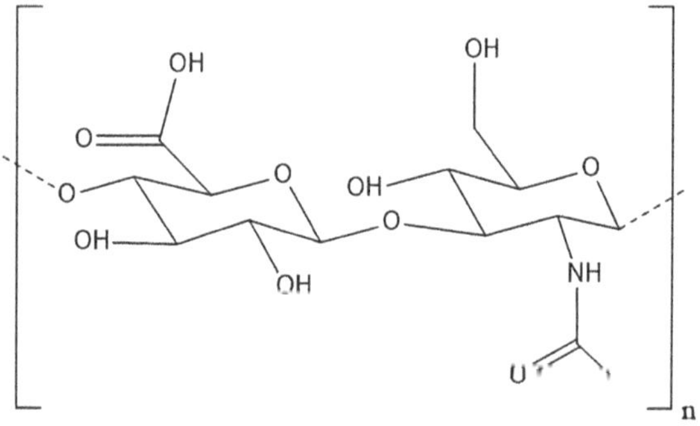

Fig. 5 Chemical structure of hyaluronic acid

3.1.7 Starch

Starch is an abundant carbohydrate present in green-leafed plants as major reserve in the form of granule in cereal grains as corn and in roots as potato and tubers. In addition, starch is a main source of energy in human diet. The structure of starch is made up of amylose (linear α-D-glucan with 1,4-linked) and amylopectin (branched α-D-glucan with 1,4-linked branches attached by 1,6-linkages to the main chain). In order to overcome the disadvantages of starch such as loss of viscosity after cooking, derivatives were developed. Their uses are not limited to the food industry but can also be used as a desiccant, dehumidifier, among others. In addition, it is widely used as biomaterial in the biomedical field such as tissue engineering, drug delivery systems, and diagnostic imaging.

3.1.8 Hyaluronic Acid

Hyaluronic acid or hyaluronan is a natural linear polysaccharide composed of repeating units of N-acetyl-D-glucosamine and β-D-glucuronic acid, alternately linked by β-1,4 and β-1,3 glycosidic bonds (Fig. 5). It is a major component of skin extracellular matrix and also is present in most connective tissues, in joints, tendons, synovial fluid, vitreous humor, umbilical cord, rooster comb, and shark skin. It can also be obtained from bacterial fermentation. Hyaluronic acid is highly hydrophilic due to the large number of hydroxyl and carboxyl groups present in its structure and it can form a viscous and elastic matrix because its chains adopt a coil configuration. Thus, it promotes cell adhesion and can absorb exudate from wounds. In this framework, it is used for producing wound dressings due to its tissue regeneration properties and its key role in the wound healing process owing to its lubricant, anti-angiogenic, analgesic, anti-inflammatory, and immunosuppressive properties.

3.1.9 Xanthan Gum

Xanthan gum is a heteropolysaccharide obtained extracellularly from bacteria of the genus *Xanthomonas* such as *X. campestris* NRRL B-1459, *X. axonopodis,*, and *X. arboricola*. The primary chemical structure is a linear (1→4) linked β-D-glucose backbone with a trisaccharide side chain on every other glucose at C-3. It is an aqueous soluble and biocompatible polymer that presents stability against heat and pH changes. Its applications cover food, pharmaceutical, and cosmetic industries as gelling, stabilizer, wetting, viscosity increasing agent, drug delivery system, and the biomedical field as wound healing and tissue engineering.

3.1.10 Guar Gum

Guar gum is a renewable polysaccharide extracted from the embryos of *Cyamopsis tetragonoloba*, a leguminous plant. It is composed of linear chains of (1–4)-β-D-mannopyranosyl connected by 1–6 linkages with α-D-galactopyranosyl. Slight modification in the guar gum structure expands its use in various fields mainly in the pharmaceutical industry for the development of different drug administration systems by several routes such as oral, buccal, ocular, and dermal, among others. Among its advantages, biocompatibility, biodegradability, stability, and non-toxicity can be highlighted.

3.2 Proteins

3.2.1 Collagen and Gelatin

Collagen, a major fraction of connective tissue in vertebrates and invertebrates, and gelatin, its partially denatured form, are biopolymers widely used as wound dressings, tissue engineering scaffolds, and in the food, pharmaceutical, and cosmetic industries. Collagen is the most abundant structural protein in mammals. The collagen molecule consists of three identical or different polypeptide α-chains depending on the types (there are 28 collagen types) and sources, each of which contains approximately 1000 amino acids. Gelatin is a heterogeneous mixture of peptides derived from collagen.

3.2.2 Silk

Silk is produced in the glands of some arthropods as spiders, scorpions, mites, bees, and silkworm. The silk produced by silkworm or *Bombyx mori* in the final stage of larval development is the most characterized and employed. Fibroin and sericin are the two major proteins of silk. Silk can be used as carrier because it is possible to incorporate drugs within or on its surface if functional groups are activated. In addition, silk can be used in tissue engineering of ligaments, cartilage, and bones and in medicine regenerative as wound dressings.

3.2.3 Keratin

Keratin is a polypeptide of different amino acids with bonding of disulfide cysteine widely found in feathers, hair, wool, nails, hooves, and horns. Among the numerous applications of keratin can be mentioned tissue engineering of bones, medicine regenerative as ocular and nerve, controlled drug delivery systems, hydrogels, and skin replacement due to amino acid sequences in keratin being similar to extracellular matrix.

3.2.4 Elastin

Elastin is a main component of the extracellular matrix of vertebrate tissues formed from tropoelastin, a soluble precursor, through crosslinking mediated by lysine. Therefore, elastin is used in tissue repair field as grafts coating, hydrogels, targeted drug delivery systems, and matrices for formation of cartilaginous tissue, among others.

4 Applications of Biopolymers

Among the multiple applications of biopolymers are food and pharmaceutical industry as packaging material, edible films, drug transport, controlled drug delivery systems, and dressing materials as well as excipients, stabilizers, thickeners, and gelling agents. In the biomedical field, they can be used for the development of implants and three-dimensional scaffolds, tissue culture, artificial grafts, bone filler materials, and in dental applications. In addition, the use of biopolymers as adhesives and cosmetics has been also reported [1, 2, 10, 14] (Fig. 6).

For example, tendon and ligament scaffolds are a very important application of biopolymers. The design involves taking into account the composition of the native tendon and ligament, cell integration, and mechanical stimulations, among other aspects. Biopolymers used in this area can be collagen, hyaluronic acid, gelatin, and silk due to their advantageous properties since they imitate the native extracellular matrix and promote cell adhesion and differentiation towards tissue formation. However, their poor tensile properties need to be improved [7]. For instance, their mechanical properties can be modulated to improve the tensile and compressive properties to withstand the applied forces [8].

Biopolymers also can be used to improve integration of artificial bones to the human body because they mimic the structure and properties of the extracellular

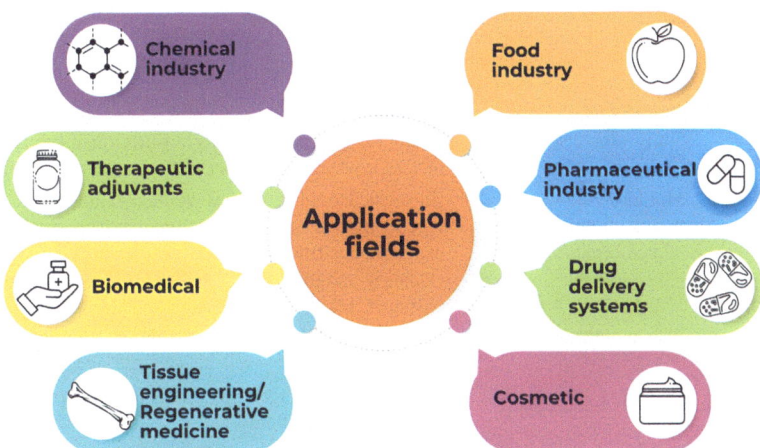

Fig. 6 Main application fields of natural polymers

matrix of the natural bone. In addition, they are biodegradable, and they can be formulated as minimally invasive implants that can be implanted with lesser tissue damage than invasive surgical procedures and, in the case of temporary implants, they do not need a second surgery to be removed [9]. Potential applications such as tissue engineering, drug delivery, and vaccine adjuvants increased over the last few years due to the manipulation of biopolymers at the nanometric scale [14]. In addition, the use of biopolymers as support materials for the immobilization of enzymes and the development of biocatalysts is being explored with growing interest in food, pharmaceutical, and biomedical applications [3].

Biopolymers such as chitosan, alginate, and other combinations could be used as adsorbents to remove surfactants, heavy metals, and dyes from wastewater reducing emerging pollutants and clarifying the wastewater due to their coagulant and flocculant properties. In this way, costs associated with commercial activated carbon commonly used for these purposes can be reduced [4]. Table 1 summarizes the main sources and applications of some key natural polymers and their derivatives.

5 Biomaterials in the Biomedical Field

Biopolymers are attracting more and more attention from researchers and industries (chemical, pharmaceutical, biomedical) due to their renewable nature. In addition, they are being increasingly studied due to their biocompatibility and biodegradability properties, necessary requirements for tissue engineering and regenerative medicine such as development of human bone substitutes and scaffolds. However, there are still some challenges to overcome such as insufficient mechanical properties. Nevertheless, certain chemical modifications in the functional groups of natural

Table 1 Sources and applications of some key natural polymers

Natural polymers	Sources	Applications
Cellulose	Green plants, algae, and oomycetes	Pharmaceutical and food industry, personal care products, oil field chemicals, adsorption of heavy metals, construction, paper, adhesives, and textiles
Chitin and chitosan	Mollusks, crustaceans, insects, fungi, and other related organisms	Biomedical applications as wound healing, tissue regeneration or controlled drug delivery, removal of heavy metals from waste waters, film forming, chelating and thickening property, food protection, cosmetic, textile, papermaking, and agriculture
Alginate	Brown algae	Biomedical applications, therapeutic adjuvant or drug carrier
Agar and agarose	Red algae	Food and gelling agent, pharmacological, cosmetic, and nutraceutical fields
Inulin	Plants	Prebiotic, fat and sugar replacer, texture modifier, functional foods
Pectin	Higher plants, gymnosperms, pteridophytes, bryophytes, and an algae	Gelling and stabilizing agent in foods
Starch	Green leafed plants	Food ingredient as source of nutrition and energy, desiccant, dehumidification of air
Hyaluronic acid	Skin extracellular matrix of animals	Tissue engineering and regenerative medicine as wound dressings
Xanthan gum	Xanthomonas bacteria	Food, cosmetic and pharmaceutic products, lotions, shampoos, wound healing, drug delivery systems, tissue engineering
Guar gum	Leguminous plant *Cyamopsis tetragonoloba*	Drug delivery systems
Collagen and gelatin	Connective tissue of mammals	Food, pharmaceutical, cosmetic industries, regenerative medicine
Silk	Arthropods as *Bombyx mori* or silkworm	Drug delivery systems, tissue engineering, regenerative medicine
Keratin	Epithelial cells of vertebrates, feathers, hair, wool, nails, hooves, beaks, and horns	Tissue engineering, drug delivery systems, regenerative medicine
Elastin	Extracellular matrix of vertebrate tissues	Tissue engineering

polymers make it possible to improve these properties. In addition, a useful strategy is the combination of natural polymers with other compounds that provide greater mechanical resistance without loss of elasticity. For example, chitosan-silk sericin/ hydrohyapatite composites containing 60–70% of organic part were prepared to promote osteoblast attachment and proliferation with suitable compressive strength and elastic modulus. However, if the tensile strength is too high it could lead to cancellous bone disruption. Therefore, the necessary characteristics of the biomedical material (compressive and tensile strength, porosity, degradability, morphology) depend on the type of tissue or organ to be repaired (cancellous bone, cortical bone, cartilage). Another challenge to be overcome by natural polymers such as chitosan, collagen, and hyaluronic acid is the degradation time in the organism, less than the 4–6 months required for biomaterials. However, the development of modified systems based on these polymers, such as nanofibers, could prolong the degradability time. Other property that can be tuned through preparation of composite materials is porosity, necessary for example in bone repair (bone growth and filling). In addition, natural polymer-based materials can be 3D printed to mimic the structure of bone in substitute applications. Currently, most of the polymeric systems used in the biomedical field are based on the combination of natural and synthetic polymers in order to take advantage of biocompatibility of the former and mechanical resistance of the latter. In addition, materials with specific properties such as promotion of regeneration, cell proliferation, osteogenesis, and antibiotic can be added to these systems [9].

6 Summary and Prospective

Biopolymers are natural eco-friendly polymers produced by living organisms such as plants, animals, or microbial biomass hence of renewable nature. They have the advantage of reducing dependence on oil reserves used for the development of synthetic polymers generally very resistant to biodegradation and limited availability. Thereby, biopolymers contribute to avoiding environmental pollution by reducing global warming. In the last decades, the interest of researchers for natural polymers as an alternative to synthetic polymers has grown due to their unique characteristics of biodegradability, biocompatibility, and relatively low toxicity. In this context, the use of these natural polymers and their semi-synthetic derivatives in the biomedical field has grown. Among their broad biomedical applications can be mentioned drug delivery systems, regenerative medicine, tissue engineering as scaffolds, tissue replacement grafts, intraocular and contact lenses, sutures, adhesives, dentistry, artificial skin, bone cement, cartilage and joint repairs, catheters, and others. In addition, they can be used alone or combined with synthetic polymers. At the same time, future research will need to address challenges such as achieving proper mechanical performance, ensuring their purity, and low inter-batch variability for which the application of standardized and validated isolation and production protocols is required.

Questions & Answers

Question 1: The following are characteristic properties of natural polymers:

(a) They are isolated from plants, animals, and/or microorganisms.
(b) They are biodegradable and biocompatible.
(c) They may contain pathogenic contaminations.
(d) All the above are correct.

Correct answer: (d). Natural polymers can be isolated from plants, animals, and/or microorganisms. They are often biodegradable and biocompatible and may contain pathogenic contaminations (e.g., bacteria, viruses). They could also be contaminated with immunogenic components (e.g., proteins) that trigger an immune response in the host.

Question 2: The main advantages of natural polymers are:

(a) Reproducibility.
(b) Biodegradability and biocompatibility.
(c) Chemical stability.
(d) Mechanical resistance.

Correct answer: (b). The main advantages of natural polymers are the biodegradability and biocompatibility. Oppositely the low reproducibility, chemical stability, and mechanical resistance are their main disadvantages.

Question 3: These are natural polymers:

(a) Polysaccharides, proteins, natural rubber, and polyhydroxyalkanoates.
(b) Polysaccharides, proteins, poly(tetrafluoroethylene), and polyhydroxyalkanoates.
(c) Polypeptides, polyhydroxyalkanoates, silicone, and starch.
(d) All the above are correct.

Correct answer: (a). Polysaccharides, proteins, natural rubber, and polyhydroxyalkanoates are natural polymers while poly(tetrafluoroethylene) and silicone are synthetic ones.

Question 4: Biopolymers and their products can be classified based on:

(a) Only the source of the isolation.
(b) Source, preparation method, and potential applications.
(c) Nature of the repeating units.
(d) (b) and (c) are correct.

Correct answer: (d). Biopolymers and their products can be classified according to their isolation source, the preparation methods, and the potential applications. In addition, they can be classified based on the nature of the repeating units.

Question 5: Cellulose is a polysaccharide that:

(a) Freely dissolves in water because it is composed of glucose.
(b) Does not freely dissolve in water.

(c) Dissolves in common organic solvents.
(d) (b) and (c) are correct.

Correct answer: (b). Cellulose is a polysaccharide that does not freely dissolve in water due to a high degree of crystallinity. The way to increase solubility is by chemically modifying it, which reduces the degree of crystallinity.

Question 6: Chitosan, obtained by partial deacetylation of chitin, is:

(a) A polycationic polysaccharide that dissolves in water at basic pH.
(b) A polyanionic polysaccharide that dissolves in water at basic pH.
(c) A polycationic polysaccharide that dissolves in water at acid pH.
(d) All the above are correct.

Correct answer: (c). Chitosan, obtained by partial deacetylation of chitin, is a polycationic polysaccharide with high charge density (one positive charge per glucosamine residue) that dissolves in dilute aqueous acid solutions. Conversely, chitosan is water-insoluble under neutral or basic pH conditions.

Question 7: Alginate is:

(a) A polycationic polysaccharide that dissolves in water at basic pH.
(b) A polyanionic polysaccharide that dissolves in water at basic pH.
(c) A polycationic polysaccharide that dissolves in water at acid pH.
(d) All the above are correct.

Correct answer: (b). Alginate is a polyanionic polysaccharide consisting of units of mannuronic acid and guluronic acid that dissolves in water under slightly basic pH conditions. Conversely, it is water-insoluble under acid pH conditions.

Question 8: Collagen is a protein that:

(a) Is used in the production of tissue engineering scaffolds.
(b) Is isolated from crustaceans and plants.
(c) Upon partial denaturation forms gelatin.
(d) (a) and (c) are correct.

Correct answer: (d). Collagen is the most abundant structural protein in mammals both vertebrates and invertebrates (approximately 30% of total proteins) and it is widely used in the production of tissue engineering scaffolds. Collagen upon partial denaturation forms gelatin.

Question 9: Agar is a linear polysaccharide which is extracted from:

(a) The extracellular matrix of vertebrate tissues.
(b) The cell walls of brown algae.
(c) Bacteria of the genus *Xanthomonas*.
(d) The cell walls of red algae.

Correct answer: (d). Agar is a linear polysaccharide which is extracted from the cell walls of red algae.

Question 10: Silk is produced by:

(a) *Xanthomonas* bacteria.
(b) *Cyamopsis tetragonoloba*.
(c) *Bombyx mori*.
(d) All the above are correct.

Correct answer: (c). Silk is produced by some arthropods as *Bombyx mori* or silkworm in the final stage of larval development.

References

1. Ashter SA (2016) Overview of biodegradable polymers. In: Introduction to bioplastics engineering. William Andrew, Amsterdam, pp 19–30
2. Baranwal J, Barse B, Fais A, Delogu GL, Kumar A (2022) Biopolymer: a sustainable material for food and medical applications. Polymers 14:983. https://doi.org/10.3390/polym14050983
3. Bilal M, Iqbal HM (2019) Naturally-derived biopolymers: potential platforms for enzyme immobilization. Int J Biol Macromol 130:462–482. https://doi.org/10.1016/j.ijbiomac.2019.02.152
4. Biswas S, Pal A (2021) Application of biopolymers as a new age sustainable material for surfactant adsorption: a brief review. Carbohydr Polym 2:100145. https://doi.org/10.1016/j.carpta.2021.100145
5. George A, Sanjay MR, Srisuk R, Parameswaranpillai J, Siengchin S (2020) A comprehensive review on chemical properties and applications of biopolymers and their composites. Int J Biol Macromol 154:329–338. https://doi.org/10.1016/j.ijbiomac.2020.03.120
6. Graça MFP, Miguel SP, Cabral CSD, Correia IJ (2020) Hyaluronic acid-based wound dressings: a review. Carbohydr Polym 241:116364. https://doi.org/10.1016/j.carbpol.2020.116364
7. Heidari BS, Ruan R, Vahabli E, Chen P, De-Juan-Pardo EM, Zheng M, Doyle B (2022) Natural, synthetic and commercially-available biopolymers used to regenerate tendons and ligaments. Bioact Mater 19:179–197. https://doi.org/10.1016/j.bioactmat.2022.04.003
8. Joyce K, Fabra GT, Bozkurt Y, Pandit A (2021) Bioactive potential of natural biomaterials: identification, retention and assessment of biological properties. Signal Transduct Target Ther 6:122. https://doi.org/10.1038/s41392-021-00512-8
9. Kashirina A, Yao Y, Liu Y, Leng J (2019) Biopolymers as bone substitutes: a review. Biomater Sci 7:3961–3983. https://doi.org/10.1039/c9bm00664h
10. Narancic T, Cerrone F, Beagan N, O'Connor KE (2020) Recent advances in bioplastics: application and biodegradation. Polymers 12:920. https://doi.org/10.3390/polym12040920
11. Olatunji O (2016) Classification of natural polymers. In: Olatunji O (ed) Natural polymers. Springer, Switzerland, pp 1–17
12. Polman EM, Gruter GM, Parsons JR, Tietema A (2021) Comparison of the aerobic biodegradation of biopolymers and the corresponding bioplastics: a review. Sci Total Environ 753:141953. https://doi.org/10.1016/j.scitotenv.2020.141953
13. Sivakanthan S, Rajendran S, Gamage A, Madhujith T, Mani S (2020) Antioxidant and antimicrobial applications of biopolymers: a review. Food Res Int 136:109327. https://doi.org/10.1016/j.foodres.2020.109327
14. Torres FG, Troncoso OP, Pisani A, Gatto F, Bardi G (2019) Natural polysaccharide nanomaterials: an overview of their immunological properties. Int J Mol Sci 20:5092. https://doi.org/10.3390/ijms20205092
15. Udayakumar GP, Muthusamy S, Selvaganesh B, Sivarajasekar N, Rambabu K, Sivamani S, Sivakumar N, Maran JP, Hosseini-Bandegharaei A (2021) Ecofriendly biopolymers and composites: preparation and their applications in water-treatment. Biotechnol Adv 52:107815. https://doi.org/10.1016/j.biotechadv.2021.107815

Fundamentals and Biomedical Applications of Smart Hydrogels

Qi Wu, Eid Nassar-Marjiya, Mofeed Elias, and Shady Farah

Abstract Hydrogels are three-dimensional elastic networks containing a large amount of water formed from crosslinked hydrophilic polymer chains, which possess tunable tissue-like physicochemical properties. This chapter thoroughly introduces the definition, classification, formation, properties, and typical representative and biomedical applications of natural and synthetic hydrogels. In detail, the essential features of hydrogels relate to swelling, mechanical strength, rheology, self-healing, and injectability. As biomaterials, biocompatibility, biodegradability, and toxicity properties should be considered preferentially. Then, we expound on several typical hydrogels, such as self-healing hydrogels, injectable hydrogels, and stimuli-responsive hydrogels, including shape memory hydrogels and hydrogel actuators. Eventually, we summarize the major attractive applications of hydrogels in biomedical fields such as contact lenses, hygiene products, drug delivery, tissue engineering, and wound healing.

Q. Wu · E. Nassar-Marjiya · M. Elias
The Laboratory for Advanced Functional/Medicinal Polymers & Smart Drug Delivery Technologies, The Wolfson Faculty of Chemical Engineering,
Technion-Israel Institute of Technology, Haifa, Israel

S. Farah (✉)
The Laboratory for Advanced Functional/Medicinal Polymers & Smart Drug Delivery Technologies, The Wolfson Faculty of Chemical Engineering,
Technion-Israel Institute of Technology, Haifa, Israel

The Russell Berrie Nanotechnology Institute, Technion-Israel Institute of Technology, Haifa, Israel
e-mail: sfarah@technion.ac.il; https://www.thefarahlab.com/

Graphical Abstract

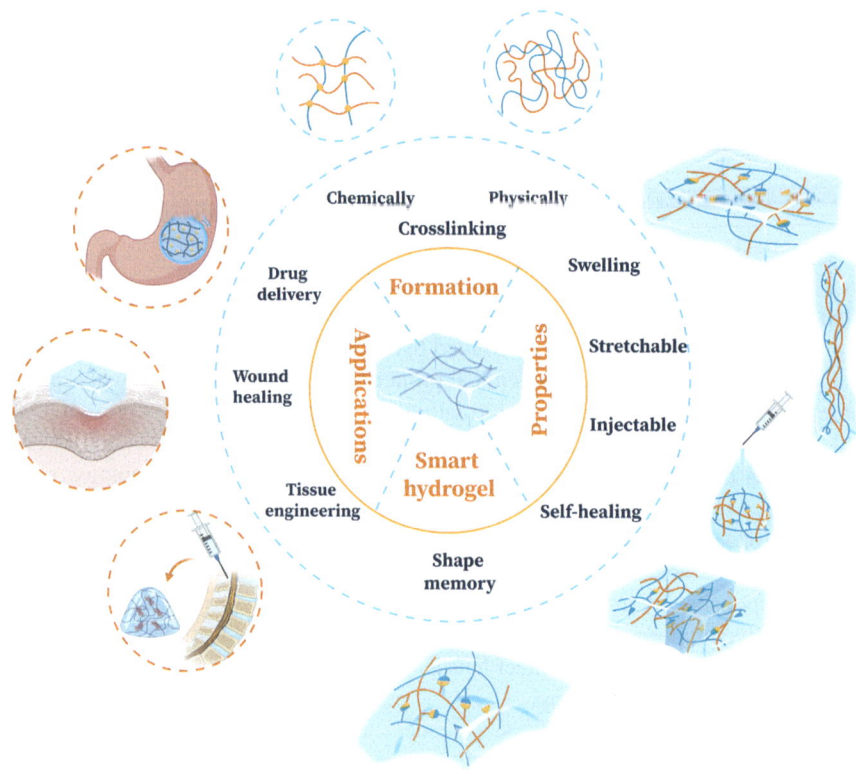

Keywords Smart hydrogels · Biomedical applications · Drug delivery · Tissue engineering · Wound healing

1 Introduction

Hydrogelsconsist of three-dimensional (3D) elastic networks formed from highly hydrophilic polymer chains with interstitial spaces that can absorb a large amount of water or biological fluids. Hydrogels represent an important class of materials possessing a watery environment and broadly tunable physicochemical properties [1]. Hydrogels have the ability to swell and hold water with their hydrophilic structures instead of being dissolved because of the crosslinking of polymeric networks, which are capable of swelling and shrinking reversibly in response to external environment changes. Generally, hydrogels can be placed into two major categories, naturally derived hydrogels and synthetic hydrogels. Naturally derived hydrogels, purified from natural sources with self-possessed advantages of biocompatibility, biodegradability, low cytotoxicity, and bioactivity, which typically formed of proteins and extracellular matrix components and resemble the native soft tissues, make them suitable for various applications, especially in the biomedical field. For synthetic hydrogels, 3D networks are prepared via chemical crosslinking, forming covalent bonds, physicochemical (noncovalent) interactions, or coordination of both, which have precisely designed structures with highly controlled chemical and physical properties. In detail, chemical crosslinked hydrogels bring together multifunctional monomers/oligomers or linear/branched polymers with crosslinkers and react to form networks. Hydrogels are found in the form of matrix, film, microsphere, and nanoparticles in multiple dimensions; even a variety of naturally derived and synthetic polymers can be processed into hydrogels being in plentiful uses. Nowadays, the versatility of the hydrogels system has endowed it with widespread applications in various fields, including soft electronics, sensors, actuators, and biomedicine such as 3D cell culture, contact lenses, blood-contacting, hemostasis bandages, wound healing, drug delivery, and tissue engineering [2].

2 Classification and Formation of Hydrogels

2.1 *Classification*

The classification of hydrogels depends on different factors, such as source, crosslinks, structures, properties, etc. There are plentiful hydrogels derived from nature that can be classified into three groups: protein-based materials such as collagen, gelatin, fibrin, elastin, and silk fibroin; polysaccharide-based materials such as chitosan and alginates; also, those derived from decellularized tissue. Based on the formation strategies of common synthetic hydrogels, they can be classified by chemical crosslinking (covalent bonding) called 'permanent' or 'chemical' gels, physical crosslinking (hydrogel bonding, hydrophobic interactions, chain entanglements, etc.) called 'reversible' or 'physical' gels and chemical-physical crosslinking dual-network hydrogels. According to the structure that forms gels, they can be divided into homopolymer hydrogels, copolymer hydrogels, multipolymer hydrogels, and interpenetrating network (IPN) hydrogels. The homo- and copolymer hydrogels are crosslinked networks with one and two/more types of hydrophilic monomer after reacting, respectively. Whereas IPN hydrogels are formed via polymerizing the monomer/comonomers within the performed crosslinked networks. Furthermore, hydrogels can be classified into conventional hydrogels and 'smart'/responsive hydrogels due to their various properties, whereas the 'smart' hydrogels show different responses under chemical (pH, glucose), physical (light, ultrasound, temperature, pressure, electric, and magnetic field), and biological (antigen, enzyme, and ligand) stimulus. Also, they can be divided broadly into stable and degradable, with the latter further categorized as hydrolytically or enzymatically degradable. Moreover, based on ionic charges on the polymeric backbone, ionic hydrogels could be classified as uncharged neutral hydrogels, positively charged cationic hydrogels, negatively charged anionic hydrogels, and ampholytic hydrogels that have both positive and negative charges, which may end up with net positive, negative, or neutral charge [3].

2.2 Formation

Gelation (gel transition) is defined as the formation of a three-dimensions network by chemical or physical crosslinking. Each molecule contributing to the crosslinking processes has a number of reactive sites, in which the normalized number per primary molecule is termed functionality. As crosslinking proceeds, molecules will develop into oligomers (e.g., dimer, trimer, etc.) with more functional groups capable of reacting. Each oligomer will progressively increase its size and aggregate up to a critical point where the 'aggregation' becomes an 'infinite network (i.e., a gel). This transition from a finite branched polymer into a network is the so-called 'gelation' or 'sol-gel transition', and the critical point is called the 'gel point'. The term 'hydrogel' involves the crosslinked 'gel' with water as a network constituent [4].

Detail descriptions of the gelation processes have been investigated through numerous techniques in diverse aspects: (1) Thermodynamic techniques (e.g., differential scanning calorimetry); (2) Spectroscopic techniques (e.g., Fourier-transform infrared spectroscopy, Raman spectroscopy); (3) Scattering methods (e.g., small-angle X-ray scattering, light scattering); (4) Electron microscopy. A valuable and direct determination of gelation could be precisely identified in terms of its rheology property. This operational definition is efficiently attributed to the quantitative correlation between mechanical response and the degree of crosslinking. Here, the details of the experiment will be described in Rheology Section. Briefly, the storage modulus G' is quantitatively related to the network's connectivity. Figure 1 is a typical plot of storage modulus G' and loss modulus G'' as a function of the gelation process. The 'gel point' is determined by the dramatic change in modulus ($G' < G''$ transit to $G' > G''$), which is the boundary between liquid-like and solid-like.

Hydrogels are formed by cross-linking molecules/polymer chains scattered in an aqueous system through the gelation process. Physical cross-linking approaches include the mechanisms of physical entanglement, ionic interactions, self-assembly (hydrogen bonding, host-guest interactions, π stacking, electrostatic interactions, and hydrophobic interactions), and crystallization, as shown in Fig. 2. Relatively, the gelation is easy and reversible but not much tunable. For many thermally driven hydrogels, a decrease or increase in temperature results in thermal gelation which comes from the temperature response of polymer chains' physical entanglement. These transition temperatures are defined as upper critical solution temperature (UCST) and lower critical solution temperature (LCST). Hydrogels show UCST behavior, such as gelatin, the gel formation occurs as the temperature drops below its UCST. Ionic interaction is a physical crosslinking that happens when a

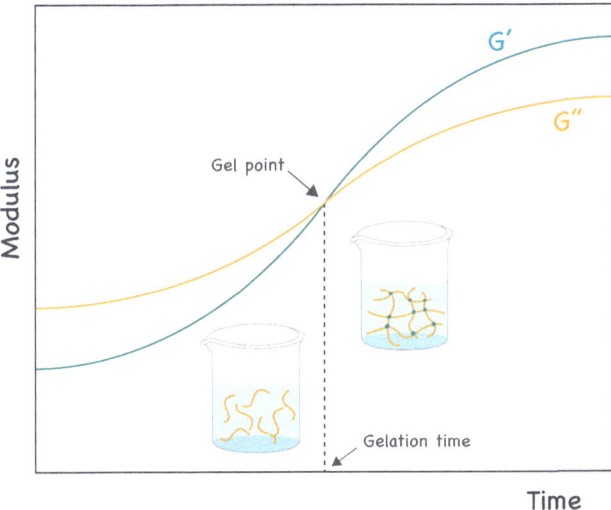

Fig. 1 Schematic of dynamic viscoelastic properties of gelation process

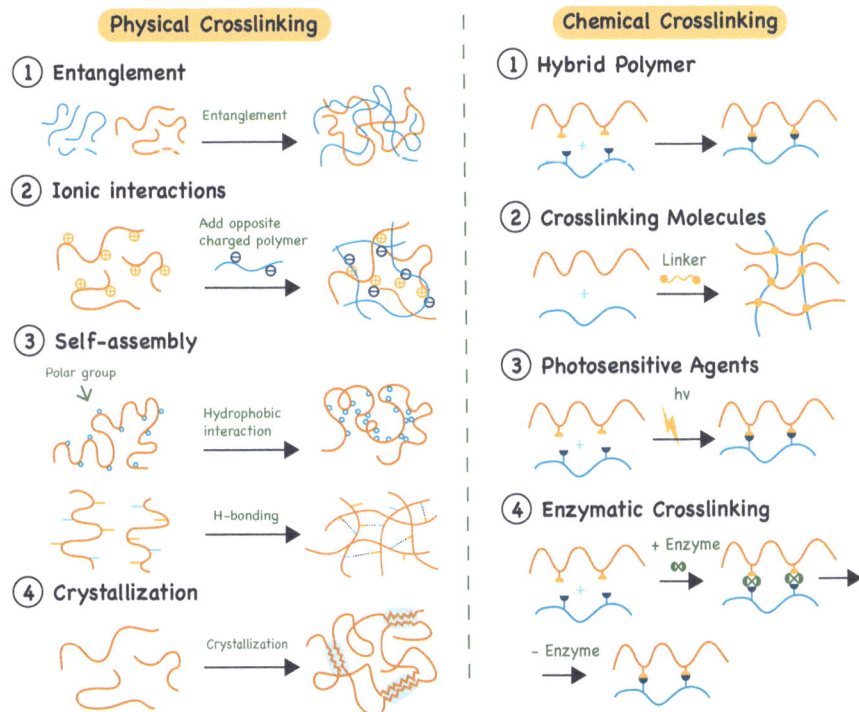

Fig. 2 The formation of hydrogels by chemical crosslinking and physical crosslinking

water-soluble and charged polymer crosslinks with an ion or another polymer chain of opposite charge. Alginate is a well-known example of a polymer that can be crosslinked by Ca^{2+} or Ba^{2+}, the hydrogel formation based on chelation. In addition, majority of protein-based hydrogels are prepared via self-assembly. Self-assembling hydrogels are formed by the spontaneous self-assembly of monomeric components into polymer-like fibrils via non-covalent interactions and the subsequent entanglement of these fibrils into an extended network that elicits gelation of the aqueous solvent. Moreover, another foremost physical cross-linking method for hydrogel formation is crystallization. The gelation phenomenon during the crystallization undergoes the gel-crystal transition, especially the competition between gel and crystal. When the temperature decreased, high solid-solution interfacial tension caused the slow growth rate of the nanometer- and micrometer-sized crystals that form a metastable-state gel system. Usually, this gel state is unstable, which will be destroyed with the change of condition and transform into a larger and regular crystal.

Chemical cross-linking gelation mainly follows four mechanisms: hybrid polymers, cross-linking molecules, photosensitive agents, and enzymatic cross-linking (Fig. 2), which significantly improves the flexibility and precision during the gelation process, showing more stable hydrogels matrix, also tunable properties from well-designed structures and controlled covalent bonds. Hybrid polymers mechanism usually happens to conjugate mixed polymer chains that possess functional groups such as –OH, –COOH, –NH$_2$, and –N$_3$ by establishing the covalent linkages from the complementary reactions. For example, azide-alkyne or Diels-Alder click chemistry, amine-carboxylic acid, isocyanate–OH/–NH$_2$ reaction, and Schiff base formation. Crosslinking molecules gelation mechanism fabricates the hydrogels by adding a crosslinker to bind bi/multi-functional molecules or pre-functionalized polymer chains under favorable conditions. For example, linking poly(vinyl alcohol) with methacrylic acid using ethylene glycol di-methacrylate as the crosslinking agent by free radical polymerization, the hydrogel is a promising system for drug delivery. Photo-crosslinking is one of the well-explored strategies of chemical cross-linking used for the synthesis of photo-crosslinkable hydrogels. The cross-linking is occurring through the exposure of a photosensitive system consisting of photosensitive functional groups, photo-initiators, and other compounds (e.g., cells and therapeutic molecules) under high-energy irradiation such as ultraviolet (200–400 nm) or visible light (400–800 nm). The photo-crosslinking gelation process mostly comes from the mechanism of free-radical photopolymerization and bio-orthogonal click reactions, which are used to bio-fabricate the photo-crosslinkable hydrogels via traditional bio-fabrication methods like inkjet 3D bioprinting and extrusion 3D bioprinting or laser-assisted bio-fabrication approaches such as stereolithography (SLA), digital light processing (DLP), and computed axial lithography (CAL). Moreover, the promising technique of four-dimensional (4D) printing allows for animated constructions that change materials' shape, function, or properties over time when exposed to specific external stimuli after fabrication. A more important thing that needs to be considered is that this technique requires the use of a secondary photosensitizer and prolonged irradiation, both of which can cause some local toxicity. Due to the substrate specificity of the enzymes, some possible cytotoxic consequences and unexpected by-products from photo-initiator- or organic solvent-mediated reactions can be avoided. Enzymatic cross-linking is another efficient way for gelation, which highly fits various enzymatic systems to form desired hydrogels with higher degrees of complexity, mimicking extracellular matrices for different biomedical purposes. Worthfully, enzymatic reactions happen with mild conditions at physiological temperatures and neutral pH in aqueous systems, whereas crosslinking under harsh chemical environments may cause the loss of bioactivities.

Drive with different molecular designs and gelation strategies, hydrogels can achieve sufficient morphologies, such as star, comb/brush, hyperbranched, and dendrimer networks. Also, hierarchical constructions include nanogels, microgels, films, and machines. Each of the physical and chemical gelation methods has its advantages and limitations. Therefore, the entire systems always combine multiple components and/or crosslinking mechanisms to prepare and functionalize hydrogels that exhibit superior physicochemical and biological properties, such as swelling/anti-swelling, stretchable, injectable, self-healing, biodegradable, biocompatible, low cytotoxicity, shape memory, and stimuli-responsive.

3 Typical Properties of Hydrogels

3.1 Swelling

The swelling behavior is a crucial parameter of hydrogels, which significantly affects their properties and is highly related to applications. When hydrogels swell, the phase transit from glassy to rubbery; some respond as volume change, and others behave in a sol-gel phase transition. The driving force for hydrogel swell mainly comes from the interactions between hydrophilic polymer chains and water molecules. Therefore, adjusting the crosslinking density and polymer structures (e.g., ionic/neutral side group) can tune the swelling behavior of hydrogels. The water capacity of hydrogels is generally described by the equilibrium swelling ratio as calculated in Eq. 1:

$$\text{SR} = \frac{W_t - W_d}{W_d} \times 100\% \tag{1}$$

where SR is the swelling ratio, W_t is the weight (g) of the swollen hydrogel at time t, and W_d is the weight (g) of the dried hydrogel.

For dried hydrogel, swelling happens when water enters and hydrates the hydrophilic groups, causing 'primary bound water'. Once the hydration of hydrophilic parts is saturated, the polymer network starts to swell, and the hydrophobic groups disclose and interact with water, resulting in 'secondary bound water'. After establishing all the interactions, the network will absorb extra water due to the osmotic pressure between polymer chains and water, which will produce a network retraction force till the swelling undergoes equilibrium. Here, this excess water is called 'free/bulk water', the 'free water' tends to pad the external space of the network, such as micropores. The swelling behavior of hydrogel is reversible, the degradation will happen while the network or crosslinks be undermined [5].

Swelling is prevalent in most hydrogel systems, which can be utilized for enzymatic reactions, tissue blockage, drug delivery, etc. However, particularly in some biomedical applications, such as soft actuating, and tissue engineering like internal wounding healing, the swelling is unexpected that may cause unfavorable oppression on the surrounding tissues. Moreover, for most hydrogels, the volume expansion during the swelling will weaken the mechanical strength. To face these situations, anti-swelling hydrogels were designed and employed in verities of fields, such as tissue grafts, tissue regeneration medicine, implantable artificial organs, etc.

3.2 Mechanical Properties

The extensive variety of materials and different chemical configurations of hydrogels endow them with tunable structures and unique interchain crosslinking networks, which achieve the materials' widely distinctive mechanical properties. Given the high percentage of water contained within a hydrogel, most hydrogels exhibit viscoelastic or poroelastic characteristics. The well-applied hydrogel often requires an optimal combination of stiffness, strength, toughness, creep, and fatigue behavior. The hydrogel's stiffness represents the material's elastic response to forces that result in small strains. Yield strength indicates the stress at which the material starts to undergo plastic deformation, while tensile strength is the maximum stress the material can withstand without failing. Toughness is defined as a material's resistance to fracture when stressed. Creep is a time-dependent deformation of materials under the influence of persistent high mechanical stresses but still below the yield strength. The fatigue behavior reflects the intrinsic ability of the materials to absorb the energy of repeated deformation.

Mechanical testing for hydrogels can be commonly classified into three categories: macroscale testing, indentation testing, and microscopy-based testing. The macroscale properties of hydrogels include uniaxial tension, compression tests, cyclic loading tests, fracture tests, and stress relation (time-dependent creep) tests, which are tested from the universal testing machine (UTM). Indentation is a multiscale method for detecting the mechanical properties of hydrogels from millimeters to nanometers, which requires minimal samples with any shape and can test the samples as prepared in containers without being attached to the instrument. It also can efficiently characterize graded or heterogeneous hydrogels with large spatial variability using scanning techniques across the sample surface. From the indentation tests, the elastic modulus of hydrogels can be measured by indenting the gel to a prescribed depth while recording the corresponding reaction force. Also ,the time-dependent behavior of hydrogels can be determined based on viscoelastic and poroelastic theories. For microscopy-based testing, atomic force microscopy (AFM) is the typical strategy that can perform nanomechanical properties of hydrogels. Also, noncontact techniques like dynamic light scattering (DLS) are widely used for the diffusivity test of hydrogels without deforming the samples.

Uniaxial tensile testing and compression testing are the most universal types of mechanical tests for hydrogels. The hydrogel strain (ε) at any particular time is calculated from the initial length of the hydrogel between the clamps (L_0) and the length at that time (L). As the strain increases, the force required to apply this strain is recorded. If we assume the hydrogel is uniform and homogeneous, the stress (σ) on the hydrogels can then be calculated from the force (F) and initial cross-sectional area of the hydrogel before stretching (A). The Young's modulus (E) can then be calculated from the linear elastic region of the stress-strain curve using Eq. 2:

$$E = \frac{\sigma}{\varepsilon} = \frac{FL_0}{A(L - L_0)} \qquad (2)$$

In addition to Young's modulus, the tensile stress-strain data can be used for yield strength, tensile strength, strain at break, and toughness of the hydrogels.

In many circumstances, some hydrogels cannot be fabricated into dumbbells or rectangle shapes, and some of them are difficult to clamp without damage for the tensile testing, whereas the compression testing will instead, in which hydrogels will be compressed between two flat plates. The compressive modulus can be easily calculated from the stress-strain curve, while it's more appropriate to calculate the equilibrium modulus or the dynamic modulus to describe the hydrogel's mechanical behavior, e.g., tissue engineering applications where the engineered tissue would be expected to undergo repeating cycles of stress in the body. For isotropic materials, the result for the tensile and compression moduli can be related using Eq. 3:

$$E = 3K(1 - 2\nu) \qquad (3)$$

where E is the tensile young's modulus, K is the bulk (compressive) modulus, and ν is the Poisson's ratio, for many hydrogels ν is assumed to be around 0.5.

The mechanical strength of hydrogels depends on network structures and crosslinks that can direct their applications. Tuning the strength of the hydrogel may follow several aspects, choosing the crosslinker types and crosslinking strategies, involving multiple crosslinking mechanisms, using the hydrogels derived from natural proteins, forming hydrogels hybrid with nanomaterials like carbon nanotubes and graphene, designing extremely stretchable hydrogels via combining molecular sliding strategy [1].

3.3 Rheological Properties

Hydrogels behave in both viscous and elastic behavior, in which the rheological properties especially viscoelasticity is the most critical parameter to learning the crosslinking degree, molecular weight, glass transition, and structural

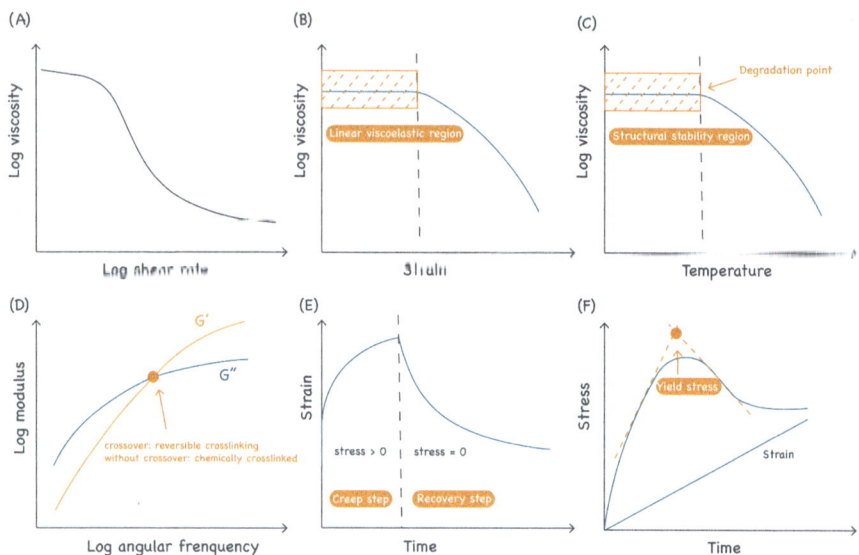

Fig. 3 The typical rheological tests of hydrogels, (**a**) steady shear flow; (**b**) strain sweep; (**c**) temperature sweep; (**d**) frequency sweep; (**e**) creep test; (**f**) stress growth test

homogeneities/heterogeneities. Generally, the rheological tests for hydrogels consist of steady shear flow, strain sweep, temperature sweep, time sweep, frequency sweep, and creep test, as shown in Fig. 3. Hydrogels are mostly non-Newtonian liquids with shear-thinning behavior; steady shear flow (Fig. 3a) detects the tendency of hydrogels to flow (e.g., injection); and the viscosity decreases as the shear rate increases. Also, zero-shear viscosity and critical strain rate can be measured by the steady shear flow test. Strain sweep also called amplitude sweep, mainly tests the linear viscoelastic region of hydrogels, showing the function of elastic modulus G′ and viscous modulus G′ with oscillatory strain under the constant frequency, as shown in Fig. 3b. Time sweep testing determines the change and stability of materials within a specific time, e.g., curing, solvent evaporation, recovery, and polymer degradation. For hydrogels, we can study the gelation process according to the time sweep results, the intersection of G′ and G′ curves is defined as the gel point, as shown in Fig. 1. The temperature sweep test (Fig. 3c) is used to evaluate the thermal stability of hydrogels, whereas the frequency sweep measurement can apply to the viscoelastic properties analysis between modulus (G′ and G′) and frequency. The selection of strain for the frequency sweep test should be within the linear viscoelastic region which is obtained by strain sweep characterization to avoid structural deformation of hydrogels happens. Also, the absence of a crossover between G′ and G″ curves means permanent chemical crosslinking, while the crossover indicates the formation of reversible crosslinking networks, as shown in Fig. 3d. Furthermore, at high angular frequency surpasses the crossover point, the elastic modulus G′ of hydrogels is higher than viscous modulus G″ corresponding to the solid-like behavior, conversely, liquid-like behavior in low frequency. The creep test provides the hydrogel's recovery capacity after the removal of constant stress, revealing and predicting the viscoelastic properties under loads for a long time, as shown in Fig. 3e. Moreover, the yield stress is another essential factor of hydrogels, which can indicate the deformation of hydrogels become to transfer from reversibly elastic to permanently plastic. For injectable hydrogels, yield stress can reveal the retained behavior of gels at the injection point. The yield stress can be determined via several methods, such as stress growth, stress ramp, creep, oscillatory techniques, and model fitting. As shown in Fig. 3f, we present the stress growth test for yield stress. When the hydrogel is sheared at a low constant rate and below critical strain showing solid-like behavior and resulting in stretchable properties; since the strain constantly increases above the critical strain, the hydrogel structure starts to break down causing shear thinning and flow. The shear stress loading when destruction happens is the yield stress, the strain defined as yield strain.

4 Several Representative Hydrogels

4.1 Biocompatible, Nontoxic, and Biodegradable Hydrogels

Hydrogels exhibit natural soft tissue-like and tunable properties that are becoming popular biomaterials for biological applications. Biocompatibility is a required property of any medical material that is used in contact with a living body. The requirements for biomaterials are nontoxicity, the ability to be sterilized, and biocompatibility. Nontoxicity means that biomaterials won't cause chronic or acute inflammation, bleeding, allergy, or cancers. Biocompatibility means that biomaterials will not disturb homeostasis, not cause unwanted rejection reactions, not cause anomalous growth, absorption, or death of organ systems, and also will adhere strongly with organs in some situations. Biocompatibility can be classified into bulk biocompatibility, which includes mechanical and design compatibility, and interfacial biocompatibility which includes adhesion and non-stimulus, such as physical non-stimulation, nonactivity, anticoagulant, and non-encapsulant. It should be emphasized that biocompatibility is a different property from nontoxicity, e.g., alumina and poly(tetrafluoroethylene) (PTFE), which are not toxic but often exhibit poor biocompatibility [6].

Biodegradable hydrogels are beneficial over nonbiodegradable hydrogels as they repeal the need to remove the 'ghost' after the useful lifetime, such as after drug release. Biodegradation leads to the conversion of polymeric networks into nontoxic, low-molecular-weight end products, evading removal from the body. Hence, the components of monomers, initiators, and crosslinking agents should be well-considered for hydrogel preparation. Hydrogels may degrade and dissolve by various mechanisms, mainly including hydrolytic, enzymatic, ionization, and photodegradation. The degradation happens with main chains, side chains, or cross-linked bonds. For example, functional groups like anhydrides, esters, and thioesters present in crosslinking agents or polymers are vulnerable to hydrolytic degradation [7]. Moreover, biodegradation via enzymatic hydrolysis is another noteworthy mechanism contributing to the degradation of hydrogels into less toxic products [8]. Lastly, most hydrogels derived from natural polymers exhibit rapid degradation upon contact with body medium. Thus, modified natural polymers and new synthetic hydrogels with better-controlled degradation properties can be utilized for more effective biomedical applications.

4.2 Injectable Hydrogels

Injectable hydrogels with high tissue-like water content and similar microenvironment to extracellular matrix (ECM) possess promising physicochemical properties that can be used as scaffolds or carriers of therapeutic agents; also they can be injected into the body under mild conditions. Injectable hydrogels attract considerable interest in the use of minimally invasive strategies for biomaterials delivery (e.g., cells, drugs, proteins, and other bioactive molecules) and tissue engineering (e.g., skin, bone, muscle, and cartilage regeneration) [9], as shown in Fig. 4a. Not only liquid biomaterials which can be injected into the body and then in situ sol-gel transition happened are regarded as injectable hydrogels, but some gels with shear thinning and self-healing features can be injected are also defined as injectable hydrogels. Besides the injectability, injectable hydrogels can especially treat irregularly shaped injured tissues as compared to conventional hydrogels, and minimally invasive procedures can highly reduce the patient's pain and cost from using the alternative pre-formed hydrogel implants. The formation of injectable hydrogels is also from chemical and physical crosslinking, usually the physical gelation process happens with some external stimuli. Moreover, remarkable parameters such as gelation kinetics, elastic rheological response, mechanical properties, stability, and injectability should be considered during the design and injection process. For example, the phase transition of gelation should take place within the restricted time interval to achieve the right injection, otherwise, the hydrogels may form inside the syringe. Furthermore, the release kinetics of therapeutic agents loaded on hydrogels can be controlled by tuning their components, porous structure, and density of cross-linking. However, several concerns do occur in injectable hydrogels system and need further exploration, such as the compatibility between hydrogels and bioactive molecules or cells during the crosslinking and proliferation process, the stability of synthetic materials to avoid the degradation and denaturation by enzymes, also cytotoxicity and inflammatory responses should under consideration.

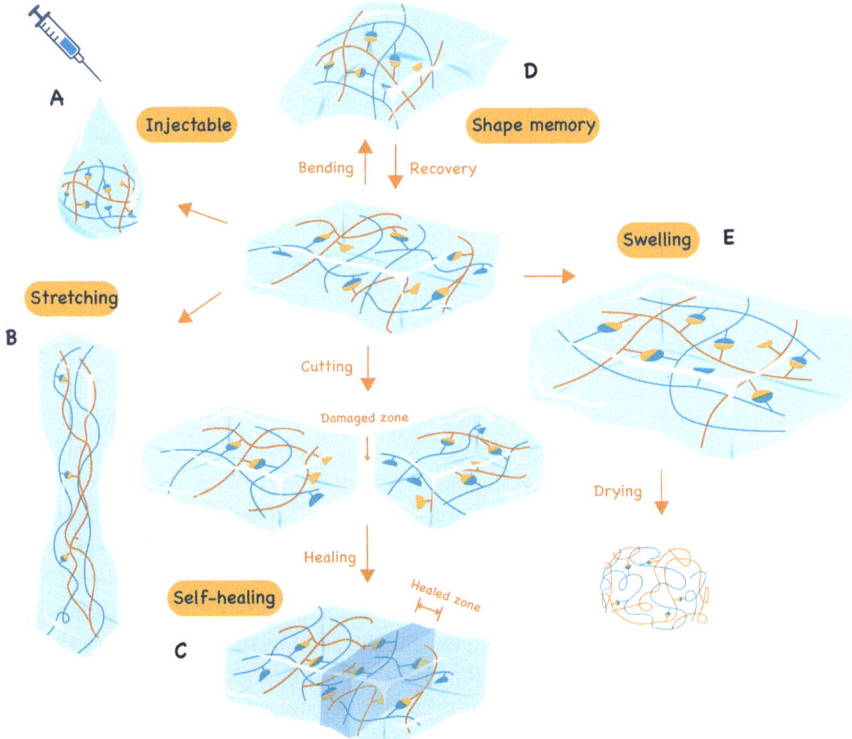

Fig. 4 Schematic of the hydrogel properties, (**a**) injectability; (**b**) stretchability; (**c**) self-healing; (**d**) shape memory; (**e**) swelling

4.3 Self-healing Hydrogels

Self-healing is one of the remarkable properties of hydrogels; self-healing hydrogels possess the recovery ability to repair self-structure after mechanical damages, such as rupture or traumatization. For hydrogels, self-healing can happen in a few seconds/minutes at room temperature, while some need a longer time with a higher temperature which depends on their structures and self-healing mechanisms (covalent and non-covalent bonding). Covalent bonding is broadly based on boronated bonds, disulfide bonds, imine bones, Diels-Alder reaction, oxime bonds, and so on, in which hydrogels are highly stable. However, there are some toxic issues that come from crosslinking agents. For non-covalent bonding, the hydrogels are highly flexible that can reconstruct efficiently after damage, which benefits from several physical interactions, such as electrostatic interactions, hydrogen bonding, host-guest interactions, and hydrophobic interactions. Especially, the novel self-healing strategy can be driven by the combination of physicochemical approaches and biological species like enzymes and microorganisms [10]. The healing process happens in the damaged zone as seen in Fig. 4c, the healing capacity is generally evaluated by mechanical and rheological recovery tests. Moreover, self-healing hydrogels with adhesive properties can be used as dressing materials for accelerating wound healing, the hydrogels loading drugs or even themselves will exhibit anti-microbial and/or anti-inflammatory functions during the healing processes. Also, promising applications like controlled drug/cell delivery can be achieved by using injectable or stimuli-responsive self-healing hydrogels. In addition, other wide uses for bone and cartilage tissue engineering, biosensors, and biomedical glues are also considerable.

4.4 Stimuli-Responsive Hydrogels

Stimuli-responsive hydrogels also called 'intelligent/smart' hydrogels, which can intelligently sense and respond to external forces or stimuli resulting in changes in stability, network structure, rheological properties, surface potential, swelling behavior, and mechanical deformation of hydrogels. 'Smart' hydrogels mean the reversible response occurs in the repeated application of stimuli. Generally, the external forces could be classified into physical stimuli (e.g., light, pressure, ultrasound, temperature, electric, and magnetic), chemical stimuli (e.g., pH and glucose), and biological stimuli (e.g., antigen, enzyme, and ligand), as shown in Fig. 5. The response mechanism depends on the functional groups and crosslinks in chemically, an ordered-disordered transition triggered by external-environmental stimuli causing reversible deformations. For example, as the pH changes in pH-responsive hydrogels, the ionization degree of ionized pendant groups changes significantly, and the sudden volume transition by electrostatic repulsion causes a high osmotic swelling stress. Temperature-responsive hydrogels are also called thermogels: chemically crosslinked thermogels undergo volume change while physically crosslinked thermo-sensitive hydrogels happen sol-gel phase transition at the critical temperature. Temperature-responsive hydrogels can be divided into positive and

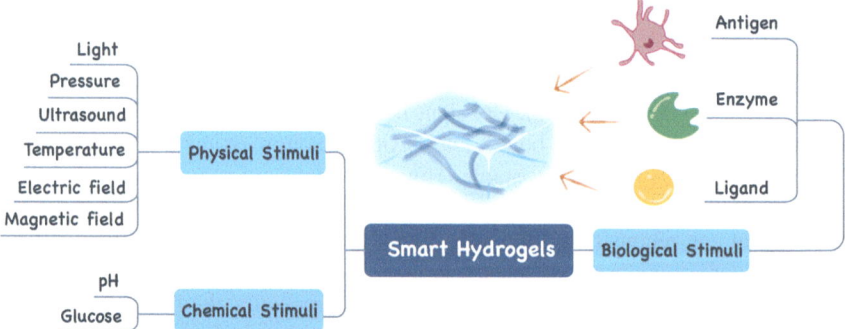

Fig. 5 Chemical, physical, and biological stimuli of stimuli-responsive hydrogels

negative systems. Polymers in positive thermogels shrink when the temperature cools below their upper critical solution temperature (UCST) while negative hydrogels shrink when the temperature heats above their lower critical solution temperature (LCST). Enzyme-responsive hydrogels can be selectively activated by specific factors, in which the polymer networks contain incorporated enzyme-detecting substrates that may access to enzymatic active center. The enzyme-sensitive hydrogels are widely tailored and applied for targeted drug delivery, the reaction between enzymes and crosslinks arouses the degeneration or morphologic phase transitions accompanied by drug release [11].

Shape-memory hydrogels and hydrogel actuators are the most typical representatives of 'smart' hydrogels developing unique and promising applications [12]. Shape memory hydrogels exhibit the ability to recover their original shapes from temporarily formed shapes, which happens in dynamic networks by building or breaking reversible crosslinks when external stimuli are applied, as shown in Fig. 4d. According to the polymer structures and crosslink number of networks, shape memory hydrogels possess dual, triple even multi-shape memory effects, which means that hydrogels can remember one, two, or more temporary shapes upon single, different stimuli even time controlled. Moreover, to meet the plentiful requirements for bio-applications, shape memory hydrogels are always designed and synthesized by combining multi-functions, such as self-healing, adhesive, antimicrobial, anti-inflammatory, antioxidant, thermoplastic, and so on. For hydrogel actuators, inhomogeneous structural polymers may swell and shrink in response to external stimulus changes performing asymmetric swelling driven by osmotic pressure, actuation stress comes from the strength and elastic modulus of the hydrogels. Typically, soft actuators prepared via photolithography and 3D printing that can mimic nature behaviors (e.g., self-folding, twisting, and bending) for microscale uses, such as biosensors, bioprobes, soft robotics, micromanipulators, tunable optical devices, and microfluidic devices. In sum, stimuli-responsive hydrogels with tunable variability in physicochemical properties as biomimetic scaffolds and carries attracting much attention in varieties of fields [13]. For example, smart hydrogels especially multi-stimuli smart hydrogels may use for responsive controlled drug release for disease therapies like anticancer and antitumor, in which the controlled release can be achieved by synergizing different treatments, such as chemotherapy, photothermal, and magnetic hyperthermia therapy [14].

5 Biomedical Applications of Hydrogels

Hydrogels with crosslinked 3D networks possess high water absorption and retention, showing tissue-like physicochemical properties applied to diverse biomedical applications. Currently, hydrogels are widely employed for the fabrication of contact lenses and many hygiene products, such as diapers, panty liners, personal care tissue paper, etc. Here, we highlight the other three critical applications of hydrogels in biomaterial fields: drug delivery, tissue engineering, and wound healing.

5.1 Drug Delivery

Hydrogels as intelligent carriers are extensively utilized in controlled drug delivery systems, which can load cells drugs, engineered, nucleic acid, and protein through highly porous 3D structures, then deliver and control release following various mechanisms. Drug loading is mainly from three strategies: swelling the hydrogel in drug solutions, designing hydrogel as reservoirs to place drug microparticles, and incorporating drugs into hydrogel during preparation. The controlled release can be achieved from different types of combinations between hydrogels and drugs, such as drugs dissolved/loaded in hydrogels, drugs encapsulated in hydrogels, drugs loaded in stimuli-responsive hydrogels, and drugs loaded in degradable hydrogels, in which drugs released from swelling equilibrium, diffusion, external stimuli (e.g., pH, temperature, enzyme), and degradation [15]. To efficiently proceed drug delivery process, several key factors should be under consideration, to control the density and degree of network crosslinking tuning hydrogel swelling behaviors to trigger drug release, enhance the intrinsic binding between hydrogel carriers and loaded drugs extending residence times for sustained delivery, design smart hydrogels obtaining well-controlled enviro-responsive drug release, and introduce molecular recognition leading precisely targeted drug delivery. As shown in Fig. 6e, drugs encapsulated in hydrogel could be fabricated as the ingested-sized pill, the drug-loaded pH-sensitive hydrogel undergoes rapid swelling till reaches the stomach, and the drug release happens when swollen and triggered by the acidic external environment. If the synthetic hydrogel possesses superior mechanical and powerful loading force with drugs, long-term release can be achieved [16]. Furthermore, the hydrogels can be designed into different porous content and multi-size like macroscopic

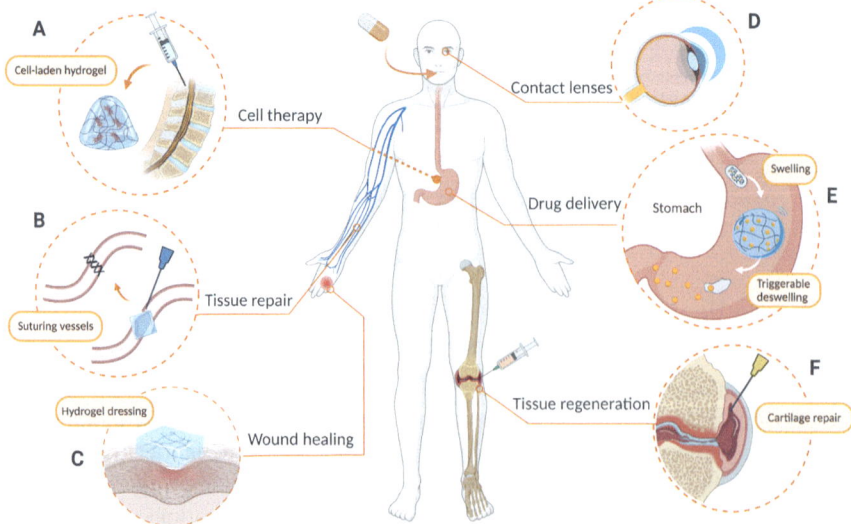

Fig. 6 Biomedical applications of hydrogels, (**a**) cell therapy; (**b**) tissue repair; (**c**) wound healing; (**d**) contact lenses; (**e**) drug delivery; (**f**) tissue regeneration

scale, mesh scale, molecular and atomistic scale. The drug delivery can proceed by surgical implantation, local needle injection, or intravenous infusion [17].

5.2 Tissue Engineering

Tissue engineering is a biomedical engineering concept that aims to develop biological substitutes to maintain, improve, repair, regenerate, or replace organs/tissues, such as skin, heart, nerve, bone, spinal cord, cartilage, blood vessels, bladder, muscle, etc. Hydrogels emerged as appealing scaffolds for tissue engineering due to their tissue-like features, which can mimic the natural extracellular matrix (ECM) and supply 3D structural templates for cell adhesion, migration, differentiation, and proliferation. As scaffold hydrogels, another desirable application is to precisely deliver bioactive molecules to the target tissues and to encapsulate secretory cells, compared with conventional drug delivery that mostly needs large doses to obtain the required dosage for aimed tissue because of the appearance of enzymatic degradation and nonspecific uptake by other tissues during the delivery process. Also, hydrogels as space-padding agents can be employed in bulking, anti-adhesive, and adhesive functional uses. Both natural and synthetic hydrogels fit the requirement of injectable, biocompatible, biodegradable, mechanical strength, and non-toxic can be considered used in tissue engineering. Hydrogels especially injectable hydrogels as scaffolds are particularly engineered for tissue regeneration and adoptive cell therapy. For tissue regeneration such as osteoarthritis cartilage repair, there are generally two strategies: (1) encapsulate autologous cells in the hydrogels first and then implant them into the defect area; (2) induce and cooperate the surrounding cells to take part in the repair, as shown in Fig. 6f. For cell therapy, the injectable cell-loaded hydrogel is an exciting realm to deliver the cells for tissue engineering, which specifically mimics the microenvironment of the targeted tissue, significantly improves the compatibility of hydrogels, and universally utilize in a non-invasive way. For example, Fig. 6a illustrates the process of injured spinal cord repair using cell-laden hydrogel. Overall, injectable hydrogels show an attractive capacity for versatile tissue engineering applications, that handle in a minimally invasive way instead of the surgical process. Furthermore, some pre-hydrogels may be difficult to load cells and then crosslink in-vitro, and most injection processes may partially destroy the cells causing the reduction of therapy efficiency, which can be improved by encapsulating the engineered cells into microspheres and binding the microsphere in the hydrogels [18].

5.3 Wound Healing

Wound dressings prevent the injured wound from external infection and provide an enabling microenvironment to promote wound healing. Basically, skin wound healing undergoes four processes: hemostasis, inflammation, proliferation, and remodeling. Hydrogel dressings attract much attention in varieties of wounds (e.g., incisional, excisional, and chronic wounds like burn, inflection, and diabetic wounds), benefit from their promising biocompatibility, degradability, removability, adhesiveness, shape adaptability, mechanical property, and bioactive multifunctionality. The schematic diagram of hydrogels for wound healing is exhibited in Fig. 6c. To preserve high efficiency and refrain infections during the healing process, the hydrogel dressings need to be replaced, while the change of dressings causes high costs and may cause a secondary injury. Therefore, hydrogel dressings' own acceptable degradability and removability are essential. The primary condition for wound healing is to stop bleeding, in which hydrogel dressings as hemostatic agents and then adhere to cover the open wound and avoid further infections. Also, the hydrogel dressings should fit the irregular wound area well and the movement like bending through rapid shape adaptability and superb mechanical properties. More importantly, promising hydrogel dressings not only provide protection but facilitate the healing process, so bioactive functionalization is the efficacious strategy to face the infection challenge and achieve efficiently wound regeneration. The multifunctionality of hydrogel dressings includes adhesive, hemostatic, anti-oxidant, anti-inflammatory, anti-bacterial, self-healing, controlled delivery, stimuli-responsive, conductive, and wound monitorable properties [19]. The challenge that should be considered is how to coordinate the multifunction into one dressing to fit mutative conditions and meet requirements of the specific regeneration process, like cells/bioactive molecules loading, pH value of wound environment, etc.

6 Conclusion and Future Perspectives

This chapter systematically explained the definition, diversity classification, and gelation process of hydrogels. The formation mechanisms of hydrogels by physical crosslinking include entanglement, ionic interactions, self-assembly, and crystallization, and chemical crosslinking consists of hybrid polymer, crosslinking molecules, photosensitive agents, and enzymatic crosslinking. Meanwhile, the properties of hydrogels like swelling, mechanical strength, rheology, biocompatibility, biodegradability, toxicity, injectability, self-healing, and stimuli-response are exhaustively introduced. In which, expound the mechanism of swelling behavior, typical mechanical tests (e.g., microscale, indentation, and microscopy-based testing), typical rheological characterization (e.g., steady shear flow, strain sweep, temperature sweep, time sweep, frequency sweep, creep test, and stress growth test). Furthermore, the chapter highlighted the essential properties and functions of injectable hydrogels, self-healing hydrogels, and stimuli-responsive hydrogels that include shape-memory hydrogels and hydrogel actuators. Finally, it emphasized the promising applications of hydrogels in the biomedical field, such as contact lenses, hygiene products, drug delivery, tissue engineering, and wound healing; thoroughly explained the principles of hydrogels for drug delivery, tissue engineering, and wound healing applications; and enumerated several practical procedures, such as cell therapy, tissue repair, tissue regeneration, and hydrogel dressing.

Hydrogels benefit from tissue-like physicochemical properties playing an attractive role in biomedical fields. However, there are still some critical challenges that should be considered and proceeded for further development. Anti-swelling hydrogels as an alternative to overcome the drawbacks occurred among swelling hydrogel applications, tuning hydrogels strength by blending with nanomaterials and involving advanced mechanisms like molecular sliding strategy. The biocompatibility is a different concept as non-toxicity, many synthetic even natural hydrogels are non-toxic but appear in poor biocompatibility; also modified natural hydrogels and/or their derivatives can achieve better-controlled degradation. Meanwhile, the compatibility between hydrogels and bioactive molecules or cells during the crosslinking and proliferation process, the stability of synthetic materials to avoid the degradation and denaturation by enzymes, also cytotoxicity and inflammatory responses in injectable hydrogel systems should be under consideration. Also, in tissue engineering, some pre-hydrogels may be challenging to load cells and then crosslink in-vitro, and most injection processes may partially destroy the cells causing the reduction of therapy efficiency, which can be improved by encapsulating the engineered cells into microspheres and binding the microsphere in the hydrogels. For drug delivery, molecular and atomistic scale hydrogels like microgels and nanogels are promising biomaterials that exhibit unique superiority. Furthermore, using hydrogel dressings should lay out multifunctional dressings to fit mutative conditions and meet the requirements of the specific regeneration process.

Quizzes

1. Which cross-linking mechanism is irreversible?

   (a) Self-assembly
   (b) Crystallization
   (c) Ionic interactions
   (d) Enzymatic crosslinking

   Answer: d.

   Explanation: The other crosslinking mechanisms are considered as physical crosslinking, which are dependent on the polymer chemical environment, while for the enzymatic crosslinking, the polymeric chains are bounded through stable covalent bound.

2. Swelling affects the mechanical properties by:

   (a) Disturbing/reducing the interactions between the polymeric chains.
   (b) Altering the chains configuration by changing the polymer network density.
   (c) Volume expansion.
   (d) All the mentioned above.

Answer: d.

Explanation: As the solvent defuse through the polymeric network, it reduces the polymer-polymer interactions and moves the chains apart, while increasing the polymer-solvent interactions, thus, imbibing water in the polymer structure. In other words, altering the chain configurations and expanding the network to a new design according to the new environment.

3. All the non-toxic hydrogels are considered as biocompatible materials.

   (a) True
   (b) False

Answer: b.

Explanation: Biocompatibility is a different property from nontoxicity, e.g., alumina and poly(tetrafluoroethylene) (PTFE), which are not toxic but often exhibit poor biocompatibility.

4. Which of the tests does not deform the sample?

   (a) Atomic Force Microscopy (AFM)
   (b) Compression test
   (c) Fracture test
   (d) Uniaxial test

Answer: a.

Explanation: The Atomic Force Microscopy (AFM) is performed by a very thin needle that scans the surface at the atomic surface providing accurate data about the atomic surface topography, where it can identify even absorbed molecules on the surface. In the compression test, the sample will be pressed between two disks. In the fracture test, the sample will be ruptured, while in the uniaxial testing, stress will be applied simultaneously in one or two directions. During all these tests, the sample will be deformed and ruptured.

5. Injectable hydrogels are considered:

   (a) Newtonian fluid
   (b) Dilatant fluid
   (c) Pseudoplastic fluid

Answer: c.

Explanation: During the injection process, pressure will be applied to the syringe to push the fluid outside the needle to its destination. When pressure is applied to the dilatant fluid, its viscosity increases, preventing it from flowing easily. In the case of pseudoplastic fluid, the viscosity decreased as the pressure is applied, allowing the fluid to flow. Polymers are considered non-Newtonian liquids since the presence of the long polymeric chains results in non-negligible friction between the chains, in other words, viscosity. Therefore, injectable polymers are considered pseudoplastic.

6. Regards the Fig. 3d, which of the following is true:

   (a) The high angular frequency in region B, corresponds with the liquid-like behavior.
   (b) The low angular frequency in region A, corresponds with the solid-like behavior.
   (c) The crossover between G' and G" confirms the existence of reversible crosslinking network.
   (d) The crosslinking between G' and G" indicates the existence of permanent chemical crosslinking.

   Answer: c.
   Explanation: High angular frequency in region B, corresponds with the solid-like behavior while the low angular frequency in region A, corresponds with liquid-like behavior. The absence of crossover between G' and G" curves indicates that there is permanent chemical crosslinking.

7. Which of the following is <u>not</u> true?

   (a) Liquid biomaterials, as well as gels with shear thinning features which can be injected, are considered injectable hydrogels.
   (b) Biocompatibility and nontoxicity are the same property, and the existence of one of them is enough.
   (c) Self-healing ability is a common property among hydrogels, and can happen in two forms, based on the bonding mechanism.
   (d) Chemically and physically crosslinked hydrogels, react in a different way to temperature changes.

   Answer: b.
   Explanation: Biocompatibility and nontoxicity are different properties, some products are nontoxic, still, not highly biocompatible.

8. Which of the following is not true?

   (a) UCST (the upper critical solution temperature), in which the hydrogel shrinks when cooled to lower than this level.
   (b) Chemically crosslinked thermo-gels are positive temperature-responsive hydrogels, while physically crosslinked hydrogels are negative responsive.
   (c) Light, pressure, ultrasound, and temperature are examples of physical stimuli for hydrogels.
   (d) Number of crosslinked networks in the polymer structure can affect to which level the polymer is considered a multi-functional.

   Answer: b.
   Explanation: Positive and negative responsive systems are two groups that the polymers are divided to, when they undergo temperature changes, regardless they are chemically or physically crosslinked (these are two different points).

9. From the written above, regards drug delivery hydrogels, one can infer:

   (a) Synthetic hydrogels, with superior mechanical strength are not compatible for drug release applications.
   (b) Density and level of network crosslinking are important keys for successful drug delivery process.
   (c) The acidic environment in the stomach inhibits the drug release process.
   (d) Hydrogels that undergo degradation are not able to deliver drugs in the human body.

   Answer: b.

   Explanation: Being a hydrogel with high mechanical strength is important for achieving long-term release, thus, it can be applicable in drug delivery. Acidic environment possesses drug release process and doesn't inhibit it. Drugs loaded in degradable hydrogels is one possible mechanism for drug delivery and release.

10. The properties of being degradable and removable in wound healing field are necessary for:

    (a) Changing hydrogel dressing.
    (b) Supplying low-cost hydrogels replacing process.
    (c) Suppling bioactive multifunctionality.
    (d) Preserving high efficiency during the healing process, as well as, refraining infections.

    Answer: d.

    Explanation: Hydrogel dressing replacement is essential for preserving high efficiency and preventing infections during the healing process, but this step is highly expensive and can cause injuries, thus, it is important to own the mentioned properties as a substitution.

References

1. Zhang YS, Khademhosseini A (2017) Advances in engineering hydrogels. Science 356(6337):eaaf3627
2. Annabi N, Tamayol A, Uquillas JA, Akbari M, Bertassoni LE, Cha C, Camci-Unal G, Dokmeci MR, Peppas NA, Khademhosseini A (2014) 25th anniversary article: Rational design and applications of hydrogels in regenerative medicine. Adv Mater 26(1):85–124
3. Peppas NA, Hoffman AS (2020) 1.3.2E – Hydrogels. In: Wagner WR, Sakiyama-Elbert SE, Zhang G, Yaszemski MJ (eds) Biomaterials science, 4th edn. Academic, London, pp 153–166
4. Rubinstein M, Colby RH (2003) Polymer physics. Oxford University Press, New York
5. Hoffman AS (2002) Hydrogels for biomedical applications. Adv Drug Deliv Rev 54(1):3–12
6. Ikada Y (2001) Section 11 – Biocompatibility of hydrogels. In: Osada Y, Kajiwara K, Fushimi T et al (eds) Gels Handbook. Academic, Burlington, pp 388–407
7. Kunduru KR, Hogerat R, Ghosal K, Shaheen-Mualim M, Farah S (2023) Renewable polyol-based biodegradable polyesters as greener plastics for industrial applications. Chem Eng J 459:141211

8. Vinchhi P, Rawal SU, Patel MM (2021) Chapter 19 – Biodegradable hydrogels. In: Chappel E (ed) Drug delivery devices and therapeutic systems. Academic, London, pp 395–419
9. Li Y, Yang HY, Lee DS (2021) Advances in biodegradable and injectable hydrogels for biomedical applications. J Control Release 330:151–160
10. Taylor DL, in het Panhuis M (2016) Self-healing hydrogels. Adv Mater 28(41):9060–9093
11. El-Husseiny HM, Mady EA, Hamabe L, Abugomaa A, Shimada K, Yoshida T, Tanaka T, Yokoi A, Elbadawy M, Tanaka R (2022) Smart/stimuli-responsive hydrogels: cutting-edge platforms for tissue engineering and other biomedical applications. Mater Today Bio 13:100186
12. Korde JM, Kandasubramanian B (2020) Naturally biomimicked smart shape memory hydrogels for biomedical functions. Chem Eng J 379:122430
13. Kutner N, Kunduru KR, Rizik L, Farah S (2021) Recent advances for improving functionality, biocompatibility, and longevity of implantable medical devices and deliverable drug delivery systems. Adv Funct Mater 31(44):2010929
14. Shang J, Le X, Zhang J, Chen T, Theato P (2019) Trends in polymeric shape memory hydrogels and hydrogel actuators. Polym Chem 10(9):1036–1055
15. Shaheen-Mualim M, Kutner N, Farah S (2022) The emerging potential of crystalline drug-polymer combination for medical applications. Polym Adv Technol 33(11):3797–3799
16. Farah S, Doloff JC, Müller P, Sadraei A, Han HJ, Olafson K, Vyas K, Tam HH, Hollister-Lock J, Kowalski PS, Griffin M, Meng A, McAvoy M, Graham AC, McGarrigle J, Oberholzer J, Weir GC, Greiner DL, Langer R, Anderson DG (2019) Long-term implant fibrosis prevention in rodents and non-human primates using crystallized drug formulations. Nat Mater 18(8):892–904
17. Li J, Mooney DJ (2016) Designing hydrogels for controlled drug delivery. Nature Rev Mater 1(12):1–17
18. Zarrintaj P, Khodadadi Yazdi M, Youssefi Azarfam M, Zare M, Ramsey JD, Seidi F, Reza Saeb M, Ramakrishna S, Mozafari M (2021) Injectable cell-laden hydrogels for tissue engineering: recent advances and future opportunities. Tissue Eng A 27(11–12):821–843
19. Liang Y, He J, Guo B (2021) Functional hydrogels as wound dressing to enhance wound healing. ACS Nano 15(8):12687–12722

Engineering Biomaterials for Nucleic Acid-Based Therapies

Parveen Kumar, Umberto Capasso Palmiero, and Piotr S. Kowalski

Abstract Nucleic acid-based therapies are emerging as a promising approach for treating a wide range of diseases, including cancer, genetic disorders, and infectious diseases. However, the delivery of nucleic acids to target cells remains a significant challenge due to their large size, negative charge, and susceptibility to degradation. Biomaterials have the potential to overcome these barriers and enable efficient and safe delivery of nucleic acids. Engineering biomaterials for nucleic acid-based therapies involves the design and optimization of materials and formulations that can protect nucleic acids from degradation, deliver them to desired target cells, facilitate cellular uptake, and promote endosomal escape. This book chapter is focused on the engineering of such materials from a chemical, formulation, and manufacturing point of view. The chapter begins by providing an introduction of nucleic acid therapies and the challenges associated with delivering them to the correct target cells. The chapter then explores various biomaterials that have been developed for nucleic acid delivery with a particular focus on lipids and polymers. The properties of these biomaterials are described, and their advantages and limitations are discussed. Next, the chapter delves into the engineering approaches used to modify their physical and chemical properties to enhance efficacy and specificity. The chapter also covers the manufacturing, characterization, and evaluation of these biomaterials for nucleic acid delivery, including physiochemical characterization, such as sizing, surface

---

Authors Parveen Kumar, Umberto Capasso Palmiero, and Piotr S. Kowalski have equally contributed to this chapter.

P. Kumar
School of Pharmacy, University College Cork, Cork, Ireland

U. Capasso Palmiero
Department of Non-Viral Delivery, RNA & Gene Therapies, Novo Nordisk A/S, Måløv, Denmark

P. S. Kowalski (✉)
School of Pharmacy, University College Cork, Cork, Ireland

APC Microbiome Ireland, University College Cork, Cork, Ireland
e-mail: piotr.kowalski@ucc.ie

charge, and pKa evaluation. In vitro and in vivo experimental set-ups to assess biocompatibility, stability, and gene transfection efficiency are also discussed. The chapter also emphasizes the importance of regulatory compliance and safety considerations in the development of clinical-grade biomaterials. Finally, the chapter concludes by highlighting the main nucleic acid therapeutics that have been already marketed and that are in development with a particular focus on siRNA, mRNA, and CRISPR applications.

Graphical Abstract

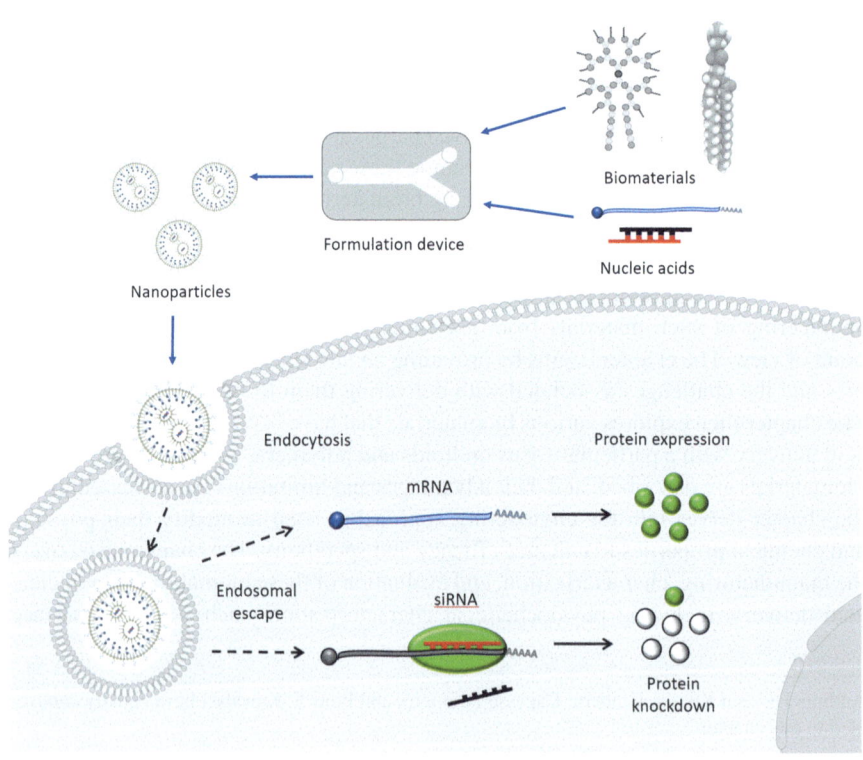

Keywords Biomaterials · Polymers · Lipids · Nucleic acid delivery · RNA therapeutics · Manufacturing

1 Introduction to Nucleic Acid-based Therapies

Gene therapy involves the delivery of therapeutic nucleic acids (DNA or RNA) into a patient's cells allowing to (i) increase target gene expression, (ii) silence the expression of the disease-causing gene, or (iii) modify a target gene by inserting, removing or entirely replacing it. Historically gene therapy refers to a gene transfer by introducing a DNA molecule into the cells but over the years it has expanded to delivery of other types of nucleic acids with different mechanisms of gene regulation, e.g. RNA interference or gene editing. Nucleic acid drugs allow specific control of gene and protein expression and are on the path to becoming a major platform in drug development alongside small molecules and other biologics. Those currently used for therapy, include plasmid DNA (pDNA), short interfering RNAs (siRNA), antisense oligonucleotides (ASOs), and messenger RNA (mRNA), as well as gene editing tools. Transient silencing of the desired gene can be achieved by neutralizing its mRNA transcript using siRNA or ASOs while the delivery of mRNA or pDNA allows producing any therapeutic protein inside the target cells. The mechanisms of action, regulation of gene and protein expression of RNA are summarized in Fig. 1. As of 2022, there were 16 drug products utilizing therapeutic delivery of RNAs (9 ASO, 5 siRNA, 2 mRNA) and 27 products

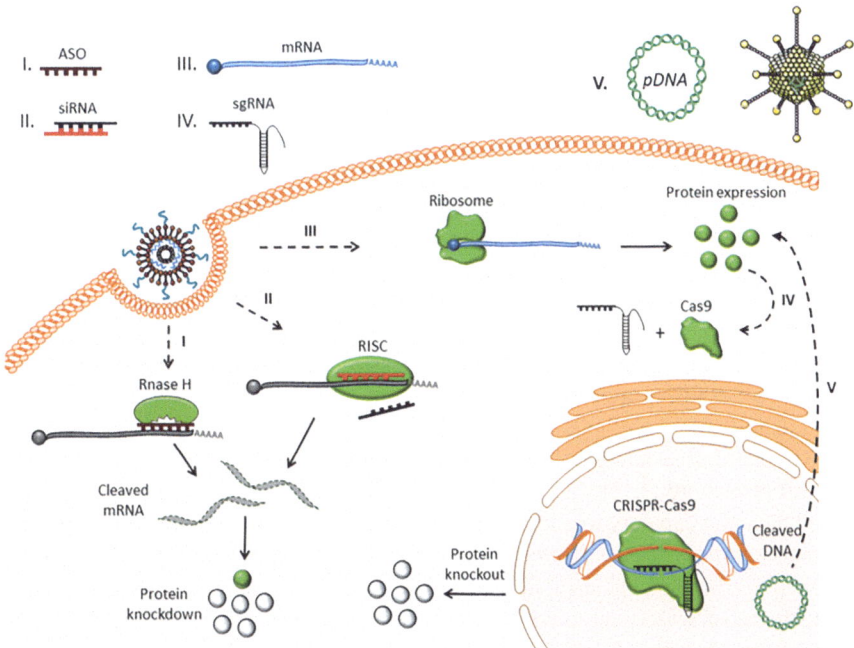

Fig. 1 Regulation of gene and protein expression using nucleic acid delivery. (Adapted from Kaczmarek et al, Genome Medicine 2017)

exploiting viral gene delivery or the use of genetically modified cells, such as chimeric antigen receptor (CAR-T) cells, that are also considered a form of gene therapy.

All types of nucleic acids share physicochemical properties which impede their uptake across biological membranes and lead to their rapid clearance from circulation, including high molecular weight (pDNA > 1000 kDa, mRNA 300–5,000 kDa, siRNA ~14 kDa, and ASOs 4–10 kDa), the hydrophilic nature, and the negative charge. These molecules are also prone to enzymatic degradation by endo and exo nucleases (e.g. ribonucleases or deoxyribonucleases) that cleave the phosphodiester bond or catalyse the removal of nucleotides from the polynucleotide chain resulting in their relatively poor half-life in the circulation (most stable: DNA > ASO > siRNA > mRNA). To exert their biological function most RNA-based drugs need to reach the cytoplasm of the cell while ASO due to their small size can penetrate through both cellular and nuclear membranes allowing them to interact with both cytoplasmic and nuclear targets. pDNA requires entry to the nucleus which creates an additional delivery barrier for this type of therapy. At the cellular level, cellular uptake and escape from the endosomal compartment are essential for their therapeutic activity and remain one of the main challenges for nucleic acid delivery. However, before even reaching their intracellular destination, nucleic acid drugs have to avoid non-specific tissue distribution and reach the diseased cells, escape renal clearance, and survive the hostile extracellular environment after systemic or local administration.

Overcoming these delivery barriers is critical to facilitate the clinical translation of nucleic acid-based therapies and requires the development of effective and safe delivery systems. The first class of vectors adopted to deliver nucleic acids was viruses since they are naturally able to transfect their genetic material (DNA or RNA) into cells. Despite a number of products based on viral vectors having been approved for clinical use, this class of carriers is facing challenges related to complex manufacturing, immunogenicity, and pre-existing immunity towards disease-causing viruses (e.g. adenovirus) which may limit their ability for repeated dosing. For some types of viral vectors potential for integration into the genome (e.g. lentivirus), and constraints on the size of DNA payload (e.g. adeno-associated virus) can also narrow the scope of possible applications. For this reason, nucleic-acid delivery approaches based on non-viral vectors have been developed which include the use of nanoparticles composed of organic and inorganic materials, extracellular vesicles, and nucleic-acid conjugates, which have been discussed in detail elsewhere [1, 2]. In general, non-viral delivery vectors are easy to scale up and have the capacity to address many limitations of viral vectors, mainly concerning manufacturing and safety. A range of biocompatible materials have been designed to complex, encapsulate and deliver therapeutic nucleic acids and among them, the most widely adopted are lipids and polymers which will be the main focus of this chapter.

2 Design of Biomaterials for Nucleic Acid Delivery

Biomaterials used in the formulation of nanocarriers for nucleic acid delivery must meet a series of important characteristics to protect and safely deliver the cargo to the correct site [3]. First, they must allow loading of nucleic acids into a delivery system capable of protecting them from the nucleases present in the environment and in the body. Preferentially, the biomaterials should be able to form nanometric colloids (i.e. nanoparticles) that can be safely injected intravenously and effectively enter the cell membranes. The nanoparticles should be stable both in simple (e.g. phosphate-buffered saline (PBS), tris(hydroxymethyl)aminomethane (Tris) buffer) and complex fluids (e.g. blood) to avoid aggregation-related toxicity and they should possess narrow size distribution to assure uniform biodistribution in the body. Moreover, the biomaterial should be able to induce the escape of the nanoparticles from the intracellular compartments such as endosomes and then, it should allow the release of the cargo once reached the target site (e.g. either the cytoplasm or the nucleus). Ideally, the nanocarrier should degrade into safe and biocompatible components to avoid accumulation and toxicity. It is also of primary importance that the nanoparticles are masked from the immune system to allow the re-dosing of the therapeutic and avoid allergic reactions. Moreover, they should be able to mainly target the desired type of cells in the desired organs to avoid off-target toxicity. In the end, the biomaterial must also be easy to produce, to characterize, and to scale-up to avoid reproducibility problems and complex quality control procedures that can affect the availability of the final product on the market (i.e. the so-called "drug shortages"). All these characteristics are difficult to obtain with only one type of biomaterial and, for this reason, the nanoparticles used in gene delivery often consist of multiple components.

In general, the priori design of nucleic acid delivery systems that possess tissue specificity, high delivery efficacy, and safety is challenging due to the many requirements and the complexity of the nanoparticle-biological environment interactions. For this reason, there is not a "one-size-fits-all" solution for every therapeutic application and optimization of the formulation composition and/or of the physiochemical characteristics of the single formulation components is generally always required. Therefore, it is crucial to identify and optimize the parameters that have the highest impact on the different NP characteristics relevant to their performance, such as NP size and surface composition. This is usually achieved by the synthesis of a chemically diverse library of biomaterials and by the optimization of the final formulation, e.g. the composition of each single component. In the contest of such multidimensional optimization problems, the use of advanced mathematical tools, such as design of experiment (DOE) and machine learning, can accelerate the discovery and selection of the lead candidate [4].

2.1 Lipids

Lipid nanoparticles represent the most advanced class of nucleic-acid delivery carriers as evidenced by the recent clinical approvals, including the first siRNA drug Onpattro in 2018 for the treatment of hereditary transthyretin-mediated amyloidosis (hATTR amyloidosis) and two mRNA-based COVID-19 vaccines (BNT 162b and mRNA-1273) in 2020/21. The lipid nanoparticles (LNPs) generally consist of four different components: a cationic or ionizable lipid, a helper lipid, cholesterol, and a PEGylated lipid (Fig. 2).

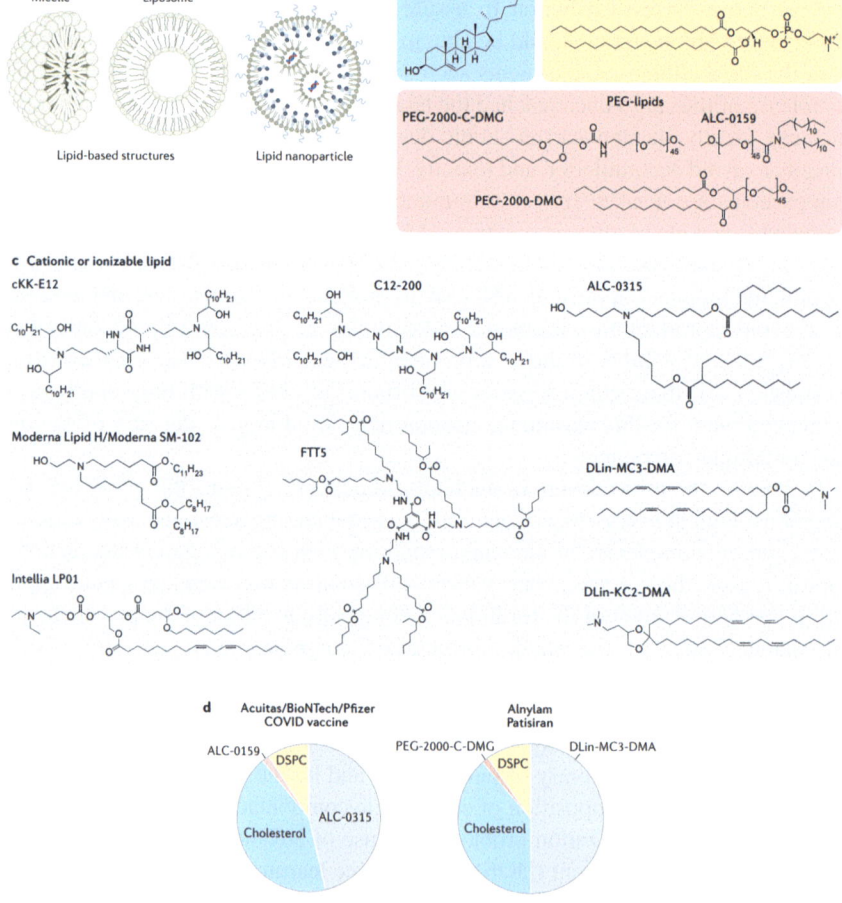

Fig. 2 (**a**) Lipid-based structures can include micelles, which consist of a lipid monolayer, or liposomes, which consist of a bilayer. Lipid nanoparticles are composed of multiple lipid layers as well as microdomains of lipid and nucleic acid. (**b**) LNPs often consist of cholesterol, a helper lipid, a PEG-lipid, and (**c**) a cationic or ionizable lipid. (**d**) The molar ratios of the four components making up the FDA-approved Acuitas/BioNTech/Pfizer COVID vaccine (BNT 162b) and Patisiran (Onpattro). (Adapted from Paunovska et al. [3])

The cationic or ionizable lipid represents one of the most important components of the formulation and it is responsible for the loading of the cargo into the NPs via ionic complexation between the positively charged polar head of the lipid and the negatively charged phosphate groups present in the backbone of the nucleic acids. In contrast, the two alkyl tails of the lipid are responsible for holding together the nanoparticle via hydrophobic effect and to interact and fuse with biological membrane bilayers to promote cellular uptake and endosomal escape. The nature of the positively charged head group and, in particular, the acid dissociation constant (pKa) of the amine groups plays a vital role in the endosomal escape, efficacy, and toxicity of the lipid nanoparticles. Lipid heads that contain chemical groups with pKa > 7.4, such as primary, or quaternary ammonium groups, confer a permanent cationic nature to these ionizable lipids in relevant physiological conditions. For this reason, although this class of cationic lipids can achieve high cargo loading efficiency, their permanent cationic nature can induce potential toxicity via perturbation of cellular membranes and hampers the release of the cargo from the nanoparticles by a strong electrostatic binding. For this reason, ionizable lipids with pKa around 5.5–6.5 have been found to be the most effective thanks to their ability to load nucleic acids during formulation at low pH (usually around pH = 3.5–5), to avoid toxicity at physiological conditions (pH = 7.4) due to their neutral charge, and then to induce endosomal escape when the nanoparticles reach the more acidic endosomal environment (pH = 5–6.5). Another important characteristic of these ionizable lipids is the presence of biodegradable chemical groups, such as esters or amides, that can be hydrolysed in the biological milieu by different enzymes (e.g. lipases or esterases) into shorter and safer compounds to avoid accumulation into the body and the associated toxicity.

While cholesterol is used to modulate the membrane fluidity and stability, the helper lipids are generally neutrally charged lipids that are primarily used to modulate the physiochemical properties of the lipid nanoparticles through the alteration of the geometry of the lipid layer, in particular its curvature and planarity. Helper lipids can be classified according to the cross-sectional area of the lipid head group and the lipid tail as cylindrical, cone, and inverse cone shapes [5].

In the end, PEGylated lipids are lipids that contain hydrophilic polyethylene glycol as hydrophilic head and are used to provide a hydrophilic surface coat able to increase colloidal stability and reduce the interaction of the LNPs with serum proteins. This hydrophilic shell is responsible for extending the LNPs circulation half-life into the blood and to avoid the clearance from the mononuclear phagocytic system. However, there is increasing evidence that PEG can induce the formation of anti-PEG antibodies. These antibodies can lower the efficacy of subsequent administered doses of the LNPs in the so-called accelerated blood clearance effect or, in the worst case scenario, they can cause allergic reactions [6]. For this reason, researchers are focused on finding PEG alternatives [7].

The nature and relative amount of all these components affect also the physiochemical properties of the LNPs, such as size, charge, and surface properties. In particular, the composition of the LNP surface plays an important role in their ability to deliver the cargo to specific sites. Once the LNPs are in contact with complex biological fluids, different biomolecules start to adsorb on their surface generating a protein coat that alters how the nanoparticles interact with the biological environment, i.e. the immune systems and cells in different tissues. This so-called protein corona is ultimately responsible for the trafficking of the LNPs into different cells and organs via the so-called "passive targeting". As an example, the adsorption of the apolipoprotein E (ApoE) on the surface of LNPs is responsible for their delivery to hepatocytes in the liver. Similarly, the biological milieu can affect the composition of the protein corona and therefore change the LNPs tropism. As an example, higher concentration or absence of specific biomolecules caused by the presence of a disease can alter the biodistribution of nanoparticles by modifying the endogenous trafficking pathways. Another common strategy (the so-called "active targeting") to improve the on-target capability is to introduce on the LNP surface ligands (e.g. antibodies) that can specifically bind to biomolecules or receptors overexpressed on the surface of specific cells.

2.2 Polymers

Polymers adopted for gene delivery applications share many traits of the ionizable lipids used in the formulation of lipid nanoparticles: (i) the presence of ionizable amine groups to promote nucleic acid complexation and to improve endosomal escape; (ii) the presence of hydrophobic groups to increase the NP colloidal stability; and (iii) the presence of hydrolysable groups to increase biodegradability and therefore reducing toxicity. In general, biomaterial hydrophobicity, pKa, and biodegradability are among the most important design parameters to obtain effective and safe gene delivery systems [8]. The biggest difference between these two classes of biomaterials relies on the fact that while ionizable lipids generally consist of a single molecule, polymers are composed of a repetition of one or more different types of subunits. This represents a double-edged sword that increases the versatility and chemical space of these biomaterials, but at the same time increases the complexity and may hamper the reproducibility of these systems. The multimeric nature of these materials allows to introduce additional parameters that can be used to optimize the nucleic acid delivery performance. Namely, the molecular weight of the polymer (i.e. the overall number of monomeric units in the polymer chain) and the charge density (i.e. the composition and distribution of different monomeric units in the polymer chain) are two important parameters that can be balanced to obtain high nucleic acid loading efficiency, nanoparticles stability, efficient endosomal escape, and reduced cytotoxicity. In a similar manner, the topological structure of the polymer (such as linear, branched, and brush-like structures) can be also tuned to obtain more stable and efficient carriers. As an example, dendrimers are orderly branched polymers with a spherical structure that are proven effective in delivering mRNA, even though they are difficult to synthesize. Since polymers consist of a distribution of different chemical species with different monomer composition, sequence, and length, the manufacturing process plays a fundamental role to control the main properties and to assure batch-to-batch reproducibility (see Sect. 3.2). As an example, it would be important to obtain polymer with narrow molecular weight distribution or, in other words, with a dispersity (Đ) close to 1. Dispersity is defined as the ratio between the polymer weight-average molecular weight (M_w) and number-average molecular weight (M_n) and it is a measure of how heterogenous is a distribution of molecules in terms of molecular weight (i.e. length in case of polymers). A polymer with a Đ = 1 is monodisperse and contains only polymer chains of the same length. In contrast, a polymer with a Đ > 1 is dispersed and contains polymer chains of different lengths.

The history of polymers resembles the history of cationic/ionizable lipids, i.e. from the use of permanently charged species, then moving to the use of ionizable amine groups, and, in the end, including biodegradable moieties to boost efficacy and safety [9]. In fact, the first generation of polymers consists of permanently charged polycations that are structurally composed of one or more type of monomer, such as polyamino acids (e.g. poly-L-lysine and poly-L-ornithine). In the second generation, ionizable amine groups were included, such as in the case of linear and branched polyethyleneimine (PEI). PEI and its derivatives are polymers with abundant amino groups with different pKa and, for this reason, they provide both high loading efficiency and strong endosomal escape. However, as in the case of the first generation, these polymers have limited biodegradability, and, for this reason, they have also poor biocompatibility and relatively high cytotoxicity.

In the latest generation, labile chemical linkages are introduced in the main backbone of the polymers to increase biodegradability. Among them, the most adopted linkages are ester bonds due to their ability to degrade in physiological conditions via hydrolysis and esterases. Example of these polymers include poly-(β-amino-esters) (PBAEs), poly(amine-co-esters) PACEs, amino-polyesters (APEs), and charge-altering releasable transporters (CARTs).

3 Examples of Polymers for Nucleic Acid Delivery

Polymers are highly valued in biology and medicine due to their versatility and unique properties resulting from their chain molecular structure. They have been widely used for the development of delivery systems for a variety of therapeutic payloads, including small molecules, proteins, and nucleic acids. Biodegradable polymers can enhance the biocompatibility of the delivery system via degradation into natural and non-toxic metabolites. Both synthetic and naturally derived biodegradable polymers hold the potential for the development of safer nucleic acid therapeutics.

3.1 Natural and Synthetic Polymers

Natural and nature-derived polymers are composed of building blocks that occur in nature and can often be extracted from natural sources, e.g. plants, shrimp shells, etc. Nature-derived polymers used for nucleic acid delivery are often composed of amino acids (polypeptides, e.g. poly-l-lysine (PLL)) or sugar molecules (polysaccharides, e.g. chitosan). These polymers are usually biodegradable. Chitosan and PLL polymers have the ability to complex nucleic acids via electrostatic interaction with their cationic amine groups, however, the design constrained to natural building blocks can contribute to their low transfection efficiency and cytotoxicity, while relatively high dispersity may complicate manufacturing these polymers.

Synthetic polymers are built from fully synthetic monomers which often do not exist in nature. These polymers are generally used to expand the design features and improve on properties of natural polymers. With the ability to select from variety of synthetic monomers they can be tailored to meet desired properties such as mechanical strength, self-assembly, degradability, stimuli responsiveness, dispersity, and drug cargo release but the use of synthetic building blocks may pose challenges with biocompatibility.

3.2 Approaches for the Synthesis of Cationic Polyesters

Polyesters are synthetic polymers widely used in drug delivery systems due to several favourable traits such as relatively low toxicity, biodegradability, biocompatibility, and most importantly they can be obtained using Generally Recognized as Safe (GRAS) molecules. Examples of clinically approved polyesters include polylactide (PLA), polyglycolide (PGA), polycaprolactone (PCL), and polylactide-co-glycolide (PLGA). However, the amine-containing polyesters are the main class of cationic polyesters that have been developed for nucleic acid delivery and the majority are synthesized using one of these three strategies: (1) poly-condensation of diol and dicarboxylic acid derivatives, (2) Michael addition of diacrylates and diamines, or (3) ring opening polymerization (ROP) of cyclic monomers (e.g. lactones) (Fig. 3).

1. Poly-condensation is a step-growth polymerization which involves the polymerization of diol and dicarboxylic acid or diester-containing monomers through the elimination of a small molecule, such as water or alcohol, to form a long-chain polymer. PACEs are cationic polyesters with low charge density usually synthesized by polycondensation reaction via enzymatic copolymerization of hydrophobic diesters and amine-containing diol monomers [10]. This process is usually carried out under vacuum with controlled temperature. Drawbacks of this method include relatively low molecular weight of the synthesized polymer with broad molecular weight distribution (Đ: 1.5–2.5) resulting in the poor control over polymerization process. This method also precludes the use of mono-

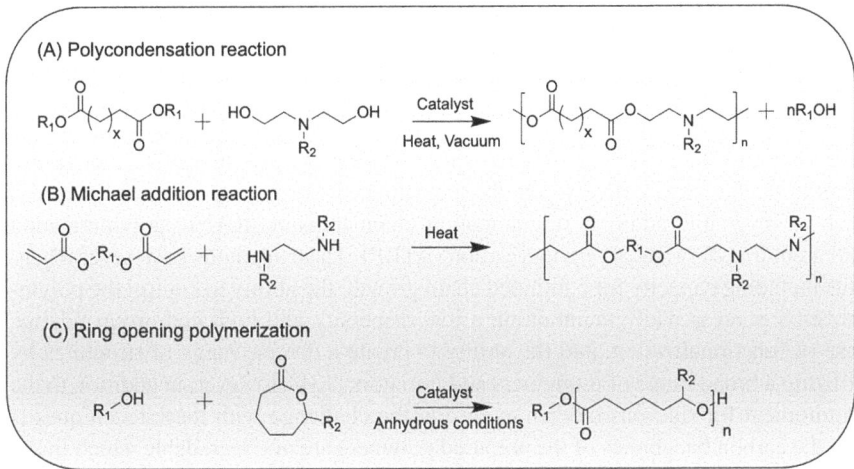

Fig. 3 Reaction scheme of (**a**) poly-condensation of diol and dicarboxylic acid derivatives, (**b**) Michael addition of diacrylates and diamines, and (**c**) ring opening polymerization of lactones

mers containing primary or secondary amines from the composition of the resulting polymer. This is because primary and secondary amines can interfere in the formation of ester bonds during the condensation process owing to the formation of amide bonds which is favoured over ester bonds.

2. Cationic polyesters can also be synthesized by Michael addition which is also a step-growth polymerization and involves the combination of a diamine with a diacrylate such as bis-acrylates or bis-methacrylates. This is a simple and one-step polymerization method which provides access to large chemical space of monomers compatible with both primary and secondary amines. This method has been used to synthesize PBAEs which endow materials with tunable properties including degradation, mechanics, hydrophilicity, and swelling [11]. One of the major drawbacks of this method is the lack of control over the polymerization process (polyaddition mechanism) especially for branched materials, resulting in a broad molecular weight distribution.

3. ROP is a versatile method for synthesizing polyesters with well-defined molecular weight and low dispersity by opening the ring of a cyclic monomer followed by transesterification reaction between monomer and a polymer chain with a reactive hydroxyl group. These reactions are often catalysed by organic or metal-based catalysts. APEs are cationic polymers synthesized via controlled ROP of lactones with tertiary amino-alcohols, offering a one-step process with the ability to control the degree of polymerization [12]. Similar to poly-condensation, this method also precludes the use of monomers containing primary or secondary amines. Charge-altering releasable transporters (CARTs) is another class of cationic polymers used for nucleic acid delivery that can be synthesized via ROP of cyclic carbonates with secondary amines and, moreover, they can be designed to be pH-sensitive for controlled release at specific pH conditions. This method

allows the use of monomers containing secondary amines but requires additional steps for amine protection and deprotection. Limitations of ROP include sensitivity to contamination with water and solvents containing hydroxyl group, limited selection of catalysis and monomers, and potential risk of racemization.

In addition to the described polymerization strategies, synthesis of cationic polymers in nucleic acid delivery can also be achieved using more advanced methods such as reversible addition fragmentation chain transfer (RAFT) polymerization and atom transfer radical polymerization (ATRP). These methods have several benefits including capacity for continued chain growth, the ability to control the polymerization process while maintaining a low dispersity and high end-group fidelity, ease of functionalization, and the ability to create a diverse range of structures by utilizing a broad range of monomers and initiators [13]. However, in addition to the requirement for rigorous oxygen exclusion the challenge with these techniques is that the carbon backbones of the prepared polymers are not degradable which limits their use for biomedical applications.

4 Manufacturing of Carriers for Nucleic Acid Delivery

4.1 The Importance of Good Manufacturing Practice (GMP)

The manufacturing of gene delivery carriers typically involves several steps such as synthesis, purification, and characterization of delivery system components (e.g. biomaterial and nucleic acids), nanoparticles formulation, quality control, sterilization, packaging, and storage. The exact process may vary depending on the type of delivery system and its intended use. All the components of the delivery systems for clinical use must be produced according to Good Manufacturing Practice (GMP) quality standards and maintained in sterile condition essential to ensure the safety and efficacy of the therapy. GMP describes the production standards that must be met to minimize risks, waste, and production losses and to ensure that medicines are manufactured and quality controlled according to set standards. All drugs intended for the European Union (EU) market must comply with GMP requirements that ensure that they meet the standards for market authorization.

4.2 Example of Manufacturing Process for mRNA Therapeutics

Production of cationic polymers and lipids for nucleic acid delivery is a key aspect of the manufacturing process. These materials should be relatively easy to synthesize at large scale (e.g. grams to kilograms), purify, and characterize, and should meet GMP standards for commercial production. Compared to cationic polymers, synthesis of lipids often requires multiple steps that can compromise product yield, but their advantage is a defined chemical structure and molecular weight, making them easier to characterize and control quality of the final product. Lipid nanoparticles are being used to manufacture siRNA-based drug Onpattro against hereditary thransthyrein amyloidosis and mRNA-based COVID-19 vaccines (Comirnaty, Spikevax) which represent the first examples of RNA-based therapeutics.

Manufacturing process of mRNA drugs involves several steps, including (1) pDNA manufacturing, (2) mRNA synthesis and purification, and (3) mRNA-LNP formulation and purification (Fig. 4).

1. pDNA template production involves cloning a gene of interest into a plasmid vector, growing the plasmid-containing bacteria, lysing the cells to release the pDNA, purification to remove impurities and contaminants, conducting quality control, formulating, and lyophilizing the pDNA. It is a complex process done under GMP guidelines.
2. mRNA molecules are synthesized using a pDNA template through an enzymatic reaction known as in vitro transcription (IVT). Purification of the synthesized mRNA usually involves tangential flow filtration (TFF) followed by sterile filtration. High-performance liquid chromatography (HPLC), anion exchangers, cellulose purification, and hydrogen bonding are other systems often used for purification of mRNA products.

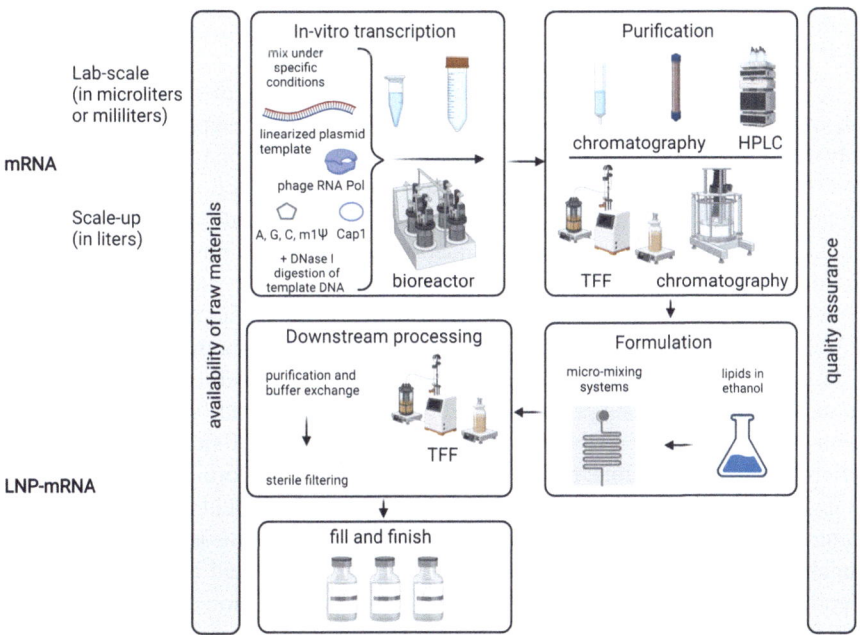

Fig. 4 Manufacturing and scale-up of nucleoside-modified mRNA vaccines. (Adapted from Szabo G et al., Mol Ther 2022).

3. The final step in the manufacturing process involves formulation of mRNA into lipid nanoparticles. This process usually involves the rapid mixing of mRNA/buffer and lipid solutions to create spontaneous self-assembly of encapsulated mRNA-LNPs formulation. The next step is diafiltration/concentration TFF which eliminates the residual solvent and concentrates the formulation, followed by sterile filtration to ensure that it is free of any contaminants before it is packaged and released for use.

4.3 Formulation Methods for Nanoparticle Manufacturing

Adequate control over the chemistry of cationic biomaterial and formulation parameters such as nucleic acid/polycations ratio, nanocarrier composition (e.g. helper lipids, PEG-lipid, cholesterol, surfactants (e.g. Pluronic)), and formulation method are critical for ensuring a high degree of reproducibility and consistent performance of nucleic acid delivery system. Optimization of the formulation requires changing the ratio of nucleic acid to ionizable lipid/polymer, modifying excipients compositions, and the appropriate selection of solvents and buffers for each of the components. To facilitate this, a reproducible method of mixing lipophilic (e.g. lipids, polymer) and hydrophilic (nucleic acid) components to control self-assembly into the uniform particles driven by electrostatic and hydrophobic interactions is necessary to ensure batch-to-batch consistency. This process should also be scalable from millilitres to hundreds of litres, to minimize the need for process re-development during translation to the clinic.

Different methods have been used for the formulation of nucleic acid complexes with cationic polymers or lipids, which include low energy methods taking advantage of simple self-assembling of nanoparticle components facilitated by mixing, vortexing or microfluidics. Ethanol injection method is one of the most frequently used techniques to produce nanocarriers by rapidly diluting ethanolic solution of lipids/polymers into an aqueous medium containing nucleic acid. In contrast, high energy methods may require additional energy input in the form of heat (e.g. electrostatic spray-drying) or pressure (e.g. dry lipid film hydration followed by extrusion or sonication to control the size of nanoparticles) to form uniform nanoparticles. Simple mixing and vortexing methods are generally used in the lab for small-scale production. However, ethanol injection method coupled with microfluidics are widely used in industry and for pre-clinical animal studies. Microfluidics is a scalable, robust, and highly reproducible formulation technique. It involves mixing of the aqueous phase containing nucleic acid and organic phase containing lipids/polymers via a micromixer (Fig. 5). Well-defined LNPs with tunable properties (e.g. different LNP size) can be produced by controlling the operating parameters which include continuous flow, diffusion distance, flow rate ratio, controlled mixing time, and temperature control, resulting in high reproducibility. The microfluidics technique allows for rapid optimization of LNP manufacturing conditions, reproducibility, and ease of scaling up, provide the

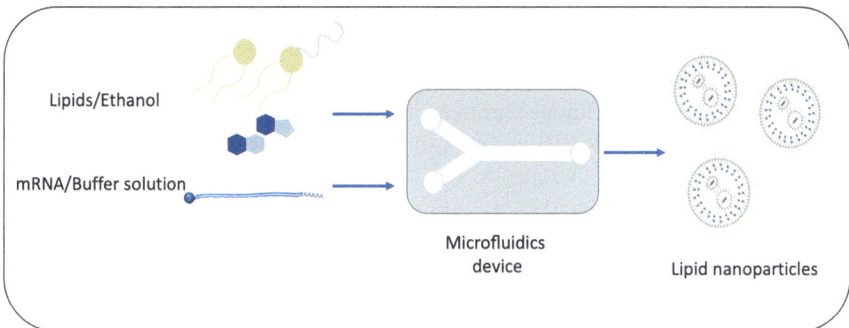

Fig. 5 Schematic illustration of the preparation method of the RNA-loaded LNP using a microfluidic device

exceptional contribution to the LNP-based nucleic acid delivery technology and their applicability of translation from laboratory to practical applications.

5 Analytical Methods to Control Quality of Nucleic-Acid Delivery Systems

5.1 *Characterization of Physicochemical Properties*

Physicochemical properties of the nucleic acid nanocarriers, such as particle size, size distribution, and surface charge, are commonly characterized using techniques such as dynamic light scattering (DLS), transmission electron microscope (TEM), nanoparticle tracking analysis (NTA), and electrophoretic light scattering (ELS). DLS measures the hydrodynamic diameter and size distribution of the nanocarriers (in the size range of 10–1000 nm), while particle tracking analyses the movement of individual particles in solution to provide information about their size, concentration, and size distribution. The particle surface charge is evaluated via ELS and it is represented by zeta potential which is a measure of the difference in potential between the bulk fluid (in which a particle is dispersed) and the layer of fluid containing the oppositely charged ions (associated with the nanoparticle surface). The electrostatic or charge attraction/repulsion between nanoparticles is one of the main parameters known to affect nanoparticle colloidal stability. Polydispersity index (PDI) reflects size distribution and PDI values between 0.05 and 0.2 indicate monodisperse particle population. Aggregation of particles leads to increase in their size and can indicate poor stability. Particle morphology can be investigated using cryogenic electron microscopy (cryo-TEM) or atomic force microscopy (AFM). Cryo-TEM captures the images of samples embedded into a thin layer of non-crystalline ice to provide the information about the nanoparticle internal structure. Monitoring those physicochemical properties is important since particles larger (>500 nm) or highly charged (> +/−20 mV) tend to be quickly cleared from circulation due to the uptake by mononuclear phagocyte system (MPS). Both concentration and nucleic acid encapsulation can be measured using nucleic acid-binding fluorescent dyes (e.g. Ribo/PicoGreen) or liquid chromatography methods (e.g. ion-pair reversed-phase chromatography (IP-RP) or size exclusion chromatography (SEC)) and are critical to determine the therapeutic dose. Nucleic acid encapsulation refers to entrapment of the nucleic acid cargo into nanoparticles and is measured after disruption of the particles, usually using a detergent (e.g. Triton X-100) or an organic solvent. Monitoring physicochemical properties allows to assess the quality of nucleic acid delivery system and ensure batch-to-batch reproducibility.

5.2 Measurement of the Acid Dissociation Constant (pKa)

pKa of polymer/lipid ionizable head groups affects the surface charge and ionization behaviour of the nanocarrier which can influence nucleic acid encapsulation, delivery efficacy and nanocarrier bio-distribution. pKa indicates the pH at which 50% of the headgroups exist in ionized (positively charged) state. The apparent pKa value is linked to the nanocarrier composition and is strongly influenced by noncovalent interactions such as dielectric constant, ionic strength, π–π stacking, hydrophobic interactions, and nature of neighbouring charges. The apparent pKa of the nanocarriers is relatively lower as compared to the calculated pKa value of individual ionizable biomaterial. Ideally, the apparent pKa value of the nanocarrier should align with the pH of the endosomal compartment to facilitate effective escape of nucleic acid cargo. The acid-base titration (or potentiometric titration method), fluorescent 2-(p-toluidino)-6-naphthalene sulfonic acid (TNS) and z-potential methods can be used for the measurement of the apparent pKa of nanocarriers (Fig. 6). In the acid-base titration method, titration of the nanocarriers (in an acidic solution) is carried out using NaOH or KOH solution to determine the pH in the middle of the two equivalence points. TNS displays strong fluorescence after binding with cationic headgroups and the pKa of the nanocarrier is determined as the pH at the half maximum value of the fluorescence. The pKa of the nanocarrier can also be calculated by measuring the zeta potential as a function of pH.

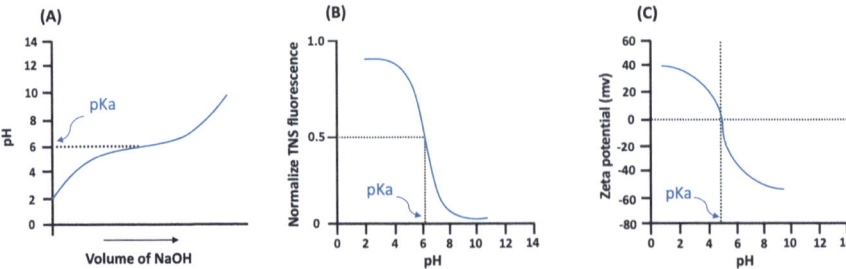

Fig. 6 Schematic diagram of the methods for pKa measurement using (**a**) potentiometric titration, (**b**) TNS fluorescent measurement, and (**c**) zeta potential

5.3 Evaluation of the Performance of Nucleic Acid Nanocarriers

To facilitate effective nucleic acid delivery, the nanocarrier must be (1) taken up by the cells, (2) escape from the endocytic compartment, and (3) release the therapeutic cargo into the cytoplasm allowing gene expression or gene silencing. The uptake is usually evaluated by fluorescence microscopy or flow cytometry which provides insight into how effectively the nanocarrier is able to cross the cell membrane and enter the cell. Nucleic acid conjugated with fluorophore (e.g. Cy5, Alexa Fluor 488) or nanocarrier labelled using lipophilic dyes (e.g. DiI, DiD) allow tracking the delivery and the fate of nanocarrier components. Flow cytometry is a quantitative method that can measure the fluorescence signal in the single cell and allows separation of different cell populations based on their size and internal complexity (e.g. granularity) rendering it particularly suitable for analysis of complex tissue samples. Intensity of the fluorescence signal reflects the extent of nanocarrier uptake providing a high-throughput and quantitative analysis but compared to fluorescence microscopy this method lacks intracellular resolution. Monitoring of the endosomal release of the nucleic acid requires the use of confocal microscopy that allows to image narrow optical section inside the cell (e.g. 0.5 um thick) combined with immuno-fluorescent staining for marker of endosomal rupture (e.g. Galectin 9) or staining of acidic compartments within a cell (e.g. LysoTracker).

Efficacy of mRNA/DNA delivery can be measured by analysing expression of a reporter gene in the transfected cell such as green fluorescent protein or firefly luciferase or desired therapeutic gene. Efficacy of gene silencing with siRNA is usually assessed by a real-time quantitative reverse transcription polymerase chain reaction (qRT-PCR) and Western Blot to determine the levels of mRNA transcript and protein knockdown in the transfected cells. Evaluation of the nucleic acid delivery in the cell culture models that lack the complexity of the whole organism offers a limited indication of the in vivo performance therefore pre-clinical efficacy evaluation in animal models such as rodents and non-human primates is needed prior to clinical trials involving human subjects. In vivo testing can also provide information about safety of the therapeutic nucleic acid delivery by assessing various parameters including organ function, body weight loss, and markers of inflammatory response.

6 Therapeutic Applications of Nucleic Acid Delivery

Whole genome sequencing enabled identifying the genetic roots of many diseases bringing us closer to developing personalized treatments based on precise control of gene and protein expression with nucleic acid-based drugs. This therapeutic strategy holds the promise to address a wide range of diseases, including genetic disorders, cancer, cardiovascular, and infectious diseases. Currently approved drug products that utilize non-viral vectors focus predominantly on systemic and local delivery of therapeutic RNA molecules (siRNA, mRNA). Lipids-based nanocarriers are presently the most clinically advanced platform for RNA delivery but the available drug products (Onpattro, Spikevax, Comirnaty) are limited to targeting cells in easily accessible tissues such as liver hepatocytes after systemic administration or muscle and dendritic cells after intramuscular administration. Polymers show the potential to facilitate RNA delivery to the lungs and gastrointestinal tract tissues, however, no products have been approved to date. Encouraging data on PBAE-mediated delivery of mRNA to lungs via inhalation of nebulized nanocarriers as well to gastrointestinal tissue via oral administration have been reported in large animals (non-human primates and pigs) [14, 15]. To fully unlock the therapeutic potential of nucleic acid base therapies, it is therefore critical to develop efficient delivery systems suitable for administration via parenteral, oral, and inhalation routes.

6.1 Short Interfering RNAs (siRNA)

Short interfering RNAs are double-stranded RNAs, 21–23 base pairs in length, that can selectively bind and degrade complementary mRNAs leading to transient silencing of protein expression. siRNAs are loaded onto the RNA-inducing silencing complex (RISC) that facilitates the cleavage of target mRNA. This conserved mechanism is known as RNA interference and its discovery was recognized with a Nobel Prize in physiology and medicine in 2006. After almost two decades of research, siRNA-based therapies represent one of the most clinically advanced platforms for RNA drugs. Onpattro was the first siRNA drug to reach the market in 2018 and was directed against the dysfunctional transthyretin gene underlying a rare genetic disease hereditary transthyretin-amyloidosis causing the buildup of amyloid in tissues and organs. Onpattro contains siRNA which silence the expression of TTR mRNA and its corresponding protein formulated into lipid nanoparticles (size range 60–100 nm) composed of a blend of four lipid excipients: DLin-MC3-DMA; PEG2000-C-DMG; DSPC; and cholesterol. These lipids protect siRNA from degradation by endo- and exo-nucleases in the circulatory system and facilitate delivery to the liver hepatocytes. Presence of ionizable lipid DLin-MC3-DMA is important for particle formation, endosomal release of the siRNA and coating of the LNP by apolipoprotein E which facilitates binding to the low-density lipoprotein receptor on hepatocytes. Following the success of Onpattro several siRNA drugs have been approved targeting liver and cardiovascular diseases, including Givlaari (acute hepatic porphyria), Oxlumo (primary hyperoxaluria type 1), Amvuttra (hATTR) and Leqvio (primary hypercholesterolaemia). Latest siRNA drugs shifted from using LNPs to N-acetylgalactosamine (GalNAc) conjugated siRNA that show high affinity binding to the asialoglycoprotein receptor (ASGR) expressed on hepatocytes. Due to their small size and good stability, siRNA conjugates allow subcutaneous administration and require less frequent dosing that improves patient compliance. However, future applications outside the liver will require the discovery of suitable cellular receptors and new types of biomaterials to design targeted nanoparticles and conjugates.

6.2 Messenger RNA

mRNA delivery allows to transiently express the desired therapeutic protein, including secreted, intracellular and transmembrane proteins, inside the host cells. Advantages of mRNA over pDNA include rapid and transient protein production allowing control of the therapeutic dose, no risk of insertional mutagenesis, and potentially greater efficacy with non-viral delivery by virtue of mRNA cytoplasmic activity. Uridine-rich mRNA sequences have been identified as a key activator of toll-like receptors and cytosolic pattern recognition receptors such as retinoic acid-inducible gene I (RIG-I) hampering the therapeutic use of mRNA. Discoveries related to nucleoside modification lead to replacing uridine with modified nucleosides (e.g. pseudouridine) which has proven effective in immune evasion without reducing mRNA translation. In addition, extensive purification of double-strand RNA (dsRNA) fragments during mRNA manufacturing, e.g. using HPLC leads to further reduction of the immunogenicity of mRNA-based drugs and improves protein production. mRNA-based drugs have a wide range of therapeutic applications, which include prophylactic and therapeutic vaccines, protein-replacement therapy aimed at rare genetic diseases, and gene editing. Naked (unformulated) mRNA has been rarely used therapeutically owing to its large size and susceptibility to degradation therefore nanoparticle-formulated mRNA has been the main method of choice for mRNA delivery. The LNPs in mRNA COVID-19 vaccines consist of four main components: a neutral phospholipid (DSPC), cholesterol, a polyethylene-glycol (PEG)-lipid, and a new generation of ionizable cationic lipid (ALC-0315 or SM-102). mRNA vaccines developed against SARS-Cov2 (Comirnaty and Spikevax) are the first mRNA-based drugs on the market that have been administered to over 5 billion people globally since 2021. The vaccine is administered by intramuscular injection and requires 2 doses 3–4 weeks apart. mRNA LNPs are captured by antigen-presenting cells at the injection site and are transported to a draining lymph node, while mRNA is translated into virus spike protein which is presented to immune cells (T-cells) and activates both cellular and humoral responses. The success of mRNA vaccines established it as a safe and effective therapeutic modality and has laid the foundation for the development of a new class of mRNA-based prophylactic and therapeutic vaccines for various indications, including cancer and infectious diseases.

6.3 CRISPR

Clustered regularly interspaced short palindromic repeats (CRISPR) technology is a simple yet powerful tool for editing genomes. It allows to alter DNA sequences and modify gene function and could potentially be used to treat or prevent diseases such as muscular dystrophy, sickle cell anaemia, transthyretin amyloidosis and familial hypercholesterolemia. CRISPR–Cas systems rely on Watson–Crick base-pairing between a single guide RNA (sgRNA) and a corresponding genomic DNA target site for binding of endonuclease CRISPR-associated protein 9 (Cas9) and cleavage of the target sequence, to introduce a double-stranded break (DSB) into a DNA molecule. DSBs can be repaired by the cells using non-homologous end joining (NHEJ) and homology-directed repair (HDR). NHEJ results in stochastic insertions and deletions ("indels") causing permanent gene knockout, whereas HDR occurs in the presence of a DNA template containing homology to regions flanking the DSB site, leading to the incorporation of desired changes encoded in the repair template into the genome. Cas9 of Streptococcus pyogenes (SpCas9) is the enzyme that is most commonly used for genome editing and genetic manipulation using CRISPR–Cas, but a growing collection of engineered RNA-guided enzymes (e.g. base and prime editing) is expanding the genome-manipulation toolbox. Therapeutic use of CRISPR-based tools focuses on genome editing leading to gene knock-out or base editing resulting in the correction of a single nucleotide mutation. The large size of Cas9 (>160 kDa) and the need for a short exposure to limit non-specific editing events pose a significant delivery challenge. Most clinically advanced approaches of in vivo gene editing utilize the co-delivery of mRNA encoding Cas9 and sgRNA formulated into liver-targeting lipid nanoparticles.

Quiz

Question 1: Which of the following nucleic acids require entry into the nucleus to exert their therapeutic function?

(a) ASOs
(b) siRNA
(c) mRNA
(d) pDNA

Correct Answer(s): (d) – Transcription machinery needed to express exogenous DNA is present in the nucleus therefore a nuclear entry is necessary

Question 2: What are the main challenges for nucleic acid delivery at the cellular level?

(a) Renal clearance
(b) Susceptibility to endo and exonucleases
(c) Cellular uptake
(d) Escape from the endosomal compartment

Correct Answer(s): (c) and (d) – Renal clearance and susceptibility to endo and exonucleases describe challenges for systemic delivery

Question 3: What is the purpose of PEGylated lipids in the formulation of lipid nanoparticles?

(a) To increase the interaction of the LNPs with serum proteins
(b) To decrease the circulation half-life of the LNPs

(c) To reduce the clearance from the mononuclear phagocytic system (MPS)
(d) To prevent the accelerated blood clearance after repeated dosing

Correct Answer(s): (c) – PEG makes nanoparticle surface more hydrophilic, masks the surface charge and decreases particle size which in turn makes them less recognizable by the MPS

Question 4: The key role(s) of ionizable/cationic lipid in the lipid nanoparticle (LNP) RNA delivery system, include:

(a) Supporting the lipid bilayer
(b) Promoting RNA encapsulation
(c) Facilitating endosomal escape
(d) Preventing LNP binding to serum proteins

Correct Answer(s): (b) and (c) – Positive charge of ionizable/cationic lipid helps attract negatively charged RNA and promote interactions with negatively charged lipids in the endosomal membrane

Question 5: What is dispersity in the context of polymers?

(a) The ratio between the polymer weight-average molecular weight (Mw) and number-average molecular weight (Mn)
(b) The number of monomeric units in the polymer chain
(c) The overall charge density of the polymer
(d) The hydrophobicity of the polymer

Correct Answer(s): (a) – dispersity is a measure of the heterogeneity of a polymer sample and is defined as the ratio of the weight-average molecular weight (Mw) to the number-average molecular weight (Mn)

Question 6: Which of the following is true about natural polymers used for nucleic acid delivery?

(a) They are built from fully synthetic monomers
(b) They can be extracted from natural sources
(c) Poly-(β-amino-esters) is an example of natural polymer
(d) They can be composed of sugar molecules

Correct Answer(s): (b) and (d) – other answers refer to synthetic polymers

Question 7: What is the major drawback of Michael Addition as a method for synthesizing cationic polyesters?

(a) Limited control over polymerization process
(b) Inability to use monomers containing primary or secondary amines
(c) Sensitivity to temperature
(d) Limited selection of catalysts and monomers

Correct Answer(s): (a) – The mechanism of Michael Addition polymerization makes it difficult to control the polymer molecular weight and reach high molecular weight

Question 8: Which of the following describes the purpose of Good Manufacturing Practice (GMP) in gene delivery carrier manufacturing?

(a) To decrease production losses and waste
(b) To maximize safety and efficacy of therapy
(c) To maintain quality control standards
(d) To ensure compliance with European Union market requirements

Correct Answer(s): (a), (b), (c), and (d) – GMP describes the production standards that must be met to minimize risks, waste, and production losses and to ensure that medicines are manufactured and quality controlled according to set standards

Question 9: Which of the following methods is used for the formulation of clinically approved RNA-based lipid nanoparticles?

(a) Self-assembly facilitated by simple mixing or vortexing
(b) Ethanol injection method coupled with microfluidics
(c) Electrostatic spray-drying
(d) Dry lipid film hydration followed by extrusion

Correct Answer(s): (b) – Ethanol injection method is often used to produce LNPs for clinical use, providing good scalability and control over particles self-assembly, size and polydispersity index.

Question 10: Which physicochemical property of the nanoparticle is reflected by the zeta potential?

(a) The size of the nanoparticles
(b) The concentration of nanoparticles
(c) Surface charge
(d) The nucleic acid encapsulation

Correct Answer(s): (c) – Zeta potential is the electric potential difference between a dispersed particle and the surrounding liquid and it is a measure of the electrical charge and stability of a colloidal system.

References

1. Elsharkasy OM et al (2020) Extracellular vesicles as drug delivery systems: why and how? Adv Drug Deliv Rev 159:332–343
2. Kulkarni JA et al (2021) The current landscape of nucleic acid therapeutics. Nat Nanotechnol 16:630–643
3. Paunovska K, Loughrey D, Dahlman JE (2022) Drug delivery systems for RNA therapeutics. Nat Rev Genet 23:265–280
4. Narayanan H et al (2021) Design of biopharmaceutical formulations accelerated by machine learning. Mol Pharm 18:3843–3853
5. Kulkarni JA, Cullis PR, van der Meel R (2018) Lipid nanoparticles enabling gene therapies: from concepts to clinical utility. Nucleic Acid Ther 28:146–157

6. Sellaturay P, Nasser S, Islam S, Gurugama P, Ewan PW (2021) Polyethylene glycol (PEG) is a cause of anaphylaxis to the Pfizer/BioNTech mRNA COVID-19 vaccine. Clin Exp Allergy 51:861–863
7. Hoang Thi TT et al (2020) The importance of poly(ethylene glycol) alternatives for overcoming peg immunogenicity in drug delivery and bioconjugation. Polymers (Basel) 12
8. Kim HJ, Kim A, Miyata K (2022) Synthetic molecule libraries for nucleic acid delivery: design parameters in cationic/ionizable lipids and polymers. Drug Metab Pharmacokinet 42:100428
9. van den Berg AIS, Yun CO, Schiffelers RM, Hennink WE (2021) Polymeric delivery systems for nucleic acid therapeutics: approaching the clinic. J Control Release 331:121–141
10. Kauffman AC et al (2018) Tunability of biodegradable poly(amine- co-ester) polymers for customized nucleic acid delivery and other biomedical applications. Biomacromolecules 19:3861–3873
11. Liu Y, Li Y, Keskin D, Shi L (2019) Poly(beta-Amino Esters): synthesis, formulations, and their biomedical applications. Adv Healthc Mater 8:e1801359
12. Kowalski PS et al (2018) Ionizable amino-polyesters synthesized via ring opening polymerization of tertiary amino-alcohols for tissue selective mRNA delivery. Adv Mater:e1801151
13. Perrier SB (2017) 50th anniversary perspective: RAFT polymerization – a user guide. Macromolecules 50:7433–7447
14. Rotolo L et al (2022) Species-agnostic polymeric formulations for inhalable messenger RNA delivery to the lung. Nat Mater
15. Abramson A et al (2022) Oral mRNA delivery using capsule-mediated gastrointestinal tissue injections. Matter

Mechanics of Biomaterials for Regenerative Medicine

Yevgeniy Kreinin, Iris Bonshtein, and Netanel Korin

Abstract The mechanical properties of biomaterials play a critical role in designing and developing medical products and selecting suitable materials for various applications. This is particularly important in regenerative medicine, where the biomaterials interact to heal tissues and restore function.

In this chapter, we define diverse types of biomaterials and describe their mechanical characteristics. Conventional methods for measurement of the mechanical properties of biomaterials will be described. The investigation of the mechanical behavior of tissues and biomaterials for regenerative medicine will be discussed, as well as functional biomechanical tests for different applications. At the end, two examples focusing on applications for biomaterials in the cardiovascular area will be presented: (1) An aneurysm embolic hydrogel; (2) A polymeric artificial heart valve.Graphical Abstract

Y. Kreinin · I. Bonshtein · N. Korin (✉)
Department of Biomedical Engineering, Technion – IIT, Haifa, Israel
e-mail: korin@bm.technion.ac.il

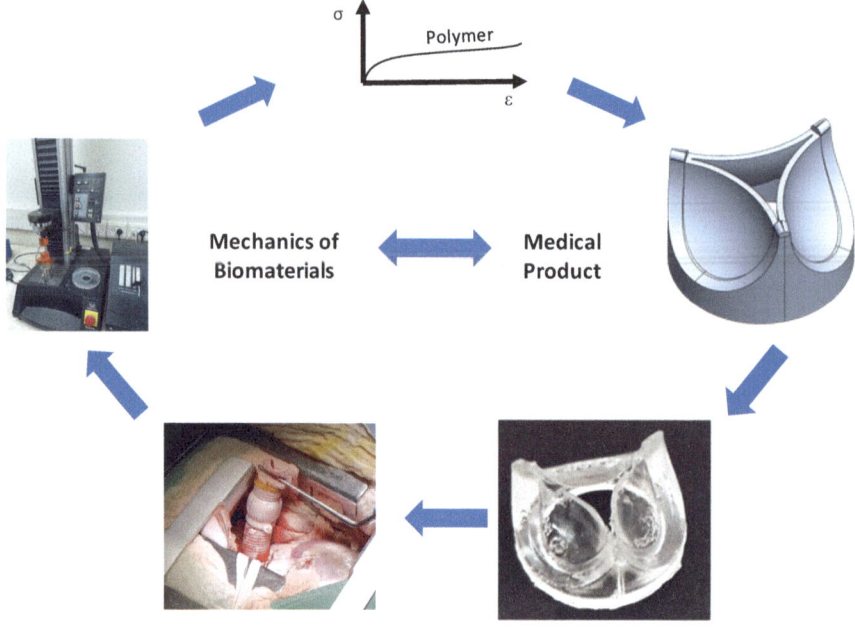

Keywords Elastic modulus · Viscoelasticity · Mechanical design · Mechanical tests · Hydrogels

1 Biomaterials: Classification and Their General Mechanical Properties

Biomaterials are materials designed for biomedical or clinical applications and thus can interact with biological systems for medical purposes. Depending on the nature of the chemical bond, biomaterials can be categorized into four main groups: (i) ceramics, (ii) metals, (iii) polymers, and (iv) hybrids. As these material structures vary by their characteristic chemical bond (covalent, ionic, or metallic), they possess different properties, and thus are utilized for different applications in the body. Another categorization is synthetic or natural biomaterials, where synthetic biomaterials include, for example, metals, ceramics, non-biodegradable polymers, and biodegradable polymers, while nature-derived biomaterials include for example, hyaluronic acid, chitin, cellulose, silk, chitosan, gelatin, and fibrin.

As mentioned above, biomaterials of different classes are characterized by different mechanical properties, *see* Fig. 1. Generally, these mechanical properties are derived by studying the mechanical behavior of materials upon subjecting them to a defined mechanical stimulation. Stress and strain are basic terms used to

Fig. 1 Material Stress-strain curve, on the left for different materials classes, on the right for different tissues

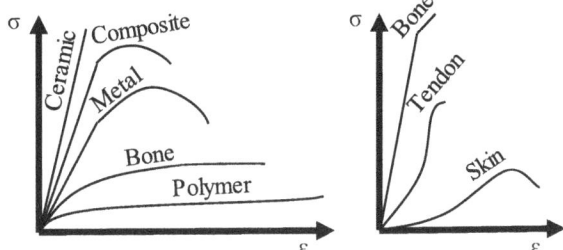

describe the behavior of a solid object to external mechanical stimulations. When a force is applied to a material (e.g., a plate made of a biomaterial) the internal forces that resist the externally applied force produce internal local stresses (area normalized forces) within the material, which also cause it to deform and thus the structure of the material undergoes geometrical changes. A measure for the deformation of a solid material is strain, which can be related to the change in a defined length compared to its original unstressed length. Biomaterials of different classes present different stress-strain relationship and are accordingly characterized by different mechanical properties, *see* Fig. 1.

Ceramics have the following mechanical characteristics: they are relatively strong (high failure stress), have high mechanical stiffness (high elastic modulus), are brittle (low strain to failure), and possess low toughness (low energy to failure). Bio-ceramics are used in medical devices as rigid materials in applications that include surgical implants for bone and cartilage regeneration or dental and hip prostheses. The main disadvantage of ceramics is their brittleness, which is the tendency of a material to fracture or break without plastic deformation when subjected to stress, particularly under rapid loading or impact. Metal biomaterials have similar high elastic modulus and yield strength; however, they are ductile, allowing them to bear a load and carry plastic deformation without rupturing. Their medical applications are similar in scope to ceramic materials and include bone regeneration and dentistry prostheses.

Natural and synthetic polymers are relativity weak (low resistance to mechanical stress), soft (low modulus of elasticity), ductile (high strain to failure), and tough (high energy to failure). Polymers have been used in bone, cartilage, tendon, and ligament regeneration, among other applications. The advantage of synthetic polymers over natural materials is that they allow to modulate their mechanical properties, and hence can be also used in cardiovascular and bone cements, suture threads, orthopedic screws, and prostheses manufacture.

Natural and synthetic hydrophilic polymers can also form hydrogels, water-soluble polymer networks that can hold a considerable amount of water (>10%) while maintaining their structure. Hydrogels are used in a variety of industries, most notably in the medical field. Although hydrogels are relatively weak, they can withstand large deformations and display complex mechanical behavior. Hydrogels can exhibit both elastic and viscous behavior, which is known as viscoelasticity. The viscoelastic properties of hydrogels are dependent on their composition, cross-linking density, and water content and can be characterized by their storage modulus, loss modulus, and complex modulus. The mechanical and transport properties of hydrogels, as well as their mass transport properties make them attractive biomaterials for various applications in tissue engineering and regenerative medicine, including drug delivery, wound healing, and hard and soft tissue regeneration [1].

Hybrid biomaterials (a composition of different biomaterials such as natural or synthetic polymers, ceramics, or metals) can have appreciable mechanical strength and hence can be used in hard and soft tissue. This class of material may offer superior mechanical properties compared to non-hybrid materials; however, their design, manufacturing, optimization, and regulations may be more challenging.

2 Mechanics of Biomaterials for Regenerative Medicine and Viscoelasticity

The mechanical properties of a biomaterial for regenerative medicine have a profound impact on the tissue to be treated and can determine the effectiveness of the tissue repair. In most cases for tissue regeneration, the general dogma is that a biomaterial that is optimal to replace a specific tissue should mimic its mechanical properties. The biomaterials need to provide temporary mechanical support as well as serve as a suitable environment for tissue regeneration. Biomaterials that are stiffer than their surrounding tissue can lead to tissue resorption. Thus, the aim is to design the biomaterial to be strong enough to prevent its mechanical failure but soft enough to avoid tissue resorption. Moreover, most tissues are complex structures that contain both liquid (water) and solid and are thus naturally viscoelastic materials. Viscoelastic materials integrate both viscous and elastic mechanical reactions to mechanical loads. The viscous reaction is time and rate dependent, while the elastic reaction is immediate. The time-dependent reaction can be observed, for example, in a creep response whereupon an immediate increase in the load to a new level of stress that remains constant, the instant change in strain (elastic response) is followed by an increase in the strain over time (creep), *see* Fig. 2a. Another effect is stress relation, in which an immediate increase in the strain results in an immediate change in the stress (elastic response) followed by a reduction in the stress over time (stress relaxation), *see* Fig. 2b. Additionally, while purely elastic materials do not dissipate energy when stress is applied and then released, viscoelastic materials do. Thus, the stress-strain curve of a viscoelastic substance shows hysteresis, which implies loss of energy during a stress cycle, *see* Fig. 2c [2]. Moreover, when cyclic sinusoidal stress is applied, a phase lag occurs between the stress and strain. The phase lag, defined by shift angle δ in the response, represents the viscoelastic damping of the material ($\delta = 0$ for elastic material and $\delta = 90$ for pure viscous liquids), *see* Fig. 2c. Additionally, the response to axial stress can be separated to the storage modulus E', which is a measure of the stored

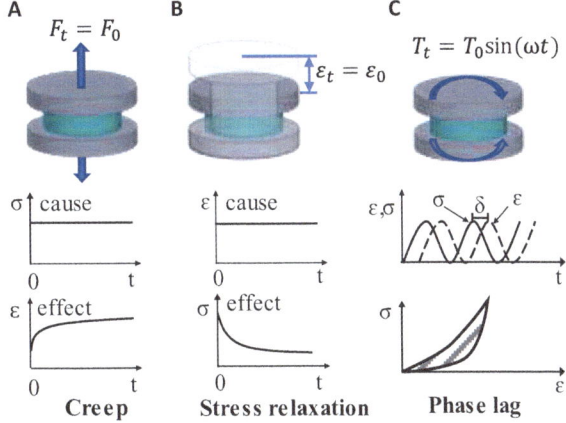

Fig. 2 Characteristics of a viscoelastic material (**a**) creep (**b**) stress relaxation (**c**) phase lag upon cyclic loading and hysteresis during a cycle (bottom graph)

energy, representing the elastic portion, and the loss modulus E'' which is a measure of the dissipated energy, representing the viscous portion. Similarly, a shear storage and a shear loss modulus, G' and G'', respectively, can be defined for the case of shear stress and strain. Altogether, viscoelastic materials are characterized by a complex modulus that can be used to describe their dynamic behavior [3].

3 Measurements of the Mechanical Properties of Viscoelastic Biomaterials

To develop biomaterial-based products for regenerative medicine, the basic mechanical properties of the material need to be defined as they critically affect the product's performance. These mechanical characterization measurements are based on applying a defined force to a sample and measuring the resulting deformation, or vice versa. The type of the applied load and deformation can differ based on the type of the instrument and measuring modality. Additionally, mathematical models describing the material's mechanics (constitutive models) are needed to extract the material properties based on the measurements. Furthermore, it is also important to define relevant standards for the measurement conditions and the measurement should be performed in a relevant physiological environment as needed (buffer, temperature, etc.). For example, most biomaterials are designed to function in an aqueous environment, and an aqueous environment may be valuable for hydrogel mechanical response. Below, we briefly describe four basic methods for measuring the mechanical properties of biomaterials, *see* Fig. 3.

Tensile Test: The tensile method is well-established and widely used for measuring materials' mechanical properties. In this method, the material is stretched by applying forces near its ends in opposite directions. The simplest form of tensile testing is uniaxial, where the sample is stretched along one axis. The material

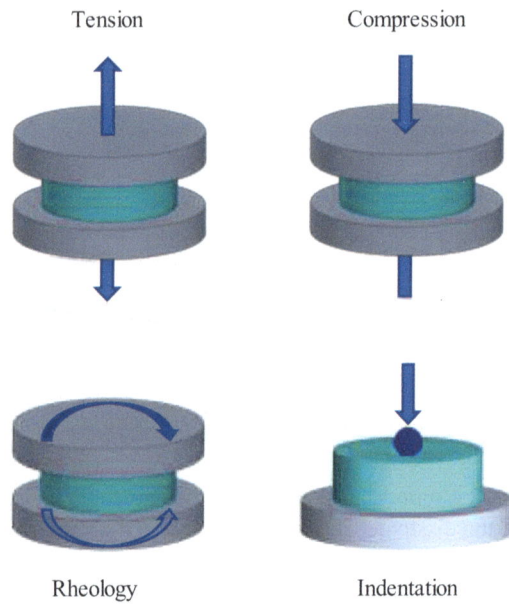

Fig. 3 Different methods for measuring mechanical properties of biomaterials

specimens are prepared in a dumbbell or dog-bone shape. Since usually, biomaterial specimens are hydrated and soft, it is not easy to grip them properly for tensile testing. Once the clamped biomaterial specimen is stretched at a uniform deformation rate up to a certain level of strain, the relationships between the deformation and stress can be extracted and its elastic modulus can be calculated (for elastic solids). Temporal changes can also be tracked to evaluate viscoelastic behavior and properties, such as in creep and stress relaxation experiments.

Compression Test: Another similar and well-established method for materials elasticity measurement is the compression test method, which is most relevant for testing biomaterial in applications where compression is the main loading modality. In this method, biomaterial samples are usually prepared in a disc form and compressed using a controlled force while their deformation is measured, *see* Fig. 3. Then the applied force and resultant biomaterial deformation are converted to compressive stress and strain. For linear elastic materials, the elastic modulus of the material specimen can be determined from the slope of the obtained stress-strain curve. Temporal changes can be tracked to evaluate viscoelastic behavior and properties.

Rotational Rheometery: A rotational rheometer is a laboratory device that is frequently used to measure how fluids or viscoelastic materials (e.g., hydrogels and polymers) "flow" or react in response to applied oscillating rotation/torque that produces shear forces, *see* Fig. 3. A biomaterial specimen is placed between the top and base plates of the rheometer, it is then slightly compressed to ensure its stable interaction and the top plate oscillates at a desired frequency and shear strain. As the specimen is twisted and undergoes shear deformation, it exerts a resistant shear force on the oscillating plate. The rheometer simultaneously measures the rotational motion and the applied torque. Based on these measurements at different frequencies and amplitudes, it is possible to measure the complex shear, shear storage, shear loss moduli, and loss angle (G, G', G', δ), *see* Fig. 2c [4].

Indentation test: In this method, a material is locally indented at a single point to a predetermined displacement depth while measuring the reaction force required to cause the indentation, *see* Fig. 3. The displacement can be measured via a tip gauge or using optical-based measurements. A force-displacement curve is used to calculate the elastic modulus of the material. This approach allows local measurements within a specimen and can be performed at different scales [4]. This method also can be done in specimens that are difficult to grip for tensile or compressive mechanical testing, allows relatively small volumes of material (micro\nanoscales), and can be performed in a specific area of interest within a specimen and on multiple different locations in the material's surface [5].

4 Functional Mechanical Tests for Different Biomaterial Applications

In addition to the standard mechanical tests of the biomaterial, every tissue/biomedical application has unique properties that must be considered and tested as per its application. Thus, functional mechanical tests are important for evaluating the performance of biomaterials in designated applications as well as for standardized testing for approval of the treatment/device. The choice of test will depend on the specific application of the biomaterial and the properties that need to be evaluated.

For example, in the case of soft-tissue adhesives, the shear adhesion can be measured by the lap shear test, which is used to evaluate the shear strength of a bond between two materials. Specimens are pulled in a direction parallel to the bond line. The maximum force that the bond can withstand before it fails is recorded as the lap shear strength. A standard test method for strength properties of tissue adhesives via lap-shear by tension loading has been defined (ASTM F2255-05). The standard is intended to provide a mean for comparison of the adhesive strengths of tissue adhesives intended for use as surgical adhesives or sealants on soft tissue [6].

Another example for functional mechanical tests is such tests performed for vascular grafts, which include burst pressure, compliance, and suture retention tests. The standard that defines these tests is the standard for cardiovascular implants – tubular vascular prostheses (ISO 7198:2016). Burst pressure is a critical parameter for vascular grafts as the graft needs to endure physiological hemodynamic pressures. The burst pressure test measures the greatest pressure before graft failure. Compliance tests measure the geometrical 3D change of a graft as a function of the change in the vessel's internal pressure. Compliance mismatch between the host vessel and vascular graft can lead to serious pathological events such as intimal hyperplasia and vessel occlusion. The graft should also have enough strength to endure the forces applied by sutures without failure. This is tested by suturing a graft that was cut in its middle, sutured back together, and then pulled at a constant rate until it fails. The maximum tensile force that it can withstand is the suture retention strength [7].

Other widely used mechanical tests also include wear tests, fatigue tests, and bending tests, among others. Wear test measures the resistance of a biomaterial to wear and tear. It is commonly used for testing the durability of materials used in joint replacements and dental restorations. Bending tests are used to measure the resistance of a biomaterial to bending forces. It is commonly used for testing the strength of orthopedic implants such as plates, screws, and rods. For orthopedic implants and dental materials, fatigue test is a common test that measures the resistance of a biomaterial to repeated loading over time. Generally, there are many other mechanical tests suitable for different applications, and thus, the choice of the biomaterial and the product design should address these tests and standards [8].

Two examples of mechanics in biomaterials application are: (1) a hydrogel embolic agent for brain aneurysms; (2) a polymeric artificial heart valve.

Study Case 1: Photopolymerizable Hydrogels for the Treatment of Brain Aneurysms [9]

Main goal: To develop injectable photopolymerizable hydrogels designed to treat brain aneurysms by selectively filling the entire aneurysm space allowing a complete separation of the aneurysm from the parent vessel.

The biomaterial: Photopolymerizable polyethylene glycol dimethacrylate (PEGDMA) hydrogels.

The mechanical requirements: Crosslink fast enough within a few minutes, swell pressure that will not damage the tissue, similar mechanics as the native tissue, fill the aneurysm, stay stable within the aneurysm.

Material Mechanical tests: Compression test, swelling pressure test.

The functional mechanical tests: Stability and fatigue tests when placed in an in vitro aneurysm model subjected to physiological flow in a perfusion system.

Background: Cerebral or intracranial aneurysms (IAs) are abnormal focal dilations of an artery in the brain caused by a weakened area in the wall of a blood vessel. The risk of IAs is that they may rupture or burst, leading to bleeding in the brain (hemorrhagic stroke). This can cause severe brain damage or even death.

Aneurysm embolization treatment is a minimally invasive procedure used to treat a cerebral aneurysm. During the embolization procedure, a catheter is used to introduce embolic material into the aneurysm. This aids in preventing blood from entering the aneurysm and reduces the risk of bleeding.

Guglielmi detachable coils (GDC) composed of a platinum alloy were the first embolization devices approved by the FDA for occlusion of aneurysms. While metallic coil embolization is minimally invasive and has replaced, in most cases, the high-risk strategy of open surgical clipping, it still has its limitations, such as partial occlusion and recurrence. Synthetic polymers like n-butyl cyanoacrylate (n-BCA) and Onyx (based on polyvinyl alcohol, PVA) and natural polymers like calcium alginate have been studied as potential embolic agents for aneurysms, however, none have yet been successfully translated into the clinic. Locally injectable hydrogels have the potential to be used as embolic agents in endovascular therapy and could provide an improved environment for vessel repair. These hydrogels can be delivered to the aneurysm in liquid form through a catheter and can solidify under various stimuli, such as temperature or pH changes. This approach may enable more accurate and targeted aneurysm blocking without causing any harm to the surrounding tissue or blood vessels. Moreover, using hydrogels, the cavity can be potently filled completely, which can be valuable for irregularly shaped aneurysms and wide-neck aneurysms, *see* Fig. 4.

An example of such an approach is a study exploring light-induced photopolymerization to form PEGDMA hydrogel in aneurysm cavities. Hydrogels belong to the soft biomaterials group and have low stiffness but offer controllable mechanical properties and swelling capacity. Swelling is defined as the ability of the material to absorb water and expand. The hydrogel swelling needs to be just enough to completely fill the aneurysm cavity, without occluding the parent vessel. Also, it must not exert too much pressure against the aneurysm wall, since aneurysms may rupture when the wall stresses exceed tissue strength. In general, the hydrogel's compliance, mechanical strength, and elasticity need to be comparable to those of the parent artery.

Since mechanical properties and swelling capacity are controlled by the molecular weight of PEGDMA and its concentrations, three different molecular weights at various concentrations of the polymer were investigated. The swelling capacity was tested in two modes: free swelling in phosphate buffered saline (PBS) or confined swelling (where the sample is exposed to the solvent just at the neck of the aneurysm in the in vitro model). After 1 month, the weight and volume swelling ratios were calculated. The results show that increasing the polymer molecular weight or concentration led to an increase in the swelling ratio, in both the free and confined swelling modes and that the free swelling was higher than the confined swelling.

The mechanical properties of the hydrogels were assessed using a compression experiment. Hydrogel samples at different swelling times in PBS were compressed to a strain of 50% under a constant speed. Load and displacement were measured, and the elastic modulus was extracted by linear regression of the stress-strain curve. The failure stress and strain were defined as the highest stress and strain that a hydrogel experienced before breaking upon being gradually compressed to a strain of 80%. The compliance of the hydrogel was evaluated under pressures between 80

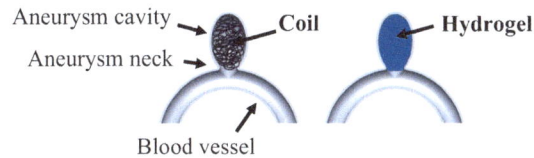

Fig. 4 Schematic showing embolization of brain aneurysms via coils (left) and hydrogel (right)

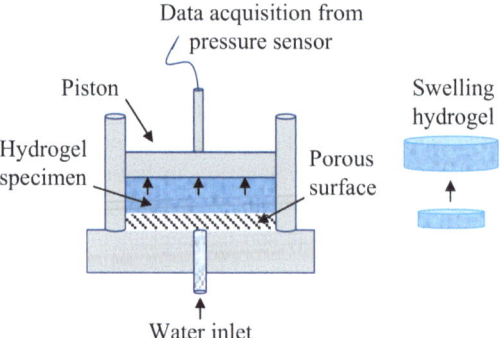

Fig. 5 Measurement tool to determine the swelling pressure under confined conditions

and 120 mm Hg to determine how easily the hydrogel may deform in response to physiological systolic blood pressures.

The swelling pressure was measured as the pressure exerted by swollen hydrogels under confined compression conditions using a piston [10], *see* Fig. 5.

To allow the hydrogel samples to expand, the chamber was filled with PBS. The load was then measured while maintaining a constant displacement until a steady state was reached. The swelling pressure was determined as the highest measured pressure during this test.

It was found that the molecular weight and concentration of the polymer affect the mechanical properties of the hydrogels. In general, increasing the molecular weight or concentration of a polymer increases both the swelling pressure and the compressive elastic modulus while decreasing its compliance under physiological pressure. Eventually, the performance of the hydrogel was evaluated using an in vitro brain aneurysm model filled with the hydrogel and connected to a flow system that applied physiological flow. Compression tests were performed on the hydrogel before and after 1 month in the model, and the surface profile of the hydrogels was studied via 3D laser scanning microscopy. The hydrogels sustained 5.5 million cycles, and no noticeable weight loss of the implant nor protrusion or migration of the polymerized hydrogel into the parent artery was observed.

These investigations revealed that photopolymerizable PEGDMA hydrogels with a molecular weight of 6 kDa and a concentration of 15% exhibited mechanical properties and compliance comparable with the natural aneurysm tissue, as well as provide a modest swelling volume and pressure and, therefore can potentially be used as a biomaterial for intracranial aneurysm treatment or repair.

Study Case 2: Aortic Polymeric Artificial Heart Valve [11]

Main goal: To develop an implantable polymeric aortic trileaflet prosthetic heart valve.

The biomaterial: Polystyrene-b-polyisobutylene-b-polystyrene triblock copolymer with about 30 wt. % polystyrene (SIBS30).

The mechanical requirements: Similar mechanical properties as a human biological aortic heart valve. To enable catheter implantation and physiological flow as well as long-term durability.

Material mechanics tests: Tensile tests, fatigue testing, and dynamic creep.

The functional mechanical tests: Hydrodynamic in vitro test under human physiological conditions (according to ISO 5840:2005 prostatic heart valve performance standards).

Background: Nowadays, approximately 4% of individuals over the age of 65 suffer from aortic stenosis (AS) resulting from calcific aortic valve disease, with an associated mortality of more than 50%. The solution to heart valve failure is implantation of prosthetic heart valves (PHVs) to replace the diseased valve. Today's PHVs are comprised of biological xenografts (tissue) valves, which are implanted via a transcatheter replacement procedure (TAVR), or mechanical valves made of pyrolytic carbon that are surgically implanted. Although these PHVs are effective, there are still unaddressed issues where biological valves deteriorate over time (<10 years) while mechanical valves require chronic anticoagulation, which may lead to bleeding complications. Polymeric heart valves can potentially offer improved durability compared to biological valves and improved hemocompatibility compared to mechanical valves [11]. Thus, there is an interest in studying and developing polymeric PHVs that could provide improved clinical outcomes and can be implanted via a transcatheter procedure.

As an example, to explore such valves, a study was performed on the design and performance of a polystyrene-b-polyisobutylene-b-polystyrene triblock copolymer with about 30 wt. % polystyrene (SIBS30) TAVR polymeric valves.

First, to study the basic mechanical properties of the polymer, such as elastic moduli, yield stress, ultimate stress, and maximal strain, tensile testing was performed by using an Instron® tensile machine according to the ASTM D 638 standard. These are essential mechanical tests for materials to be used in catheter implantation. To find the viscoelastic dynamic modulus, SIBS30 specimens were subjected to a stress-controlled sinusoidal oscillation with varied cyclic loading frequencies of the stress levels. An important mechanical test for materials to be used in PHV is a fatigue test which measures the material durability when subjected to intense cyclic loading. Fatigue testing on SIBS30 was carried out using an Instron® instrument. Once the material's properties were defined, the design of the PHV was done in an iterative process that combined simulations followed by functional experiments [12], *see* Fig. 6.

Based on anatomical data, a TAVR 3D computer-aided design (CAD) model was created using CAD software, *see* Fig. 6 left upper picture. To study the functionality of the design, a dynamic computer simulation was performed using the Finite Element Analysis (FEA) method, *see* Fig. 6 right upper picture. For this purpose, the mechanical data of the tested SIBS30 polymer was integrated into the simulation data. Finally, the valve's stress and hemodynamic parameters were simulated under normal physiologic aortic pressures to show its functionality in silico. Following this analysis step, a functional polymeric TAVR model was produced via a molding method. To test the produced SIBS30 TAVR prototype under physiological

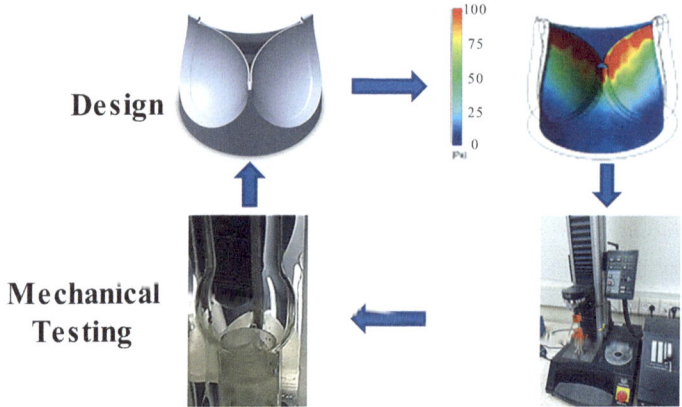

Fig. 6 Schematic representing the polymeric PHVs development process which includes mechanical analysis and polymer mechanical testing, CAD modeling, FEA simulations and in vitro functional tests

conditions and compare it to the computer simulation results, in vitro hydrodynamic testing was conducted in a Vivitro left heart simulator (LHS), as per ISO 5840:2005 prostatic heart valve performance standards. The LHS can simulate precise physiological pressure and flow waveforms, *see* Fig. 6 bottom section. The transvalvular pressure gradient, regurgitation, energy loss, and effective orifice area were recorded and analyzed optically and by an electromagnetic flow meter [11].

Results of the study showed that proper design and leveraging advances in material science can be instrumental in developing a functional polymeric valve that may potentially serve as a PHV.

5 Summary and Future Prospective

In this chapter, we briefly reviewed the mechanical properties of biomaterials, presented measurement techniques, and discussed the pivotal role these properties play in the design of new medical products. We succinctly also presented two examples highlighting the integrative process in designing new medical products/ therapeutic procedures while integrating material science and material mechanics.

We believe that recent advances in manufacturing, which include bioprinting, will offer new opportunities to improve medical products as well as produce complex structures that would possess better mechanical properties. Moreover, bioprinting and new fabrication modalities require a better understating of the biomaterials' mechanical properties, both for optimizing the fabrication process and developing a new product with superior mechanical properties. Additionally, the ability to integrate cells in printed tissue-engineered products would require further expanding our understanding of biomechanical processes and materials that combine living and artificial components and allow us to propose new approaches for disease treatment.

Questions

Question 1: Which of the following materials is not mechanically suitable for use as a bone tissue repair and regeneration:

(a) Synthetic polymers
(b) Ceramics
(c) Natural polymers
(d) Metal

Explanation: Natural polymer are relativity weak (low resistance to mechanical stress), soft (low modulus of elasticity) while bone tissue repair requires stronger materials.

Question 2: For tissue repair, it is advised that the biomaterial will be:

(a) Much stiffer than the target tissue
(b) Much softer than the target tissue
(c) With similar stiffness to the target tissue
(d) The stiffness of the biomaterials is not important

Explanation: In most cases for tissue regeneration, the general guideline is that a biomaterial that is optimal to replace a specific tissue should mimic its mechanical properties. Biomaterials that are stiffer than their surrounding tissue can lead to tissue resorption. Thus the biomaterial needs to be strong enough to prevent its mechanical failure but soft enough to avoid tissue resorption.

Question 3: Which test can be used to directly derive the elastic Modulus of a material?

(a) Fatigue test
(b) Tensile test
(c) Stress-controlled sinusoidal oscillation test
(d) Aging test

Explanation: The elastic modulus of a material can be derived from the slope of the stress-strain curve obtained via a tensile test.

Question 4: Which sentence about viscoelastic materials is false:

(a) Viscoelastic materials show hysteresis while elastic materials do not
(b) Viscoelastic materials show creep behavior
(c) Viscoelastic materials show stress relaxation
(d) Elastic materials are stiffer than viscoelastic materials

Explanation: Viscoelastic materials show hysteresis, creep, and stress relaxation. However, viscoelastic material can be also stiffer than soft elastic materials.

Question 5: Creep experiment is an experiment where:

(a) Constant strain is applied, and the force changes over time are measured.
(b) Cyclic force is applied, and the strain is measured over time
(c) Constant force is applied, and the strain is measured over time
(d) Cyclic strain is applied, and the force is measured over time

Explanation: A creep response occurs when a material that is subjected to a force exhibits a continuous increase in the strain over time (creep). Thus in creep experiments a force is applied and the strain is measured over time.

Question 6: The potential advantage of using hydrogels over coils for the treatment of intracranial aneurysms according to study case 1 is:

(a) Hydrogels have high stiffness
(b) Its feasibility to be delivered through a catheter
(c) Its ability to completely fill the aneurysm cavity
(d) Being a synthetic material

Explanation: While coils have high mechanical stiffness, they cannot completely fill the aneurysm cavity whereas hydrogels can be used to completely fill the cavity.

Question 7: The mechanical properties of the hydrogels can be controlled by their concentrations, according to study case 1 with the increasing of their concentration:

(a) Both the elastic modulus and the compliance increased
(b) The elastic modulus increased while the compliance decreased
(c) The elastic modulus decreased while the compliance increased
(d) Both the elastic modulus and the compliance decreased

Explanation: Result in the experiments performed in the study showed that as the hydrogel concentration increased (there is more polymer in it) the hydrogel became stiffer.

Question 8: Compared to mechanical prosthetic heart valves and biological valves, polymeric valves may offer:

(a) Improved durability compared to the biological valve and improved hemocompatibility compared to mechanical valves
(b) Improved durability compared to the mechanical valve and improved hemocompatibility compared to biological valves
(c) Improved durability compared to both of them
(d) Improved hemocompatibility compared to both of them

Explanation: Biological valves deteriorate over time whereas mechanical valves require chronic anticoagulation. Polymeric heart valves can potentially offer improved durability compared to biological valves and improved hemocompatibility compared to mechanical valves.

Question 9: Functional mechanical tests for vascular grafts include:

(a) Creep experiments of the material
(b) Rotational rheometer tests
(c) Burst pressure, compliance, and suture retention tests
(d) A nano-indentation tests

Explanation: Creep, rotational rheometery, and indentation experiments can be used to characterize the material properties but are not functional mechanical tests. Burst pressure, compliance, and suture retention tests are functional mechanical tests used for testing vascular grafts.

Question 10: According to the material stress-strain curve, which type of the biomaterial suit for skin regeneration:

(a) Ceramic
(b) Polymer
(c) Metal
(d) Composite

Explanation: The skin is a highly deformable tissue whereas ceramics; metals and composite materials are not ductile enough to serve for skin regeneration. Polymers and polymeric hydrogels are ductile materials that can deform substantially.

References

1. Dolcimascolo A, Calabrese G, Conoci S, Parenti R (2019) Innovative biomaterials for tissue engineering. In Biomaterial-supported tissue reconstruction or regeneration. IntechOpen. https://doi.org/10.5772/INTECHOPEN.83839
2. Meyers MA, Chawla KA (1999) Viscoelasticity. Prentice Hall, Upper Saddle River, pp 98–103
3. Lakes RS (2009) Viscoelastic materials. Cambridge University Press
4. Lee D, Zhang H, Ryu S (2018) Elastic modulus measurement of hydrogels. Cellulose-Based Superabsorbent Hydrogels:1–21. https://doi.org/10.1007/978-3-319-76573-0_60-1
5. Oyen ML (2014) Mechanical characterisation of hydrogel materials. Int Mater Rev 59(1):44–59. https://doi.org/10.1179/1743280413Y.0000000022
6. Wang M, Kornfield JA (2012) Measuring shear strength of soft-tissue adhesives. J Biomed Mater Res Part B Appl Biomater 100B(3):618–623. https://doi.org/10.1002/JBM.B.31981
7. Rodríguez-Soto MA, Polanía-Sandoval CA, Aragón-Rivera AM, Buitrago D, Ayala-Velásquez M, Velandia-Sánchez A, Peralta Peluffo G, Cruz JC, Muñoz Camargo C, Camacho-Mackenzie J, Barrera-Carvajal JG (2022) Small diameter cell-free tissue-engineered vascular grafts: biomaterials and manufacture techniques to reach suitable mechanical properties. Polymers (Basel) 14(17). https://doi.org/10.3390/POLYM14173440/S1
8. Niinomi M (2007) Fatigue characteristics of metallic biomaterials. Int J Fatigue 29(6):992–1000. https://doi.org/10.1016/J.IJFATIGUE.2006.09.021
9. Poupart O, Conti R, Schmocker A, Pancaldi L, Moser C, Nuss KM, Sakar MS, Dobrocky T, Grützmacher H, Mosimann PJ, Pioletti DP (2021) Pulsatile flow-induced fatigue-resistant

photopolymerizable hydrogels for the treatment of intracranial aneurysms. Front Bioeng Biotechnol 8:1567. https://doi.org/10.3389/FBIOE.2020.619858/BIBTEX
10. Schmocker A, Khoushabi A, Frauchiger DA, Gantenbein B, Schizas C, Moser C, Bourban PE, Pioletti DP (2016) A photopolymerized composite hydrogel and surgical implanting tool for a nucleus pulposus replacement. Biomaterials 88:110–119. https://doi.org/10.1016/J.BIOMATERIALS.2016.02.015
11. Claiborne TE, Sheriff J, Kuetting M, Steinseifer U, Slepian MJ, Bluestein D (2013) In vitro evaluation of a novel hemodynamically optimized trileaflet polymeric prosthetic heart valve. J Biomech Eng 135(2).1–9. https://doi.org/10.1115/1.4023235
12. Fray ME, Prowans P, Puskas JE, Altstädt V (2006) Biocompatibility and fatigue properties of polystyrene-polyisobutylene-polystyrene, an emerging thermoplastic elastomeric biomaterial. Biomacromolecules 7(3):844–850. https://doi.org/10.1021/bm050971c

Biomaterials for Controlled Drug Delivery Applications

Krishanu Ghosal, Merna Shaheen-Mualim, Edwar Odeh, Nagham Moallem Safuri, and Shady Farah ⓘ

Abstract A drug delivery system (DDS) is by definition a platform that can transfer a therapeutic agent to the disease site in order to elicit and improve the desired therapeutic response. Currently, the traditional DDSs include tablets, syrups, capsules, creams, ointments, etc. The efficiency of the aforementioned systems is limited by their bioavailability level and their stability at the administration stage, which determines their ability to control the drug's dosage level and frequency. Researchers and clinicians aimed to overcome the current limitations by exploiting novel biomaterials for controlled drug delivery applications. The new biomaterials exhibit improved bioavailability and the ability to control drug release kinetics in terms of steady and long-term drug release within the therapeutic window, with minimum side effects.

This chapter overviews the concept of DDSs, from understanding its basics up to describing the role of biomaterials in different drug delivery applications. Initially, the chapter starts by introducing the fundamentals of drug delivery systems including classification of drugs based on drug delivery systems, why there is a need of controlled drug delivery, different routes of drug administration, pharmacokinetics of drug delivery systems, and different release kinetics of drugs. These discussions provide a brief understanding for a particular type of drug and disease model which

---

Authors Krishanu Ghosal and Merna Shaheen-Mualim have equally contributed to this chapter.

K. Ghosal · M. Shaheen-Mualim · E. Odeh · N. M. Safuri
The Laboratory for Advanced Functional/Medicinal Polymers & Smart Drug Delivery Technologies, The Wolfson Faculty of Chemical Engineering, Technion-Israel Institute of Technology, Haifa, Israel

S. Farah (✉)
The Laboratory for Advanced Functional/Medicinal Polymers & Smart Drug Delivery Technologies, The Wolfson Faculty of Chemical Engineering, Technion-Israel Institute of Technology, Haifa, Israel

The Russell Berrie Nanotechnology Institute, Technion-Israel Institute of Technology, Haifa, Israel
e-mail: sfarah@technion.ac.il; https://www.thefarahlab.com/

type of biomaterials should be designed. Whereas in the second part of the review, we have focused on the design considerations for controlled drug delivery systems, role of biomaterials for controlled drug delivery applications, and different biomaterials for drug delivery applications.

Keywords Drug delivery systems · Biomaterials · Drug release kinetics · Polymers · Drug crystals

1 Introduction

According to the United States Food and Drug Administration (US FDA), a drug is defined as an active pharmaceutical ingredient (API) which is intended for use in diagnosis, treatment, or prevention of disease [1]. Drug delivery is a technique by which a drug can be delivered in such a way that the concentration of that drug is localized in some parts/organs of the patient's body compared to others. To achieve this, researchers and clinicians are continuously trying to develop novel materials which can deliver the drugs in an extended, controlled, and targeted fashion so that the drug can reach the disease site with protected packaging. For successful treatment of any disease, the delivery of drugs to the disease site, their on-demand release, and their dosage play a very critical role. Each dosage form is an amalgamation of drugs/APIs with its non-drug counterpart excipients. Excipients are substances added in a pharmaceutical dosage form to facilitate the manufacturing process, and preserve, maintain, or increase stability, bioavailability, or improve patients' acceptability. The conventional forms for drug delivery are tablets, capsules, lozenges, pills, granules, bulk solid dosage forms, ointments, lotions, syrups, etc. These conventional drug delivery systems have few drawbacks; for example, in most cases, the drug release is too fast, as a result, most of the drugs are eradicated from the body very quickly, which might demand taking the dosage multiple times after short interval. Another major problem is that as most of the drugs are eliminated within a short time it is not sustained within a particular therapeutic window. Therefore, after a single conventional dose, the drug metabolizes very quickly, and the drug level rises, followed by a rapid decline. The period may be insufficient to produce a significant therapeutic effect, resulting in a subtherapeutic response.

Figure 1 demonstrates the fluctuations of drug concentration within blood plasma after conventional drug administration. Several strategies have been exploited to keep the plasma drug concentration above the minimum effective concentration (MEC) but below the toxic level

One of the strategies is to give a single dose that is more than what is necessary, which results in a longer duration above the MEC but increases the chance of more side effects. Another more prevalent strategy is prescribing several doses at regular intervals. This is a better option than giving the total amount at once. However, fluctuations of drug concentration in plasma still takes place and frequently causes the drug concentration to fall below effective levels and then rise again over toxic

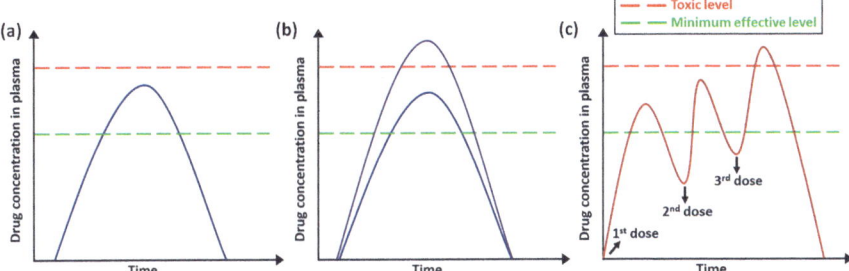

Fig. 1 Plasma drug level after conventional drug administration (**a**) Single, convenient dose, (**b**) Increased single dose, (**c**) Multiple doses

level. Moreover, patient compliance suffers when there are multiple dosages taken throughout the day. To keep plasma drug levels constant within the therapeutic window and provide the desired therapeutic impact for a prolonged period of time, controlled-release drug delivery system is crucial. There are several advantages of controlled drug delivery systems, such as improved bioavailability, safeguard from metabolism by enzymes/chemicals, targeted specificity, and better patient compliance. In this chapter, we have discussed the biomaterials used in controlled drug delivery applications along with some important aspects related to it, such as biopharmaceutics classification system of drugs, routes of drug administration, the pharmacokinetics of drug delivery systems, drug release kinetics, and design considerations for preparation of controlled drug delivery systems. These aspects play a critical role in the biomaterials design for controlled drug delivery systems.

2 Classification of Drugs Based on Biopharmaceutics

Based on the permeability and solubility, drugs are classified into four subcategories (shown in Fig. 2).

The first subclass includes drugs with high solubility as well as high permeability. This type of drugs possesses a superior absorption rate compared to the excretion rate, e.g., paracetamol, metoprolol. The second subclass includes drugs with high solubility but low permeability. In this case, the drugs solvate very fast while absorption into the blood is limited (e.g., cimetidine). Therefore, the drug formulations should be adjusted appropriately with their counterpart excipients. The third subclass includes drugs with low solubility but high permeability (e.g., aceclofenac, glibenclamide). The bioavailability of these types of drugs is very restricted. Finally, the last subclass includes drugs with low solubility and low permeability (e.g., bifonazole), leading to poor bioavailability and high inconsistency. Accordingly, these considerations should be specifically addressed during the design of any biomaterial for controlled drug delivery applications.

Fig. 2 Classification of drugs based on biopharmaceutics

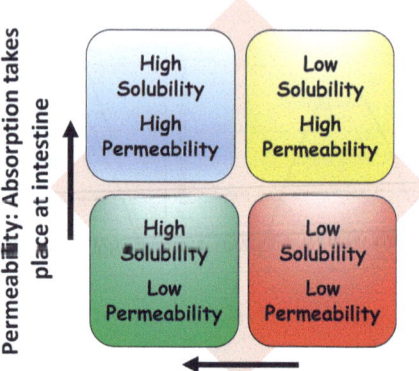

3 Routes of Drug Administration

In addition to the type of drug, the administration route also significantly impacts the designing of biomaterial for controlled drug delivery applications. Drugs can be administrated through different routes. The target organ/part of the body, how the drugs act within the body, the drugs' permeability, and solubility, dictate the optimal route of administration. For example, insulin is a well-established drug for the treatment of diabetes. However, if administered orally it is prone to degradation and loss of activity within the gastric acidic environment resulting in a very low bioavailability. Bioavailability refers to the percentage of the administered drug in its active form that enters the systemic circulation and is available for the target site. Thus, the choice of drug administration route is crucial. The different drug administration routes are (a) Intravenous route, (b) Intramuscular route, (c) Subcutaneous route, (d) Rectal route, (e) Vaginal route, (f) Inhaled route, (e) Transdermal route, (f) Ocular route, (g) Otic route, (h) Oral route, (i) Nasal route, (j) Inhalation route, (k) Nebulization route, (l) Cutaneous route, and (m) Surgical implantation. Figure 3 demonstrates the different routes of drug administration.

4 Pharmacokinetics of Drug Delivery Systems

Pharmacokinetics is the branch of pharmacology where the movement of drugs is studied within the body. It takes place in four distinct stages: (a) drug absorption, (b) distribution, (c) metabolism, and (d) excretion.

(a) *Absorption*

Absorption is the first stage of pharmacokinetics where drug is absorbed from the drug administration site to the bloodstream. The percentage of drug absorbed

Biomaterials for Controlled Drug Delivery Applications

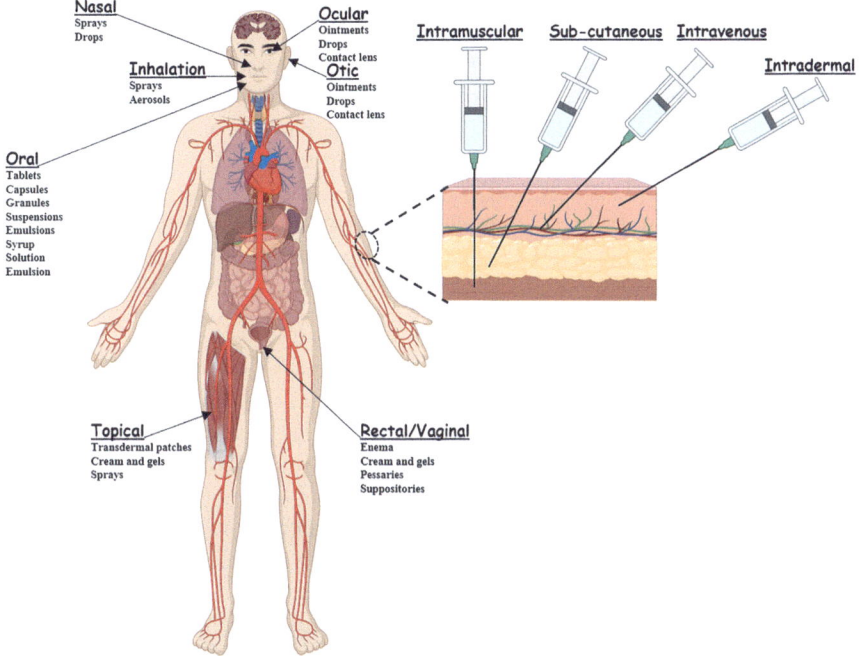

Fig. 3 Different routes of drug administration

within the body depends upon several factors such as selection of administration route, physicochemical properties of drug administrated, interaction of the drugs with enzymes and other in vivo factors, and type of formulations. In this regard, intravenous administration offers 100% bioavailability of drugs as in this case the dosage form is directly injected into the bloodstream. However, it should be noted that it is not possible to deliver all drugs directly to the bloodstream due to their different working mechanisms, dosage factor, and other physicochemical properties.

Absorption of drugs takes place through the plasma membrane *via* either active or passive transport [2]. In case of active transport of drug molecules, there is an energy requirement to enable drug molecules transport against a concentration gradient, which typically occurs at particular sites in the small intestine. Most of the drug molecules which are absorbed by this route share a similar mechanism with endogenous elements that are present in blood such as metal ions, vitamins, sugars, and amino acids. Whereas passive transport includes the transfer of drug molecules across a cell membrane from an area of high drug concentration to an area of low drug concentration. Unlike active transport, energy is not a requisite, and the drug diffusion rate is proportional to the concentration gradient.

(b) *Distribution*

In the distribution stage of pharmacokinetics, the reversible transfer of drug molecules takes place in between blood, extravascular fluids, different body organs, tissue, and ultimately to the cells. The distribution of drug regulates the amount of drug that reaches target sites in comparison to the other body part, hence it plays a decisive role in determining drug's efficacy as well as toxicity. The prime factors that significantly influence the biodistribution of any drug include the hydrophilic/hydrophobic balance, the molecular size of the drug, blood flow, and binding affinity of the drug molecules with plasma proteins. Furthermore, for specific targeting site, certain anatomical barriers restrict the biodistribution of drug molecules. Such as, for the successful delivery of any drug molecules which specifically used for treating brain diseases, it should pass the blood-brain barrier, which acts as an additional blockade for drug molecules.

(c) *Metabolism*

Metabolism of drugs takes place by various processes such as oxidation, reduction, hydrolysis, hydration, conjugation, condensation, or isomerization; by these processes, drug molecules are converted into either less active or inactive forms, which makes them easier to absorb into the body as well as excrete from the body. For example, in the case of oral administration, before the drug reaches the bloodstream, its concentration is dramatically reduced. Although the enzymes necessary for metabolism are found in various tissues, the liver typically contains most of them.

For several drugs, metabolism takes place in 2 phases. In the 1st phase, formation of a new or modified functional group or cleavage of drugs occurs *via* oxidation, reduction, hydrolysis, etc.; these reactions are non-synthetic. In the 2nd phase, conjugation of endogenous substance takes place; these reactions are purely synthetic in nature. Synthetic metabolites are more polar, and thus more easily excreted by the kidneys (in urine) and the liver (in bile) compared to the non-synthetic metabolites.

(d) *Excretion*

In the last stage of pharmacokinetics, the unmetabolized drug molecules and their metabolized products are removed from the body; this process is called drug excretion. There are many routes of excretion of drugs such as sweat, urine, bile, saliva, tears, and stool.

5 Drug Release Kinetics

The drug release kinetics is generally represented as a plot of plasma-drug concentration with time. In Fig. 1a, we have already represented the minimum effective concentration (shown with green dashed line) below which the drug is ineffective and the toxic concentration (shown with red dashed line) above which drug molecules can cause adverse effects. For successful therapeutic effectiveness without any adverse effect, there is constant need to maintain a certain concentration which is greater than the minimum effective concentration but less than toxic concentration. There are certain drug release models which are well-established in this field and most of the drug release profiles follow one of them. In the below section we briefly review these models.

5.1 Zero-Order Release

In case of zero-order release kinetics, a constant amount of drug is released per unit time. The rate of release is independent of the drug concentration. In mathematical term, zero-order release kinetics is represented by Eq. 1:

$$Q = Q_0 + Kt \qquad (1)$$

where Q represents the amount of drug released, Q_0 is the amount of drug at initial time (it is usually zero), t is the time, and K is zero-order constant. Zero-order release profile is usually a straight line with a gradient of K.

5.2 First-Order Release

For first-order release the amount of drug release rate is directly proportional to the concentration gradient and is a function of the amount of drug remaining in the dosage form. In term of equation the first-order release is represented by Eq. 2:

$$\log C_t = \log C_0 - Kt/2.303 \qquad (2)$$

where C_t is the drug concentration at a time point t, C_0 is the initial drug concentration, K is the first-order release kinetics constant which is expressed by unit of time^{-1}, and t is the time point. The zero-order and first-order release kinetics are most primary release kinetics. However, in real life it was observed that in most of the cases these kinetics models failed to describe the release pattern of drug molecules. Later, researchers and clinicians proposed several other drug release models which are discussed below.

5.3 Higuchi Model of Drug Release

Higuchi model of drug release was proposed in 1963 by Higuchi. This model is primarily applicable for water soluble as well as low soluble drugs which is incorporated in semisolid and solid matrices. This model is expressed with Eq. 3:

$$Q = X\left[Y(2C - C_s)C_s \cdot t\right]^{1/2} \qquad (3)$$

where Q is the amount of drug released at a time point "t" per unit area X, C is the initial drug concentration, C_s is the solubility of the drug in the release media, and Y is the diffusion coefficient of drug molecules in matrix.

5.4 Hixson-Crowell Cube Root Model of Drug Release

The cube root model of drug release is jointly proposed by Hixson and Crowell. Their model is applicable for drug delivery systems in which there is an alteration in surface area and diameter of drug particles/tablets/crystals. As a suspended solid dissolves, its surface drops by a factor of 2/3rd of its weight, assuming that there is no change in its shape. The Hixon-Crowell model is represented by Eq. 4:

$$Q_t^{1/3} = Q_0^{1/3} - K_{HC} t \tag{4}$$

where Q_t represents the remaining weight of the solid after a time t, Q_0 is the starting weight of the solid when time (t) = 0, and K_{HC} denotes the Hixson-Crowell constant also known as the dissolution rate constant. This model is preliminarily used taking into consideration that the release rate is limited by the drug particles/tablets/crystals dissolution rate and not by the diffusion. Additionally, Hixson and Crowell's model demands certain consideration for the law's validity which are summarized below:

(i) This law is applicable for those systems which are usually monodispersed and spherical in nature with identical properties in terms of size, morphology, surface, and volume.
(ii) Drug dissolution usually occurs close to the surface. In addition to that, the rates at various crystal faces should differ very little from one another, and the agitation of the liquid against all areas of the surface has the same result.
(iii) The liquid should be vigorously stirred to avoid the stagnation of the drug molecules close to the drug particles/tablets/crystals, which can slow down the rate of diffusion.

5.5 Korsmeyer-Peppas Model of Drug Release

In 1983, Korsmeyer proposed a model for drug release from a polymeric system. Later Ritger and Peppas, as well as Korsmeyer and Peppas, derived an equation which can rationalize both Fickian/non-Fickian release of drug release from swelling/non-swelling polymer-based drug delivery systems.

To determine the mechanism of drug release, 60% of the drug release data was initially fitted in the Korsmeyer-Peppas model, which is represented by Eq. 5:

Biomaterials for Controlled Drug Delivery Applications

$$M_t / M_0 = Kt^n \qquad (5)$$

where M_t is the amount of drug released at time t, M_0 is the amount of drug at the initial time, and K is the rate constant which relates structural and geometrical characteristics of the drug delivery system. For example, "n" denotes the mechanism of drug transport through the polymer, also known as the release exponent. According to Ritger-Peppas model, the value of "n" lies in between 0.45 and 0.89 when a non-Fickian release takes place from non-swelling cylinders whereas the value of "n" ranges in between 0.43 and 0.85 when non-Fickian release happens from non-swelling spherical particles [3, 4]. But it should be noted that this value of "n" is applicable when M_t/M_0 is <0.6.

5.6 Baker-Lonsdale Model of Drug Release

This is an extension of Higuchi model and it deals with the drug release from spherical matrices according to Eq. 6:

$$f = 3/2\left[1 - \left(1 - M_t / M_0\right)^{2/3}\right] - M_t / M_0 = Kt \qquad (6)$$

Here also M_t represents the amount of drug release at time t, M_0 drug at initial time, and K is the release constant. This model is usually applicable for linearization of the release data from drug encapsulated within microcapsules [5].

5.7 Weibull Model of Drug Release

In the year of 1951, Weibull proposed a generalized empirical equation for the dissolution/release process of drug [6]. The Weibull model of drug release is represented by Eq. 7:

$$M = 1 - \exp\left[-\left(t - T_i\right)^{b/a}\right] \qquad (7)$$

Here M is the fraction of drug released at time point "t", "a" represents scale parameter, which defines the time scale of the process. T_i is the location parameter, which signifies the time lag before the actual onset of the dissolution process which, in most cases, will be equal to zero. "b" denotes the shape parameter, which describes the shape of dissolution curve progression. When $b = 1$, the curve is exponential; while when $b > 1$ the curve is sigmoidal in nature with a turning point, on the other hand when the value of $b < 1$ the curve should be parabolic in nature with a steeper initial slope.

5.8 Hopfenberg Model of Drug Release

Hopfenberg drug release model is applicable for drug release from surface eroding polymers as long as the surface area does not change during the deterioration process. This model is expressed by the mathematical Eq. 8:

$$M_t / M_\infty = 1 - \left[1 \quad k_0 t / C_L \alpha \right]^n \tag{8}$$

Here M_t/M_∞ is the cumulative fraction of drug released at a time point "t". "k_0" is the zero-order rate constant relating the surface erosion degradation process. C_L is the total drug loading within the polymer, "α" is the half thickness of the system, and "n" is an exponent that changes with different geometry. For example, $n = 1$ for flat/slab geometry, $n = 2$ for cylinder shaped geometry, and $n = 3$ for spherical geometry.

5.9 Gompertz Model of Drug Release

Gompertz model of drug release primarily used to compare the release profile of drugs which own good solubility and intermediate release profile. The Gompertz model of drug release is expressed by Eq. 9:

$$X(t) = X_{max} \exp\left[-\alpha \bullet e^{\beta \log t}\right] \tag{9}$$

Here $X(t)$ represents the percent of drug dissolved after time t, X_{max} denotes maximum dissolution of drug, α represents undissolved fraction of drug at time $t = 1$, which is also known as location or scale parameter, and β expresses the dissolution rate of drug per unit time also known as shape parameter. The drug release profile of this model usually has a sharp rise at early stage and later it reaches slowly a maximal dissolution.

5.10 Gallagher-Corrigan Model of Drug Release

Gallagher-Corrigan model of drug release is applicable for biodegradable polymeric drug delivery systems. For this model, drug release takes place in a combination of degradation as well as diffusion. The drug release profile for this model is commonly sigmoidal in nature. The Gallagher-Corrigan equation relates the kinetic profile of a drug that is not bound to the drug matrix, and initially shows a burst release followed by a controlled release governed by matrix erosion. The Gallagher-Corrigan model of drug release is expressed by Eq. 10:

$$F(t) = b + Y_1\left(1 - e^{-k_1 t}\right) + Y_2\left(\frac{e^{k_2(t - kt_{max})}}{1 + e^{k_2(t - kt_{max})}}\right) \tag{10}$$

Here $F(t)$ is the fraction of drug released at a time point t, Y_1 is the fraction of the drug released during the first stage, Y_2 is the fraction of drug released during the second stage, k_1 denotes the first-order kinetic constant (first stage of release), k_2 signifies the kinetic constant for the second phase of the drug release when matrix degradation takes place, and t_{max} denotes the time of the maximum fraction of the drug release rate.

5.11 Cooney Model of Drug Release

This model is based on the hypothesis that a single zero-order kinetics process takes place on the surface of the drug delivery system. Cooney model of drug release provides a detailed analysis of cylindrical and spherical drug delivery systems which undergoes surface erosion. For a cylindrical drug delivery system with an initial length of L_0 and dia D_0, the following equation (Eq. 11) is developed to calculate the drug release rate "f" as a function of time "t".

$$f = \left[(D_0 - 2kt)^2 + 2(D_0 - kT)(L_0 - 2kT)\right] / \left(D_0^2 + 2D_0 L_0\right) \tag{11}$$

In this equation k is a constant. It should be considered that when L_0/D_0 approaches to zero the release profile turns into a horizontal line with a constant drug release rate of 1, whereas when $L_0/D_0 < 1$, the drug release rate is finite until the drug is completely released from the system. On the other hand, if the $L_0/D_0 > 1$ (for cylinders) the drug release rates reach to zero at extended time points.

5.12 Sequential Model of Drug Release

The sequential model of drug release is applicable for drug release system in which swelling of matrix takes place. It is utilized to determine the swelling and release behavior of hydrophilic matrix tablets, as well as to understand the influence of device shape on the release profile of drugs. For the sequential model of drug release, the tablet system is assumed as a given number of single layers penetrated by water, and the model is run in a computational grid with a modified grid structure for numerical analysis. One major advantage of this model is that the use of computational grid permits modeling of inhomogeneous swelling also. Additionally, in this model it was considered that swelling takes place *via* layer-by-layer, with the outermost layer swelling first and then the nearby interior layers. The following physicochemical events take place during the course of drug release from hydrophilic tablet matrix. (i) At the initial stage, formation of water concentration gradient takes place at the water/matrix interphase resulting in water penetration into the system. (ii) Due to the penetration of water molecules the drug carrier swells, which leads to significant changes in its shape as well as drug concentrations. (iii) When the drug molecules come to close contact with water molecules, the drug dissolves into it and diffuses out from the carrier. The drug's diffusion coefficient increases significantly as the water content increases. (iv) In case of poor water soluble drugs, the soluble as well as insoluble part drug co-exist within the polymer matrix, but the insoluble part does not take part into diffusion. (v) In case of high drug loading, the structure of hydrophilic polymer matrix changes significantly. In that case, it will be more porous and less restrictive for drug dissolution. (vi) Based on the polymer chain length and degree of substitution of hydrophilic polymer employed to formulate the drug carrier matrix the degree of polymer dissolution takes place. A dissolution rate constant, k_{diss}, was taken into consideration based on the reptation theory to describe the polymer mass loss velocity normalized to the system's real surface area. The sequential model of drug release is expressed by Eq. 12:

$$M_{pt} = M_{po} - k_{dis} A_t t \qquad (12)$$

Here M_{pt} denotes the dry polymer matrix mass at time t, and M_{po} is the dry polymer matrix mass at time $t = 0$, A_t represents the surface area of the drug carrier system at time point t.

6 Controlled Drug Delivery Systems

Controlled drug delivery systems are those systems which maintain a certain level of drug concentration within the range between the minimum effective concentration and minimum toxic concentration in blood, tissue, or site of interest, for an extended period of time so that there is no requirement of multiple dosage at a certain interval. The pharmacokinetics profile of conventional drug delivery systems and controlled drug delivery systems are demonstrated in Fig. 4. In case of conventional drug delivery systems (Fig. 4a), the drug level fluctuates continuously and in few instances; it could also exceed the minimum toxic concentrations which may lead to hazardous side effects, whereas in case of a controlled drug delivery system the drug is released for a prolonged duration while maintaining a certain concentration within the range between minimum effective concentration and minimum toxic concentration.

More evidently, controlled drug delivery systems continuously maintain drug plasma levels by dispensing the exact dose of the drug at specific time for a certain duration, which helps to minimize the dosing frequency and improve patience compliance (shown in Fig. 4b).

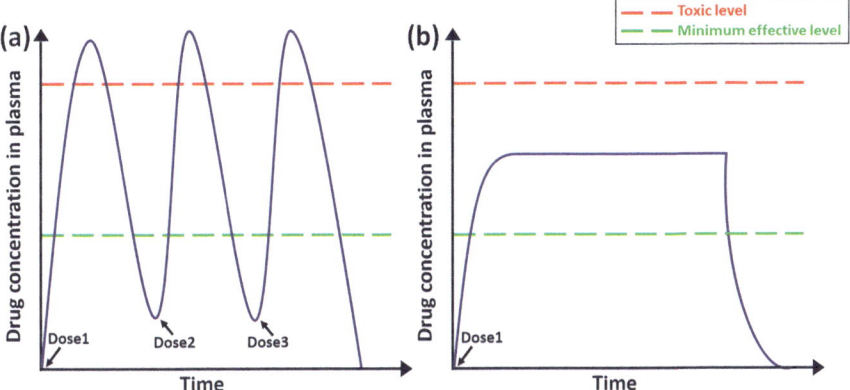

Fig. 4 Typical drug release profile of (**a**) Conventional drug delivery system (**b**) Controlled drug delivery system

6.1 Design Aspects of Controlled Drug Delivery Systems

Several considerations must be kept in mind while designing a controlled drug delivery system for a specific medical application. The prime and major aspect is choice of material (excipient) to prepare the delivery system. The excipient should be biocompatible so that there is no host response within the patient body. Other material properties that play significant role in determining suitability for good, controlled drug delivery system are surface chemistry, hydrophilic/hydrophobic properties, degradation profile and products, and mechanical and rheological properties. Moreover, delivery routes, pharmacokinetics behavior, stability of the carrier system (for example, in certain levels of pH, temperature, or in presence of specific enzyme carrier system can degrade and the drug will not reach to the target site), targeting capability to the disease site, and capability to surpass biological barriers also play key roles in the design of a suitable material for drug delivery applications.

6.2 Different Mechanisms of Controlled Drug Delivery Systems

Based on the release mechanisms of drug, controlled drug delivery systems are classified into six different categories. These categories are (a) diffusion-controlled systems, (b) dissolution-controlled systems, (c) swelling-controlled systems, (d) water penetration-controlled systems, (e) osmotic pressure-controlled systems, and (f) chemically controlled systems.

6.3 Source of Materials for Drug Delivery: Natural vs Synthetic

Source of materials also plays a major role in determining the effectiveness of drug delivery systems. Materials for drug delivery applications should be biocompatible, biodegradable, and non-immunogenic in nature. In this regard, both natural and synthetic resourced materials are exploited for drug delivery applications. Each of them has a few advantages as well as few disadvantages. In the case of naturally resourced materials, the prime advantage is their excellent biocompatibility, biodegradability, and non-immunogenicity. Although good biodegradability is an advantage for naturally resourced materials, in some instances, the rapid biodegradability of naturally resourced materials leads to the burst release of drugs which leads to poor bioavailability. Typically, three types of naturally resourced materials are employed for controlled drug delivery applications: protein-based biomaterials (e.g., gelatin, silk), polysaccharide-based materials (e.g., cellulose, chitosan), and decellularized tissue-derived materials (e.g., exosomes).

For synthetic resourced materials, there is more control over design parameters. Compared to natural ones, the biodegradability and mechanical properties of the delivery system can be fine-tuned precisely. However, if the material is not chosen correctly, there is a chance of immune response with synthetic resourced drug delivery system. So, the selection of material is very crucial, in case of synthetic resourced drug delivery system. Example of synthetic resourced materials for drug delivery include nanomaterials, synthetic polymers, liposomes, etc.

7 Role of Biomaterials in Controlled Drug Delivery

Biomaterial is defined as an engineered substance that can interact with biological systems for a medical need, either a diagnostic or therapeutic one [7]. Biomaterials can play a critical role in controlled drug delivery by modulating the pharmacokinetics behavior of the drug. Biomaterials which have exploited so far in the field of controlled drug delivery applications are natural and synthetic polymers in different forms, proteins, lipids, peptides, metallic and non-metallic nanomaterials, and drug crystals. Most of these materials are typically in the range of nano to micrometer scale. Hydrogels have also emerged as an attractive set of candidates for controlled drug delivery applications. The different biomaterials used in controlled drug delivery applications are depicted in Fig. 5. The main advantage of these sub-micron-sized forms is that they offer higher loading or dosing per unit volume due to a large surface area. Together with enhanced bioavailability, efficient navigation in the remote sites of tissue, flexibility in formulations, and improved intracellular trafficking of drugs motivate researchers and clinicians to make these formulations in nano/micro form.

Biomaterials for Controlled Drug Delivery Applications 149

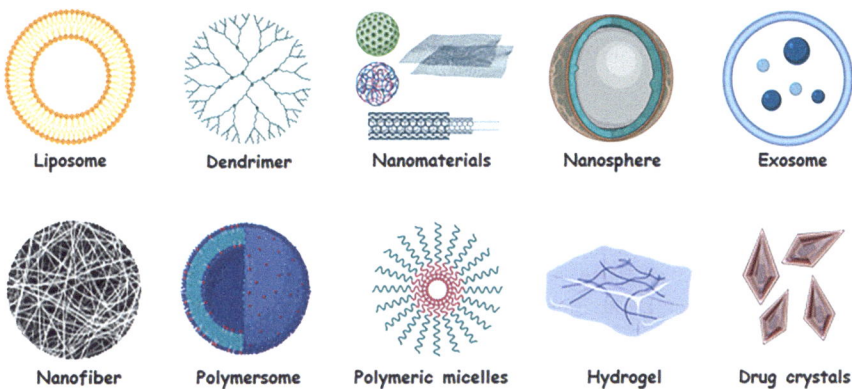

Fig. 5 Different biomaterials used for controlled for drug delivery applications

In the below section we have presented short overview about these different biomaterials used in drug delivery applications.

7.1 Liposomes

Liposomes are one of the most popular drug delivery vehicles for controlled drug delivery applications. They are amphiphilic phospholipid-based colloidal particles that surround an aqueous compartment with lipid bilayers. The alignment of the closed bilayer structure takes place due to the hydrophobic effect of the amphiphilic molecules where it is organized in such a fashion that the unfavorable interactions between hydrophobic part of the molecules and the aqueous environment are minimum. The size of the liposomes is usually within the range of 25–200 nm, although liposomes with size greater than 200 nm also exist. However, due to their large size they are excreted by the reticuloendothelial system within a short period of time, and as a result they have short circulation time in the blood. Due to enhanced permeability and retention (EPR) effect, liposomes are primarily used for drug delivery to cancer cells. One commercialized example of liposomal-based controlled drug delivery system is amphotericin B liposomal injection used in the treatment of fungal infections.

7.2 Dendrimers

Dendrimers are highly ordered polymeric materials. They are highly symmetric around a core with three-dimensional architecture. Due to the free space between its branched structure, it can carry a large number of drugs within it and can release them in a controlled fashion with time. Polyamidoamines (PAMAM) dendrimer [8] is the most common dendrimer used in biomedical applications.

7.3 Nanomaterials

Nanomaterials can be defined as materials having, at minimum, one external dimension within a range of 1–100 nm. Nanomaterials can be zero dimensional (nanoparticles, quantum dots), one dimensional (nanowires, nanorods), or two dimensional (MXenes). Nanomaterials can load drugs in both physical (Van-Der Waals interaction, H-bonding, etc.) as well as chemical bonding (drugs covalently bounded to nanomaterials), although it was observed that chemically bound drugs are physically more stable and release more sustained fashion due to greater stability of covalently bound drugs compared to physically bound drugs. The primary advantage of the nanomaterials is that it can deliver drugs to a difficult-to-reach site such as in brain, in addition to that it can be administered *via* different routes.

7.4 Nanosphere and Nanocapsules

Nanosphere and nanocapsules are especially designed core-shell type drug delivery carriers where the inner core acts as a reservoir for the drug and the shell protects the drug and releases it in a controlled manner. Nanosphere and nanocapsules are usually made of biodegradable polymers with a size of 5–200 nm.

7.5 Exosomes

Exosomes are nano-sized, cell-derived, membrane-bound vesicles with a diameter of 30–100 nm that participate in the intercellular transportation of foreign and endogenous chemicals. Due to the excellent internalization capability into cells, exosomes are primarily used for targeted delivery of small proteins, mRNA, or nucleic acid drugs into cells.

7.6 Polymeric Nanofibers

Nanofibers are usually solid or hollow fibers with a diameter less than 100 nm. Due to its high surface-to-volume ratio, it is well acknowledged for drug delivery applications. The features of the nanofiber such as diameter, surface morphology, and porosity can be fine-tuned to control the drug release kinetics. Usually, nanofibers are synthesized using electrospinning technique where patterning can be also implemented to control the drug release. In addition, nanofiber can be prepared in a core-shell model, where two different drugs can be loaded into the system, one into core and another one into shell; after degradation of shell and release of the drug present in shell the second drug will be released from the core. Therefore, sequential release of two drugs also can be achieved using nanofiber.

7.7 Polymersomes

Polymersomes are synthetic microscopic vesicles that are usually made of diblock copolymers/polymer-lipid composites which offer improved stability, enhanced drug encapsulation efficacy, and membrane characteristics. Polymersomes are more stable compared to liposomes and have a lower toxicity in the body. It can encapsulate both hydrophilic as well as hydrophobic drugs.

7.8 Polymeric Micelles

Polymeric micelles are made of amphiphilic polymers with a hydrophilic and another hydrophobic block that spontaneously self-assemble to form micelles. The hydrophobic block of the polymer forms the core while the hydrophilic unit of the polymer forms the shell. Generally, the size of polymeric micelles ranges in between 20 and 250 nm. Polymeric micelles are very popular for delivery of hydrophobic drugs such as Doxorubicin [9]. Polymeric micelles are primarily used for tumor cell targeting and delivery of hydrophobic anticancer drugs to the cancer cells. The hydrophobic core of the micelles protects the drug from premature degradation while due to the EPR effect it selectively targeted toward tumor cells compared to healthy normal cells.

7.9 Nanoemulsions

Nanoemulsions for drug delivery applications usually are made of heterogeneous oil-in-water system stabilized by surfactants or emulsifiers. They can load hydrophobic drugs and administrated *via* different routes. Compared to conventional emulsions, nanoemulsions are more stabilized and have better creaming property. In addition to that, due to the larger surface area, it can cover more area on the skin than conventional emulsion. Drugs through nanoemulsion are usually administrated *via* the dermal route.

7.10 Hydrogels

Hydrogels are three-dimensional (3D) chemically/physically cross-linked polymer networks that can absorb and hold a substantial amount of water [10]. Within hydrogel, the drug is distributed in a glassy polymer, and when it comes in contact with water the hydrogels swell and subsequently release the drug. The release profile of the drug is controlled by the penetration ability of water molecules into the hydrogel, the solubility of the drug in water, and the swelling behavior of the hydrogel. It was observed that temperature, pH, and ionic strength in single term or in combination may be used to manipulate the swelling of the hydrogel. Depending on the design considerations, hydrogels can offer excellent spatio-temporal control over the release profile of therapeutic agents including small molecule drugs, nucleic acids, proteins, growth factors, and cells. Additionally, due to their tunable physicochemical properties, biodegradability, and protecting capability of drugs, they serve as an ideal candidate for controlled delivery of many drugs.

7.11 Drug Crystals

Drug crystals, in opposite to amorphous form, are formulations in which the drug molecules are arranged in a defined structure. The crystal is composed of a repeating block called the unit cell that builds the crystal lattice. Some drugs have the ability to exhibit different molecules arrangement, resulting in different crystalline phase; this phenomenon is called polymorphism [11].

Crystalline drug formulations can stand themselves as a carrier-free DDSs, or be a part from a carrier-based DDSs [12]. Many advantages can be counted for the first platform; the entire DDS is composed 100% from the therapeutic agent without a drug content limitation as well as it decreases the body immune response. Furthermore, due to the crystal's compact structure, the drug molecules are trapped within the lattice, making the drug molecules release mainly *via* surface erosion. However, the drugs releasing rate depends on the crystal morphology and size, which can be controlled by the crystallization conditions.

Drug crystallization from a solvent include the solvent/anti-solvent, temperature-induced, and solvent evaporation techniques. In addition to single-component drug crystals, multi-component crystals are being investigated in order to enhance the physicochemical properties of the individual components. Multi-component crystals include drug-drug and drug-coformer combinations, such as co-crystals, as well as crystal salts and solvates/hydrates.

Drug crystals primal administration as carrier-free DDS is *via* implantation locally at the disease site. However, as stated before, drug crystals can be incorporated in a carrier-based platform, such as liposomes that can provide protection from enzymatic degradation and impact the drug bioavailability and functionality.

7.12 Stimuli-Responsive Smart Biomaterials

Another interesting class of substances for controlled drug delivery application is stimuli-responsive biomaterial. These types of biomaterials are very sensitive toward stimuli such as temperature, pH, light, redox, enzyme, magnetic field, etc., and depending on design factor, they can be sensitive to single or multiple stimuli. These types of drug carriers are specially designed for very particular applications. Stimuli-responsive biomaterials can offer on demand drug delivery so that, in presence of particular stimuli the drug will be released. For example, in cancer cells the concentration of glutathione is much higher compared to normal healthy cells. Based on this fact, researchers have developed glutathione-responsive drug delivery system containing sulfur-sulfur bond (S-S) [13], that cleave in the presence of excess glutathione at cancer cells, and the drug can be released.

8 Conclusion

Formulation, which is a combination of drugs and excipients, plays most important part in successful treatment of any disease. Excipients are used to provide structure, improve stability of the drugs, and mask the unusual flavor of drugs. A traditional dosage includes solid, semisolid, and liquid forms, which suffer from fluctuations in plasma drug concentration, impose high dosing, as well as multiple dosing frequency which leads to poor patient compliance. For any drug, bioavailability is very crucial to attaining the intended effect from any dosage form. In this regard, controlled drug delivery systems have emerged as an improved alternative of traditional drug delivery systems in order to achieve improved bioavailability, extended drug release, and to maintain drug-plasma levels within the therapeutic window with minimum side effects. Biomaterials are the prime candidate for controlled drug delivery applications as they can provide good biocompatibility, tunable biodegradability, excellent bioavailability, and extended drug delivery time within the therapeutic window without compromising the patient compliance. Till now, several biomaterials including liposomes, dendrimers, nanomaterials, nanosphere and nanocapsules, exosomes, polymeric nanofibers, polymersomes, polymeric micelles, nanoemulsions, hydrogels, and drug crystals are exploited in controlled drug delivery applications. Recently stimuli-responsive drug delivery systems became popular in controlled delivery applications for their targeted and on-demand drug delivery capability. The recent trends of emerging biomaterials include microfluidic-based carriers, personalized biomaterials using additive manufacturing, and CRISPR/Cas9-based controlled delivery systems. However, the major challenge is that while several studies focused on biomaterials for controlled delivery applications, till now only very few are translated into real-life clinical applications. To overcome this, long-term preclinical studies and clinical trials are needed more frequently following the preliminary in vitro and in vivo studies.

Acknowledgments The Neubauer Family Foundation is thanked for their generous funding and support. S.F. was supported by MAOF Fellowship from the Council for Higher Education, Israel. This work was supported also by the Technion's president grant, and we thank them for that.

Questions

1. How the physicochemical properties of a drug can affect its functionality?
    - Drug functionality depends mainly on its solubility that can be adjusted by using appropriate counterpart excipients.
    - Drug functionality depends mainly on its permeability, which also dictates the drug administration form.
    - The drug's solubility and permeability affect the drug's functionality, and therefore, should be considered in the way of drug administration as well as the DDS designing.
    - The drug's physicochemical properties don't affect the drug's functionality.

    Explanation: both drug solubility and permeability do affect the drug functionality. For effective drug functionality, it should be first dissolved followed by its absorption in the blood. For different drugs classifications, different solutions are used to overcome lower drug solubility/permeability.

2. What is the need for a DDS utilization?
    - Modulating the pharmacokinetics behavior of a drug
    - Targeted delivery of a drug
    - Stabilizing the drug
    - All of the above

    Explanation: DDS can alter the drug release rate, where it can maintain the drug level in between the minimum effective and minimum toxic levels. Moreover, DDS can enable long-term drug release, as well as enabling its delivery to a certain organ/tissue, for targeted treatment. For less stable drugs, a suitable DDS can stabilize it to ensure its effective effect.

3. What are the principle considerations in DDS election?
    - DDS pharmacokinetics and administration route
    - The medical application
    - Drug's classification and release kinetics
    - All of the above

    Explanation: DDS election and designing depends first on the medical problem that should be addressed, as well as the relevant drug classification, which supposed to be adjusted by the DDS pharmacokinetics and administration.

4. What are the main material properties that dictate their utilization in a DDS?
    - Biocompatibility, biodegradability, and non-immunogenicity
    - Biocompatibility, toughness, and non-immunogenicity

- Biocompatibility, biodegradability, and flexibility
- Biodegradability, toughness, and non-immunogenicity

Explanation: the main material properties that dictate their utilization in a DDS are biocompatibility, biodegradability, and non-immunogenicity. These material properties can insure their enhanced functionality with minimal toxicity to the body, as well as minimal immunological response. The material biodegradability is also important to insure their elimination in a safe way. For specific applications, the material's mechanical properties do also matter.

5. Nano-formulation of DDS are preferred because of their:

- Low surface-to-area ratio
- Efficient navigation within the tissue
- Mechanical properties
- None of the above

Explanation: Nano-formulation of DDS are preferred because of their higher drug loading per unit volume due to a large surface to area ratio. Together with enhanced bioavailability, efficient navigation in the remote sites of tissue, flexibility in formulations, and improved intracellular trafficking of drugs.

6. Drug release from crystals is controlled by?

- Crystal size
- Crystal purity
- Crystal Morphology
- All of the above

Explanation: Drug release from crystals can be manipulated by all of the above parameters, where they can help to achieve a slower or faster drug release to fit a specific application.

7. Drug release from crystals versus amorphous formulation is expected to be?

- Faster
- Similar
- Slower
- Either similar or faster

Explanation: Drug release from crystals is always slower than amorphous as the crystalline structure represents the most stable form compared to the randomly organized amorphous form.

8. Drug crystals of hydrophobic drugs versus drug crystals of hydrophilic drugs are expected to exhibit?

- Faster release
- Similar release
- Slower release
- Either similar or faster release

Explanation: Although both are crystals and should exhibit a slower release profile from the amorphous version but since the drug is released from hydrophobic-based crystals the release is anticipated to significantly exhibit a slower release from the hydrophilic-based crystals.

9. Conventional drug delivery system versus controlled drug delivery system?

- Increases the chances to range between low-effective dose to toxic dose
- Can be used for long-term release
- Favored for all of the applications
- None of the above

Explanation: Conventional drug delivery system allows drug concentration in the plasma following the administration to range from very low effective dose to high concentration (might be toxic both in the short or long term).

10. Which of the following is best describing classification of Type 4 in drugs based on biopharmaceutics?

- High solubility & High permeability
- Low solubility & Low permeability
- High solubility & Low permeability
- Low solubility & High permeability

Explanation: Drugs-based biopharmaceutics that classified Type 4 are identified with both low solubility and low permeability.

References

1. Langer R (1998) Drug delivery and targeting. Nature 392(6679 Suppl):5–10
2. Hedaya MA (2012) Basic pharmacokinetics. CRC Press, Boca Raton
3. Ritger PL, Peppas NA (1987) A simple equation for description of solute release I. Fickian and non-fickian release from non-swellable devices in the form of slabs, spheres, cylinders or discs. J Control Release 5(1):23–36
4. Ritger PL, Peppas NA (1987) A simple equation for description of solute release II. Fickian and anomalous release from swellable devices. J Control Release 5(1):37–42
5. Shukla AJ, Price JC (1991) Effect of drug loading and molecular weight of cellulose acetate propionate on the release characteristics of theophylline microspheres. Pharm Res 8(11):1396–1400
6. Langenbucher F (1972) Letters to the editor: linearization of dissolution rate curves by the Weibull distribution. J Pharm Pharmacol 24(12):979–981
7. Hudecki A, Kiryczyński G, Łos MJ (2019) Chapter 7 – Biomaterials, definition, overview. In: Łos MJ, Hudecki A, Wiecheć E (eds) Stem cells and biomaterials for regenerative medicine. Academic, London, pp 85–98
8. Wang G, Fu L, Walker A, Chen X, Lovejoy DB, Hao M, Lee A, Chung R, Rizos H, Irvine M, Zheng M, Liu X, Lu Y, Shi B (2019) Label-free fluorescent poly(amidoamine) dendrimer for traceable and controlled drug delivery. Biomacromolecules 20(5):2148–2158
9. Zhang C-G, Zhu W-J, Liu Y, Yuan Z-Q, Yang S-D, Chen W-I, Li J-Z, Zhou X-F, Liu C, Zhang X-N (2016) Novel polymer micelle mediated co-delivery of doxorubicin and P-glycoprotein siRNA for reversal of multidrug resistance and synergistic tumor therapy. Sci Rep 6(1):23859

10. Ghosal K, Chakraborty D, Roychowdhury V, Ghosh S, Dutta S (2022) Recent advancement of functional hydrogels toward diabetic wound management. ACS Omega 7(48):43364–43380
11. Beckmann W (2013) Crystallization: basic concepts and industrial applications. Wiley, Weinheim
12. Farah S, Doloff JC, Müller P, Sadraei A, Han HJ, Olafson K, Vyas K, Tam HH, Hollister-Lock J, Kowalski PS, Griffin M, Meng A, McAvoy M, Graham AC, McGarrigle J, Oberholzer J, Weir GC, Greiner DL, Langer R, Anderson DG (2019) Long-term implant fibrosis prevention in rodents and non-human primates using crystallized drug formulations. Nat Mater 18(8):892–904
13. Quinn JF, Whittaker MR, Davis TP (2017) Glutathione responsive polymers and their application in drug delivery systems. Polym Chem 8(1):97–126

Biomaterials Application: Implants

Aditya Ruikar, Chase Bonin, Gauri S. Kumbar,
Yeshavanth Kumar Banasavadi-Siddegowda, and Sangamesh G. Kumbar

Abstract Biomaterials are a class of materials used for a wide range of therapeutic and diagnostic applications inside the body cavity without severe side effects. Structural components derived from biomaterials can successfully mimic the function of tissue and integrate with a biological system. Biomaterials and their implants are widely used in a variety of applications, but this chapter will focus on those used in skeletal, skin, cardiac, neuronal, and ocular implants. Skeletal implants have been designed and used to enable locomotion for centuries. The first-generation implants were derived from metals primarily for load-bearing without any bioactivity. However, the generation implants focused on the use of a variety of biomaterials including polymers, bioceramics, and their composites, as well as bioactive factors to improve their performance and integration with host tissue. Skin implants are typically made of natural and synthetic polymers as well as their composites in the form of wound coverings, bioactive bandages, and scaffolds to deliver bioactive agents and cells to promote wound healing. Cardiac implants are designed to mimic the electroactive nature of the tissue and are derived from a decellularized extracellular matrix (ECM) and conductive materials to enable electrical recording and

---

A. Ruikar
Department of Orthopaedic Surgery, University of Connecticut Health, Farmington, CT, USA

Department of Biomedical Engineering, University of Connecticut, Storrs, CT, USA

C. Bonin · G. S. Kumbar
Department of Biomedical Engineering, University of Connecticut, Storrs, CT, USA

Y. K. Banasavadi-Siddegowda
Surgical Neurology Branch, National Institute of Neurological Disorders and Stroke, National Institutes of Health, Bethesda, MD, USA

S. G. Kumbar (✉)
Department of Orthopaedic Surgery, University of Connecticut Health, Farmington, CT, USA

Surgical Neurology Branch, National Institute of Neurological Disorders and Stroke, National Institutes of Health, Bethesda, MD, USA

Department of Materials Science and Engineering, University of Connecticut, Storrs, CT, USA
e-mail: kumbar@uchc.edu

stimulation. Similarly, neural tissue implants also focus on the use of electroactive scaffolds along with factors and cells to promote neural tissue and axonal regeneration. A wide variety of ocular implants, including contact lenses, keratoprostheses, and intraocular lenses have been developed and are mainly used for vision correction. The chapter will also provide an overview of biomaterial and implant applications in cosmetic surgeries, medical devices, and equipment. The chapter will also cover the more recent progress in biomaterials and implants that can truly mimic the function of specific tissues to improve healing outcomes.

Graphical Abstract

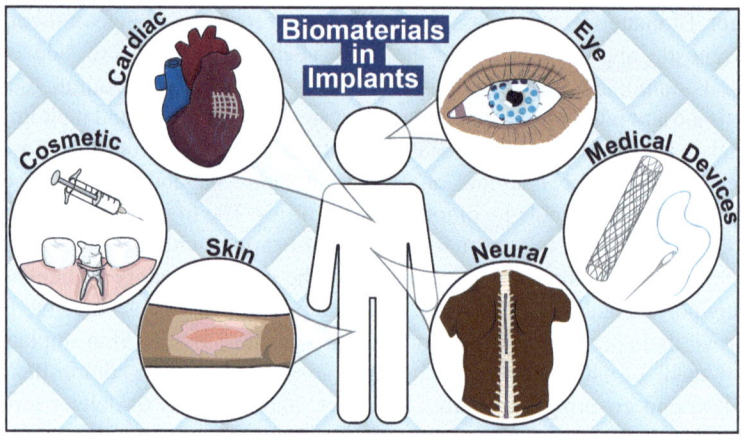

Keywords Biomaterials · Implants · Skeletal implants · Skin implants · Cardiac implants · Neural tissue implants · Ocular implants

1 Introduction

The term "biomaterial" refers to any material that is used to replace or repair damaged tissue in the human body. Throughout history, people have made sutures from a variety of materials, including gold wires in ancient Greece (~100–200 AD). Since 600 A.D., when the Mayans fashioned their teeth out of seashells, which mostly contained calcium as its main chemical constituent, biomaterials have been used as implants. Biomaterials include ceramic, polymer, metal, or their composites find applications as implants in organ or tissue systems (cardiac, nervous, soft tissue, or skeletal). However, the most common way biomaterials are classified is by their source of origin, i.e., natural biomaterial or synthetic biomaterial. Natural biomaterials are materials that occur naturally, such as polysaccharides, proteins, lipids, or decellularized materials. Because these materials are abundant in nature, they are considered to be highly biocompatible and mimic the human extracellular matrix (ECM) compositionally and structural shaping which makes them ideal "scaffolds" for tissue repair and regeneration. Although natural biomaterials are a popular choice for scaffold fabrication, they have drawbacks such as low strength, poor degradation, and physical properties. Synthetic polymers, on the other hand, are simple to manufacture and have excellent physical properties, and low cost; however, as their acidic degradation byproducts are known to compromise biocompatibility of the product, these are generally not used alone in production of scaffolds. Poly(lactic acid) (PLA), poly(lactic-co-glycolic acid) (PLGA), polycaprolactone (PCL), poly(vinyl alcohol) (PVA), and poly(glycerol sebacate) (PGS) are synthetic materials that are commonly used as biomaterials [1]. Semi-synthetic polymers combine the best features of natural and synthetic materials to mimic both physical and biological requirements for implant applications. Therefore, often natural and synthetic materials are combined to create composite biomaterials that exhibit ideal properties of both natural and synthetic materials.

Biomaterials are used to create implants that substitute or replace a damaged or diseased body part's function. Implants, like biomaterials, are classified in a variety of ways, but one of the most common is based on the system in which the implant is used, such as skeletal, cardiac, neuronal, skin, or ocular. Knowing the system in which the implant will be used is important when selecting the biomaterials that will create the implant, as every system has unique requirements (Fig. 1).

We will learn about the use of biomaterials in various medical implants as well as their application in cosmetic surgeries and other medical needs in this book chapter. This chapter will also describe how material science and engineering have progressed in implant fabrication and discuss the future scope of implant research. The chapter is divided into three sections: application of biomaterials in implants, which covers the use of biomaterials in skeletal, skin, cardiac, neuronal, and ocular implants; application of biomaterials in cosmetic surgeries, which covers materials used for cosmetic enhancement; and biomaterials used in other medical needs, which covers biomaterials used in the manufacturing of other medical devices.

2 Application of Biomaterials in Implants

2.1 Bone and Skeletal Implants

Bone is a mineralized tissue that is composed primarily of type I collagen and hydroxyapatite, which is produced by osteoblasts. Since it is a mineralized tissue, bone is prone to fracture due to trauma or erosion caused by a variety of factors such as osteoporosis and extensive wear and tear.

Every year, around 280,000 hip fractures, 700,000 vertebral fractures, and 250,000 wrist fractures are reported in the United States, with an estimated cost of $10 billion. Surgeries are critical when the loss of bony tissue exceeds the body's regenerative capacity, which leads to around 4,000,000 bone grafting or bone substitute operations performed globally every year. Back pain, diseased or eroded bone/joints that are beyond repair, and bone tumors have all been treated with

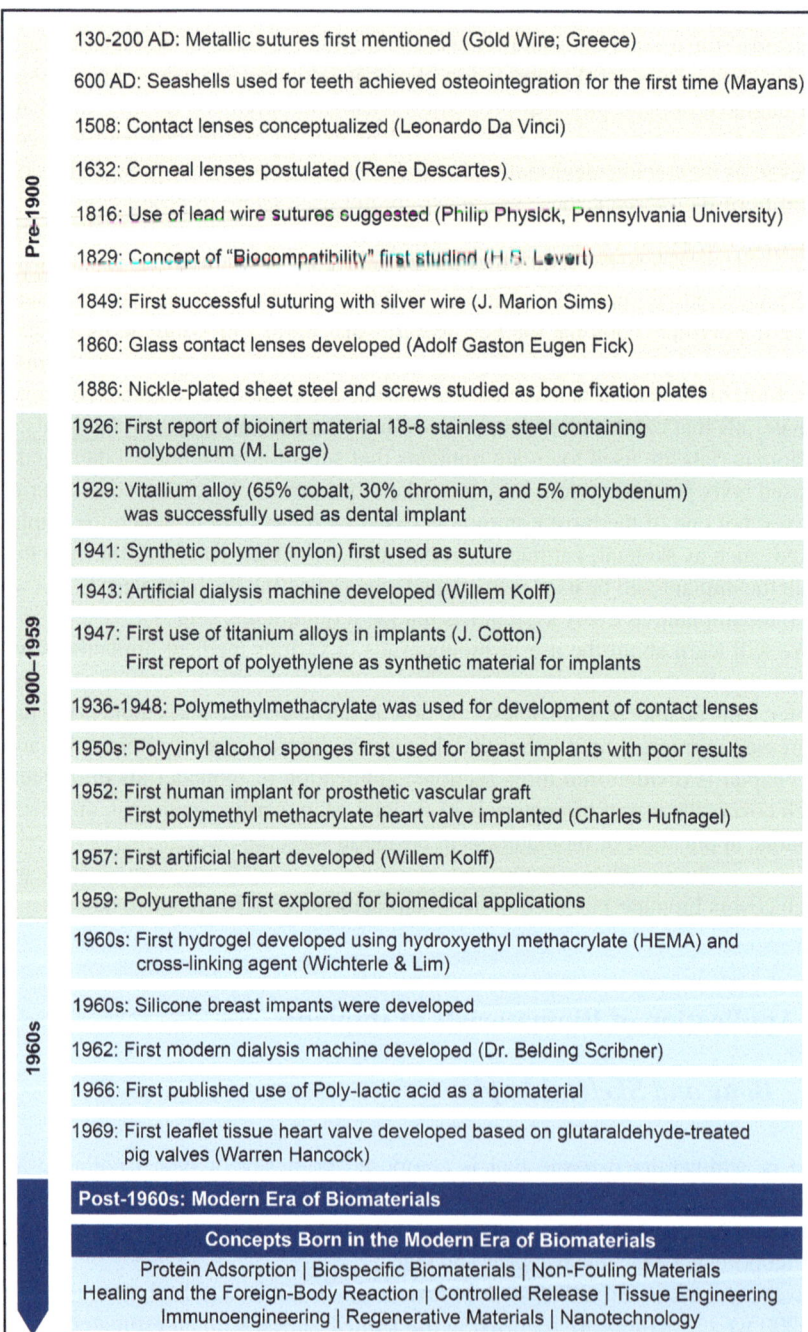

Fig. 1 History of biomaterials

surgical interventions such as bone replacement or grafting. According to reports, about half of all autologous bone grafting surgeries performed in the United States are for spinal fusions, with only 15% of those achieving a successful union. Joint replacement therapy for corroded joints, on the other hand, has increased in recent years, with hip arthroplasty and total knee replacement surgeries being the most common. Although hip and knee replacement surgeries are common, they have a revision rate of around 15% for hip replacement and 7% for knee replacement surgeries in the United States. As a result, improvements in surgical procedures and implants used in these procedures are needed to improve the success rate of these surgeries.

Cells with osteogenic potential (osteoblasts), an osteoconductive matrix, stimulus, and a mechanically stable environment are essential for bone healing or regeneration following such trauma. Grafting, or the procedure of replacing a damaged part with a substitute, has long been considered the gold standard in surgeries after traumatic injuries. Autografts (tissue transplants from one's own body) and allografts (tissue transplants from a donor body) were once thought to be cutting-edge implantation techniques and are still used in developing countries when the amount of bone damage is minimal. Autografts are generally harvested from the iliac crest, while cortical and fibular grafts are also used for maintaining the structural and functional integrity of the bone. Although autografting is the safest method of implantation since the tissue will not be rejected by the immune system, the limited availability of tissue that can be used for grafts is a disadvantage. Allografts on the other hand are available as massive bone pieces, demineralized bone matrix (DBM), processed bone chips, or cortical struts. Massive allografts are used in tumor resurrection surgeries, whereas DBM is produced from cadaveric bone and is commercially available as allograft bone paste (Osteofil™, Medtronic Sofamor Danek, Memphis, Tennessee), putty (Grafton™, Osteotech Inc., Eatontown, New Jersey), gel, or cement. Cortical struts lack the osteogenic potential and hence are mainly used as load-bearing materials in the graft. Although the availability of allografts is greater than autografts, a major disadvantage of allografts is that they are susceptible to the host-versus-graft immune response and pose an inherent risk of disease transmission. Due to the above disadvantages associated with grafting procedures, numerous alternative biomaterials were engineered to improve bone repair and regeneration.

Since the early 1900s, biomaterials have revolutionized the process of bone repair and regeneration by providing improved mechanical stability and low host-graft immune reactions. Metals such as stainless steel, cobalt-chromium alloys, and titanium or titanium alloys were the first biomaterials used as bone substitutes in the first generation of implants. These materials were used as load-bearing prostheses or fracture-fixing materials due to their remarkable load-bearing capacity and low host-versus-graft immune response. Stainless steel and titanium alloys have been well appreciated for their biocompatible nature, good load-bearing ability, and low inflammatory response; therefore, they are still used as fixing materials or bone substitutes in developing countries. However, one of the major disadvantages associated with first-generation implants is their non-biodegradable nature which necessitates revision surgery for removing the implant after complete healing. Also, low flexibility and leaching of metal toxic products into the bloodstream are some of the added disadvantages associated with first-generation implants. Therefore, second-generation implants were focused on the improvement of their biocompatibility and biodegradability over first-generation implants.

Second-generation implants mainly consisted of biocompatible and biodegradable polymers which mimicked the exact structure of bone and also provided a suitable environment for osteoblast growth and mineralization, thereby leading to rapid bone healing and regeneration. Natural polymers such as chitosan, alginate, carrageenan, silk fibroin, collagen, gelatin, hyaluronic acid, and cellulose, and synthetic polymers such as polylactic acid (PLA), polyglycolic acid (PGA), polycaprolactone (PCL), polymethyl methacrylate (PMMA), and polyurethane are the major biomaterials used in second-generation biomaterial implants. These polymers are either used individually or in a mixture (composite form) to mimic the structure of the bone. Natural polymers have greater biocompatibility and biodegradability, whereas synthetic materials tend to have better mechanical strength. Hence, the composite use of these polymers is advantageous for modulating mechanical strength as well as the biocompatibility of the final product. Bioceramics such as calcium phosphate, hydroxyapatite, calcium silicate, and calcium sulfate in conjugation with the polymers supported osteoconduction and induction to support bone regeneration. Furthermore, 3D printing of polymeric materials has given an added benefit in the manufacture of implants such as orthopedic screws, resulting in the production of biocompatible and biodegradable alternatives to metal implants [2, 3]. Although these materials appear to be excellent bone substitutes and provide the necessary environment for bone regeneration, bone healing requires a stimulus to initiate osteogenesis, which was not sufficient for the second-generation implants.

To improve second-generation implant osteogenic potential, third-generation bone implants were developed with osteogenic-stimulating substances embedded in second-generation biopolymers. BMP-2 (Bone Morphogenic Protein – 2) and BMP-7 (Bone Morphogenic Protein – 7) are common growth factors that are embedded in second-generation implants and provide the necessary stimulus for bone regeneration. BMP-7 embedded in a collagen carrier was investigated in posterolateral spinal fusions and produced a 55% fusion rate as compared to the gold standard iliac crest which produced only a 40% fusion rate, indicating its improved efficiency in bone regeneration. In addition, antibacterial molecules can be embedded into polymers to prevent bacterial infection following transplant surgery. Although third-generation biomaterial implants are currently a promising alternative, they also carry the risk of osteosarcoma (pathogenic bone overgrowth) due to the presence of stimulation agents [4].

The technology of skeletal implants has come a long way since its beginning, yet there has not been a single commercialized implant that can support bone growth and lead to rapid healing. A variety of osteoinductive molecules in the form of small molecules and growth factor alternatives are being evaluated to identify those that will provide the necessary environment for bone growth and repair. Furthermore, osteoinductive molecule eluting formulations using novel polymers and scaffolds are being explored to control the release to improve bone healing.

2.2 Skin

Skin is the largest organ in the human body, and it performs many functions including being a barrier to outside materials, thermoregulation, moisture retention, immune protection, imparting sensation, and self-healing response. Skin is made up of three layers: epidermis, dermis, and hypodermis, also known as the outer, middle, and inner layers. It also has sweat and sebaceous glands, mechanoreceptors, hair follicles, vasculature, and nerve endings in addition to these layers.

Factors such as burns, acute trauma, chronic wounds, surgeries, and infections are mainly responsible for skin damage. According to the World Health Organization (WHO), approximately 180,000 deaths occur annually due to fatal injuries from burns, and the global wound care market has risen from 18.35 billion USD to 22.81 billion USD between 2017 and 2022. Hence, wound management has a huge socioeconomic impact on the world.

As with skeletal implants, autologous split-thickness skin grafts have been considered the gold standard implant for repairing damaged skin tissue since 1874. However, the technique of autologous grafting comes with certain drawbacks such as limited availability of donor skin, failure to treat full-thickness wounds, and scarring on both donors as well as recipient sites. These drawbacks limit their clinical use but have paved the way for developing biomaterials for skin tissue engineering. Since Jacques-Louis Reverdin pioneered the work of skin tissue engineering through the application of skin allografts on the damaged site in 1870, the science of skin regeneration and replacement has come a long way: now, biocompatible materials along with cultured cells are used as substitutes for the original skin.

To generate functional skin substitutes, biomaterials used for skin are required to possess some key features such as biocompatibility, non-toxicity, no immunogenicity, biodegradability, ability to retain moisture, optimal elasticity, and porosity with good interconnectivity for a free exchange of gases and nutrients. Also, for commercialization and to increase the global impact, the biomaterial substitutes should be economic, scalable, have a considerable shelf life, and should be available in every customer outlet.

Natural materials are widely used for formulation of skin implants. Since ancient times, silk has been used as a wound dressing due to its properties such as good biodegradability, ability to retain moisture, and permeability for air and oxygen. Silk fibroin, a protein made by silkworms and a core protein of silk fiber, is widely used as a skin tissue scaffold. The molecular weight of silk fibroin affects its wound healing capacity: silk fibroin protein with a narrow molecular weight distribution accelerates healing with less scar formation and lower immune response compared to silk fibroin with wide molecular weight distribution. Silk fibroin is widely used in 3D bioprinting applications; however, the use of silk fibroin alone in 3D bioprinting leads to certain disadvantages such as clogging needles, inappropriate mechanical properties, and the lithium bromide (LiBr) dissolution process often used in 3D bioprinting leads to degradation of the silk fibroin protein chain. Hence, to avoid these hindrances, silk fibroin is combined with other natural polymers such as alginate, cellulose, pectin, collagen, polylactic-co-glycolic acid (PLGA), keratin, and gelatin to form a composite that exhibits improved properties compared to silk fibroin alone. Collagen is another common protein that is used in skin tissue engineering which not only has excellent biocompatibility but also yields high mechanical strength and can provide the necessary amino acids required for wound healing after its degradation. Collagen plays a hemostatic role by promoting platelet aggregation and has high hydrophilicity like most natural biomaterials. Collagen is used as a cross-linked scaffold by using low-toxicity cross-linking agents and is now commercially available as Apligraf and Integra for skin regeneration. Proteins such as gelatin – a denatured product of collagen, keratin, and fibrin also are extensively used as natural biomaterials for the formulation of skin implants.

Polysaccharides are also used as natural biomaterials for the production of skin implants. Out of all the polysaccharides, chitosan, cellulose, and alginate have gained immense popularity for skin tissue engineering. Chitosan is a deacetylated product of chitin which is commonly found in crustacean cell walls. It possesses good gel and film-forming properties and, due to the presence of cations in the chitosan structure, it also has an additional antibacterial property which is necessary for preventing wound infections. The hydrophilic nature and moisture retention feature make it attractive for skin cell growth and skin tissue engineering. Although chitosan is an excellent biomaterial for fabrication, it is highly brittle and naturally lacks the necessary mechanical strength; therefore, chemically modified chitosan is used in commercially available products. Cellulose on the other hand is the most abundant polysaccharide available and is used widely in skin implants due to its outstanding mechanical strength and biocompatibility. Chemically modified derivatives of cellulose are available wherein cellulose acetate and hydroxyethyl cellulose are the most commonly used derivatives for scaffolding. Alginate is another marine biopolymer that is used in skin implants. It is usually obtained from brown seaweeds and is composed of mannuronic and glucuronic acid subunits. Alginate is widely used as a hydrogel scaffold due to its excellent gelling ability in the presence of calcium ions. Alginate is not only used as a biocompatible scaffold but is also used as a carrier for delivering drugs and cells in minimally invasive treatments. One commercially available form of alginate is sodium alginate. Sodium alginate possesses good biocompatibility, hydrophilicity, and degradability making it attractive for tissue engineering applications.

Synthetic materials are also being used as biomaterials for skin implants. Polylactic acid (PLA) is commonly used in skin tissue scaffold which is obtained by polymerizing lactic acid, which in turn is a fermentation product of corn and wheat. Although PLA possesses no cytotoxicity, it is hydrophilic and has poor flexibility which limits its use in the fabrication of scaffolds alone. One study showed that 3D PLA nanofibrous mats with embedded bone marrow stromal/stem cells (BMSC) accelerated the healing of full-thickness skin wounds in rats, proving that PLA may be a synthetic alternative to natural biomaterials. [5, 6]

Skin tissue engineering is being revolutionized due to novel technologies such as 3D bioprinting and novel biomaterials. Now, autologous cells from one's skin can be harvested, cultivated, and grown to an extent where the damaged skin can be fully replaced with in vitro-grown autologous cells. Therefore, the field of skin tissue engineering is growing at a rapid pace, which will hopefully lead to painless skin replacement with scarless tissue soon (Fig. 2).

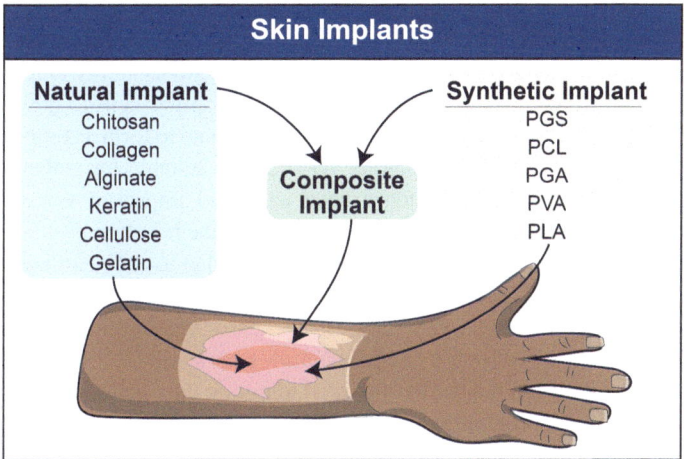

Fig. 2 Biomaterials in skin implants

2.3 Cardiac

The heart is a vital organ in the human body, thus inefficiencies in cardiac functioning can lead to serious consequences and even death. Cardiovascular diseases are one of the major contributors to total global mortality and loss of productivity, wherein over 92 million adults in the United States are affected by cardiovascular disease. The global impact of cardiovascular disease is also alarming, as about 30% of medical expenditures – which is equivalent to $149 billion – are spent on cardiovascular diseases annually worldwide.

Current treatment options are limited to the use of autologous grafts for diseases such as myocardial infarction and coronary artery disease. Although much like other autografts, limited availability and difficult surgical procedures have hindered their usage in cardiac repair and regeneration. This need has stimulated research in the field of cardiac tissue engineering, leading to the development of novel biomaterials which can be used as cardiac tissue replacement and to restore normal cardiac function. The process of selecting a graft material is generally based on the specific function of diseased or damaged cardiac tissue and its complexity. A variety of natural and synthetic biomaterials have been developed to repair and regenerate cardiac tissue.

Natural materials mimic the environment of cardiac tissues, hence they are now widely used to make scaffolds for cardiac repair. These materials mainly include well-characterized polymers such as collagen, chitosan, alginate, and decellularized ECM.

Collagen and its processed product gelatin are some of the most commonly used biomaterials in cardiac tissue engineering. These materials have excellent resorption, making them an ideal candidate for the fabrication of cardiac scaffolds. Collagen and gelatin patches with embedded autologous cardiomyocytes have shown promising results in pre-clinical animal studies wherein treatment with the scaffold prevented dilation of the infarcted area and induction of systolic wall thickening in a rat model.

Alginate is also used in cardiac tissue engineering due to its ability to form porous hydrogels, thereby creating a suitable environment for cardiac cells to grow and proliferate. Furthermore, chitosan can also be added to alginate to form a composite polyelectrolyte complex with a highly porous 3D structure suitable for embedded cells. These polyelectrolyte complex patches have exhibited improved vascularization, attenuated fibrosis, and have effectively integrated into the host tissue.

Other polymers such as fibrin and silk fibroin are also used in cardiac implants. Fibrin is thought to be an ideal cardiac tissue engineering candidate due to its precise and controlled degradation, easy processing, ease in chemical modification, injectability, and ability to form a complete autologous scaffold. Silk fibroin, on the other hand, can easily be chemically modified to mimic the cardiac niche and hence is used as a biomaterial for cardiac repair.

Decellularized tissue and organs are one of the emerging sources of biomaterials for a variety of tissue engineering applications. Decellularized ECM is one of the best scaffolding techniques for cardiac repair and regeneration, as it has a very low risk of immunological reaction due to the absence of cells. It mainly depends on the isolation of ECM from the tissues with minimal loss, damage, or disruption to the structure and maximized cell removal. Removal of cells from a cadaver heart is achieved using either physical, chemical, or enzymatic methods to obtain a full-sized ECM heart, in which autologous cells can be cultured so that a full-grown autologous heart can be created. These decellularized ECM scaffolds have immense potential to overcome the limitation of biocompatibility in organ transplantation and open up a new clinical possibility shortly.

Synthetic biomaterials are also used for the production of cardiac scaffolds due to their superior mechanical strength, tuneability, diversity, and optimal degradation rates. Also, unlike natural materials, synthetic materials can be used in a variety of techniques such as electrospinning, conventional casting, and 3D bioprinting for the fabrication of scaffolds. PLA and PGA are some of the widely investigated synthetic biomaterials for cardiac tissue engineering. Specifically, researchers have extensively used PGA for surgical applications and as a scaffold. Current synthetic grafts are mainly composed of polytetrafluoroethylene (PTFE) or Dacron® which is a non-biodegradable material. Other materials are also used in the fabrication of scaffolds such as polycaprolactone, which exhibits rapid endothelization and ECM production in comparison to PTFE grafts, and polyglycerol sebacate (PGS) which displays excellent biocompatibility and elastomeric properties in comparison to other materials.

Composite or hybrid scaffolds are now used extensively since they combine the properties of both natural and synthetic polymers. The properties of synthetic polymers such as mechanical strength, tuneable degradation rates, and other physical properties are combined with those of natural polymers to improve biocompatibility and cell adhesion [7].

Cardiac tissue has an intrinsic electrical activity which is essential for proper heart function. To achieve this proper heart function, implants fabricated using conductive biomaterials can facilitate the conduction of electrical signals through the heart. Materials such as gold nanoparticles (spheres, rods, and wires), silicon

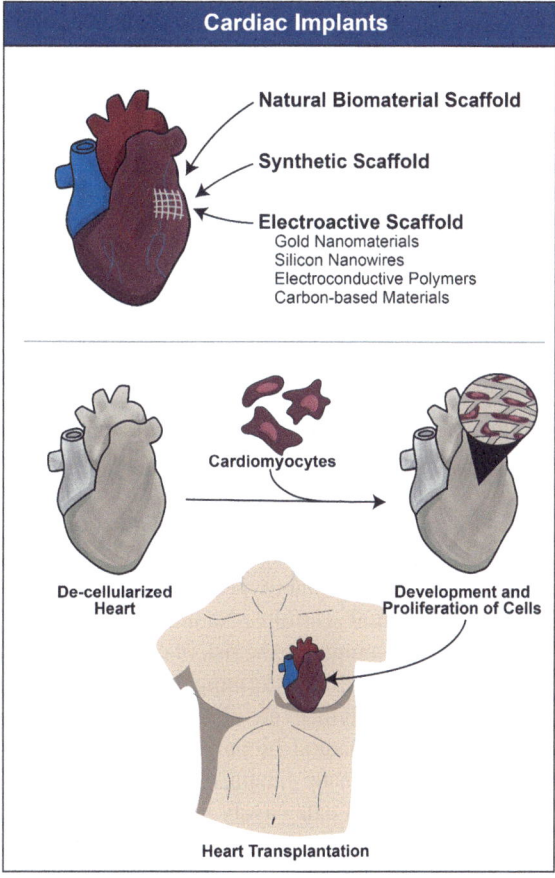

Fig. 3 Summary of cardiac implants

nanowires, carbon-based nanomaterials (carbon nanotubes, graphene nanosheets, nanohorns, nanofibers, and nanowires), and electroconductive polymers like polypyrrole, polyaniline, and poly(3,4-ethylene dioxythiophene) polystyrene sulfonate (PEDOT:PSS) are commonly used in the fabrication of electroconductive cardiac implants. However, before these conductive biomaterials can improve patient outcomes, these materials must improve cardiac function and electrical activity both in vitro as well as in vivo [8] (Fig. 3).

2.4 Neuronal

The nervous system is one of the most intricate and complex systems present in mammals. Due to the limited regeneration capacity of the nervous system, it is challenging to repair and regenerate this intricate system after an injury. The nervous system is mainly divided into the central nervous system (CNS) and peripheral nervous system (PNS). CNS impairments can occur due to various factors such as trauma due to falls, accidents, assault, sport-related injuries, or neurodegenerative diseases such as Alzheimer's disease, Parkinson's syndrome, Amyotrophic lateral sclerosis, etc. Stroke and tumors are additional factors that can cause CNS impairments. On the other hand, PNS impairments are also observed due to trauma caused by car accidents, penetrating injuries due to violence, falls, and occupational accidents. These nervous injuries to either central or peripheral nervous systems are devastating to the quality of life of the patients and can lead to permanent sensory or motor defects and excruciating neuropathic pain.

End-to-end nervous replacement with the help of autologous grafting was considered to be a gold standard treatment of nervous injuries until now. However, the scarce availability of neural grafts, the possibility of neuroma formation, and severe immunological reactions after grafting hinder the practical applications of grafting.

These limitations have now been overcome by the use of biomaterials, wherein the biomaterial scaffold provides a suitable environment for neural growth and proliferation which can lead to rapid repair of the injured nerve. Although there are challenges in nerve regeneration and repair, neural tissue engineering has set a benchmark using a variety of natural, synthetic, and novel materials that can completely repair the injury.

As with the other implant types, the natural polymers remain to be a good choice for fabricating neural scaffolds. Natural polymers not only have chemically tunable properties but also minimize the risk of cytotoxicity and immunological reaction after implantation. In addition to forming 3D scaffolds that fit a variety of physiological geometries, these polymers are also used for matrix formation, gelling agents, drug carriers, etc. Some natural biopolymers that are used in neural tissue engineering include collagen, alginate, gelatin, elastin, hyaluronic acid, chitosan, keratin, and silk.

Collagen has been extensively studied for neural regeneration and has also displayed promising results in regenerating rat sciatic nerves, dog sciatic nerves, and cat neural conduits. Collagen is highly biocompatible and can be used alone or combined with other natural or synthetic polymers to form scaffolds. Commercial collagen products are available, such as NeuraGen, which proved to be effective in neural reconstruction in around 43% of patients, and another product named Neuromaix, which displayed outstanding results in its first clinical trials. Studies using these two collagen products suggest that collagen can be effectively used in nervous system repair and regeneration. Although most research is done using bovine collagen, recently, fish collagen has gained more attention due to its excellent biocompatibility, cell adherence, and biodegradability. Even though fish collagen is an interesting alternative to its bovine counterpart, its use in neural regeneration remains unexplored. Gelatin is a processed product of collagen and is widely in tissue engineering applications. In neural tissue engineering, electrospun fibrous scaffolds with various combinations of gelatin and other natural or synthetic polymers are widely used. The electrospinning process allows the optimization and manipulation of biological and kinetic properties, thereby creating an effective environment for nerve regeneration. Gelatin-PCL blends act as a positive stimulus for neurite growth and for Schwann cell proliferation in vitro. When implanted in a rat model with a 10 mm sciatic nerve gap, a guide conduit made from genipin cross-linked gelatin and containing tri-calcium phosphate exhibited improved motor functionality than silicone tubes alone. Together, these results indicate that composite gelatin biomaterials may improve the functional healing of nerves.

Hyaluronic acid (HA), a glycosaminoglycan that performs a crucial role in lubrication in the human body, has been identified as an effective biomaterial for neural tissue engineering. HA has been successful in neural tissue engineering due to its tunable properties such as biodegradability, biocompatibility, degradability, and hydrogel-forming ability. Additionally, it supports neurite outgrowth, differentiation, and proliferation on a variety of substrates. HA is primarily used as a hydrogel and has shown promising results in CNS as well as PNS injury repair and regeneration due to its suitable mechanical properties which influence the differentiation of neural progenitors. HA can be combined with various natural and synthetic polymers. For example, HA/collagen conduits show improved regeneration of facial nerves in rabbits, HA/chitosan injectable hydrogels are used for the repair of PNS injuries, and HA/PLGA blends display great potential for controlled drug delivery in spinal cord regeneration. Currently, HA nanoparticles are being used to improve cell adhesion to electroconductive scaffolds. Chitosan has also been successful in neural tissue engineering, wherein chitosan hydrogels used as neural scaffolds exhibit improved cell adhesion, interaction, and survival. Porous chitosan scaffolds containing appropriate neural growth factors (NGF) have synergistic effects on neural stem cells, which leads to the effective repair of CNS and PNS injuries. Chitosan composite biomaterials are often made with PVA, PCL, and PLGA to improve the biocompatibility and cell adhesion of the overall product. Chitosan is also used as a drug delivery system, wherein chitosan micro/nanocarriers deliver stem cells or drugs directly at the site of injury like the chitosan neural scaffold developed for controlled delivery of 4-aminopyridine which has exhibited promising result in guiding the peripheral nerve conduit leading to complete repair and regeneration of nerve. 3D printing, a sophisticated technique for the preparation of scaffolds, also uses chitosan composite bioink to print with human neural stem cells.

Other natural polymers such as alginate, keratin, elastin, and silk fibroin have also been used to fabricate neural implants. Keratin and its hydrogels promote nerve regeneration by enhancing Schwann cell activity, attachment, and proliferation. Alginate is a commonly used marine biopolymer in tissue engineering, but its use in neural implants is limited due to the presence of impurities such as heavy metals, endotoxins, proteins, and polyphenolic compounds in natural alginate. Currently, alginate exhibits great potential when combined with other biopolymers to make neural implants. Silk is also one of the materials which can be easily molded into films, hydrogels, fibers, or particulate scaffolds, and used for neural regeneration. Silk hydrogels, silk fibroin, and electrospun silk scaffolds have displayed excellent results for neural growth and regeneration both in vitro and in vivo.

Synthetic polymers are also used in neural implants due to their mechanical strength and flexibility combined with their ease of modification. Synthetic polymers are compatible and can be modified for any fabrication technique such as casting, electrospinning, or 3D bioprinting. These polymers are usually combined with natural materials to balance biocompatibility and mechanical properties in the final composite material. The functionalization of synthetic polymers with neural growth factors has increased their usage in drug delivery and gene delivery applications.

PLA, PLGA, and PGA are some of the common synthetic polymers that are employed in the fabrication of neural scaffolds. PLA is successfully used in scaffolds that provide support to Schwann cells, which allow elongation of axons and promote vascular growth. However, the structural instability of PLA leads to shattering and crumpling which limits using PLA alone in the fabrication of scaffolds. On the other hand, PGA exhibits excellent mechanical strength in the scaffold but has been shown to lose its strength in only 1–2 months after implantation. Therefore, researchers usually combine PLA and PGA to form PLGA which provides sufficient mechanical stability to the scaffolds during their lifetime. Alteration in the ratio of PLA:PGA can lead to alteration of properties such as permeability, swelling, deformation, and degradation rate. This property of PLGA is useful in the preparation of scaffolds suitable for a variety of applications such as nerve conduit, drug delivery, gene delivery, and/or as support. Since PLGA is very effective in crossing the blood-brain barrier, it is now commonly used as a carrier molecule for carrying drugs across the blood-brain barrier to treat CNS injuries. Polyethylene glycol (PEG) is a non-active, non-toxic synthetic polymer that can be used in neural tissue engineering. PEG is generally combined with other polymers to form a hydrogel that can be used for implantation and to support neuronal cell growth and differentiation. PEG is being examined preclinically as an injectable polymer scaffold to treat head injuries. These studies show that after a traumatic brain injury, intravenous application of PEG hydrogel slows down the degeneration of injured axons to the point that the treated brains closely resemble the uninjured brains.

Similar to cardiac tissues, neurons in the nervous system have innate conductivity and communicate through electrical signaling. Scaffolds with some electrical activity can mimic the natural neuronal environment and help neurons regenerate faster; therefore, electroconductive polymers which mainly contain loosely held electrons in their structural backbones have been developed. Purposeful manipulation of these electroconductive polymers is done by a process called "doping" in which chemicals are added to oxidize or reduce. This pushes the electrons of the conductive polymers into the conducting orbit, leading to their activation. Some of the materials used as electroconductive polymers are polypyrrole, polyaniline, poly(3,4-ethylenedioxythiophene) (PEDOT), and Indium phosphide. Polypyrrole is an organic polymer and is mainly used in combination with other biodegradable polymers such as PLA, PLGA, and PCL to improve its biocompatibility. When used in combination with these polymers, polypyrrole has shown excellent results in neuronal growth and development. Also, its combination with natural polymers, such as hyaluronic acid, has been investigated for constructing electroconductive hydrogels to improve recovery from traumatic brain injuries and stroke. Polypyrrole is also used as an electrode material for chronically implanted neuroprosthetic devices. Polyaniline is another electroconductive polymer with high conductivity, low cost, ease of synthesis, and wide availability. Even with all these advantages, a critical disadvantage of using polyaniline is its suboptimal biocompatibility; therefore, it is used in composite materials with other biocompatible polymers. Polyaniline hydrogels have shown promising results in neural tissue engineering for peripheral nerve regeneration and as biosensing electronic patches that can be integrated into electro-responsive tissues to record their activity. Poly(3,4-ethylenedioxythiophene) or PEDOT is one of the most commonly used electroconductive materials in neural tissue engineering and shows excellent results in neural stem cell differentiation, longer neurite growth, and neural stimulation. Indium phosphide nanowire scaffolds were recently developed and exhibit influence on neuronal and cell morphology, circuit formation, and function.

Recently, carbon-based nanomaterials that have exclusive electrical, mechanical, and biological properties which make them an ideal biomaterial for the fabrication of neural scaffolds have been developed. Carbon nanotubes (CNTs) are carbon-based biomaterial that has a tube-like cylindrical structure and displays excellent thermal, optical, and electrical properties. They are biocompatible, conductive, and non-biodegradable which makes them an ideal candidate for neural tissue engineering. Both single-walled carbon nanotubes (SWCNTs) and multi-walled carbon nanotubes (MWCNTs) are commonly used in neural tissue engineering and have displayed rapid healing of neurological injuries and promotion of neurite outgrowth, leading to the establishment of neural circuits. Graphene, a carbon allotrope, is also used in neural scaffolds mainly in the form of foams and graphene nanogrids. Graphene foams are electrically conductive 3D scaffolds that stimulate neural growth and differentiation leading to rapid healing of injuries. Graphene nanogrids can also increase neural to glial cell ratio due to their excellent biocompatibility. Graphene has also been applied to neural probes to improve the quality of the neural-device interface [9, 10]. Thus, biomaterials that have electroconductive properties have been used in a wide range of neural and bio-instrumentation applications (Fig. 4).

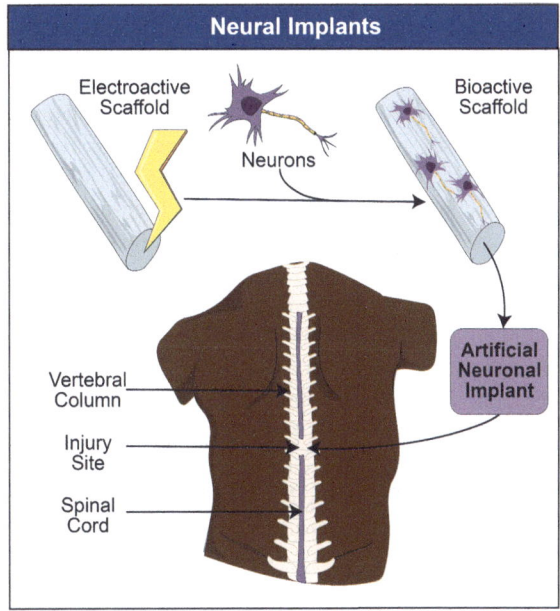

Fig. 4 Production and application of neural implants

2.5 Eye

The human eye is a complex organ and is vital for everyday life. The eye consists of various parts joined together and fitted in a socket in the skull known as the orbit. The main anatomical structure of the eye from the exterior to the interior consists of the cornea, iris, anterior chamber, lens, posterior chamber, and retina. Several biomaterials are used to make implants for the correction of deficiencies caused by disease, age, or ocular trauma.

Orbital implants are cosmetic implants mainly used after the removal of some contents (evisceration) or removal of the complete eyeball (enucleation). Numerous problems can lead to evisceration or enucleation of the eye such as retinoblastoma, trauma, uveitis, and rubeotic glaucoma. In these cases, the socket is mainly replaced with a spherical biomaterial implant to fill in the gap of the eyeball. Throughout history, a variety of materials such as glass, cork, ivory, and aluminum were used as orbital implants, but due to the non-biological nature of these implants, they lacked biocompatibility which led to unnecessary immunological reactions. Modern orbital implants generally consist of materials such as hydroxyapatite or porous polyethylene with embedded antibiotics to surpass all the limitations of previous implants. Although modern scientific implants surpass the earlier implants, these implants fail in 10% of patients due to frequent extrusion or a high infection rate.

Contact lenses are the most commonly used ocular implants for the correction of mild ametropia (when distant points are no longer visible by the eye) in people who prefer to not wear spectacles. The idea of contact lenses was first recorded by Leonardo da Vinci in 1508, but their first clinical use was not until the 1880s. In the early days, contact lenses were made from glass shells and were molded using rabbit and cadaver eyes. Although these lenses provided some correction to the vision, they were difficult to manufacture and uncomfortable to wear which led to the discovery of polymer-based contact lenses. The discovery of poly(methyl methacrylate) (PMMA) in the 1940s revolutionized the process of manufacturing contact lenses and provided an opportunity for using other materials for the fabrication of ocular devices. Contact lenses are mainly classified into two types, i.e., hard contact lenses and soft contact lenses. As the name suggests, hard contact lenses are rigid and non-elastic polymeric lenses that are mainly prepared from PMMA. PMMA is a synthetic biomaterial that is obtained by free-radical polymerization of methyl methacrylate and is said to have good optical properties, is lightweight, has good surface wettability, and is durable. However, PMMA has a low oxygen permeability which limits the long-term use of PMMA lenses. To overcome this problem, the first rigid, gas-permeable lenses were manufactured by Polycon Laboratories in 1970 by copolymerization of methyl methacrylate (MMA) with methacryloxypropyl tris (trimethyl siloxy silane) (TRIS). One of the key advantages of using this MMA/TRIS combination is that oxygen permeability, modulus of elasticity, hardness, and wettability can be modulated by changing the ratio of MMA and TRIS. Today, copolymerization of fluoromethacrylates like hexafluoroisopropyl methacrylate (HFIM) with TRIS, MMA, crosslinkers, and wetting agents is done to produce modern-day hard lenses. Soft contact lenses are mainly hydrogels or silicon-based elastomers. The first soft contact lens material was developed by Otto Wichterle in 1961 which was poly(hydroxyethyl methacrylate) (PHEMA). The originally developed PHEMA contained 38% water and had excellent wettability which provided instant wearer comfort. Subsequent development of soft lens materials led to the usage of hydrophilic monomers such as N-vinyl pyrrolidinone (NVP) and glyceryl methacrylate (GMA) in the daily use of soft lenses. Another material used for the preparation of soft lenses is the silicon elastomer known as polydimethylsiloxane (PDMS). This material has excellent optical properties, tear resistance, and high oxygen permeability which makes it suitable for long-term usage. PDMS comes with the disadvantage of low tear wettability which makes it difficult to wear; therefore, currently companies graft hydrophilic polymers, such as polyethylene glycol (PEG), on the lens surface.

When an alteration of the refractive state of the eye requires direct intervention surgery, it involves the implantation of permanent devices such as keratoprostheses or intracorneal implants. Keratoprostheses are devices that involve full-thickness penetration of the cornea resulting in a complete substitution of corneal function, whereas intracorneal implants are devices that are implanted to augment the natural function of the cornea. Keratoprostheses were first designed by Cardona and coworkers in the 1960s which was a "nut and bolt" design type manufactured from high-quality PMMA. As a result of tissue erosion followed by extrusion of the implant, this design had a severe failure rate of over 30%. Recently, twin plate keratoprostheses from PMMA were designed to use in dry eye conditions, but it was observed that, over time, PMMA keratoprostheses were not able to achieve large-scale success. An alternative to PMMA keratoprostheses was developed in Italy by Strampelli in 1963 called osteo-odonto-keratoprostheses (OOKP). This keratoprosthesis contains a lamina from a single-root tooth extracted from a patient on which a PMMA optic is cemented in the center and the device is implanted in a subcutaneous pocket.

Other commonly used ocular implants include scleral buckles which indent the sclera thereby bringing the choroid in contact with the retina. These are mainly made up of absorbable material, which early in their development were autogenous tendons or fascia lata from the patient. Today, scleral buckles are made with nonabsorbable materials such as silicones and hydrogels of poly(glyceryl methacrylate) (PGMA), poly(2-hydroxy-ethyl acrylate) (PHEA), and co-poly(methyl acrylate-2-hydroxyethyl acrylate) (MAI), which provide advantages over absorbable materials because of non-eroding nature thereby providing a tough fibrous capsule for scleral buckles.

Intraocular lenses are a special type of intraocular device which are inserted in the eye, usually after cataract surgery, as a substitute to the natural lens. Three types of intraocular lenses are commonly available: they are anterior intraocular lens (sits in front of the iris but behind the cornea), iris clip lens, and posterior chamber intraocular lens (behind the iris and on the capsular bag). PMMA has been used as a standard material for the production of intraocular lenses but has a disadvantage of postoperative inflammatory reactions which, in turn, lead to uveitis and cystoid macular edema that ultimately result in vision loss. To increase the biocompatibility of intraocular lenses, highly polished surfaces of NVP, HEMA, and perfluoropropane with a layer of heparin or hyaluronic acid on the outer surface have been developed. These lenses show reduced protein adsorption, cellular adhesion, and neutrophil activation. Recently, foldable intraocular lenses made up of silicone elastomers, collagen copolymers, PHEMA hydrogels, and acrylates have been developed [11] (Fig. 5).

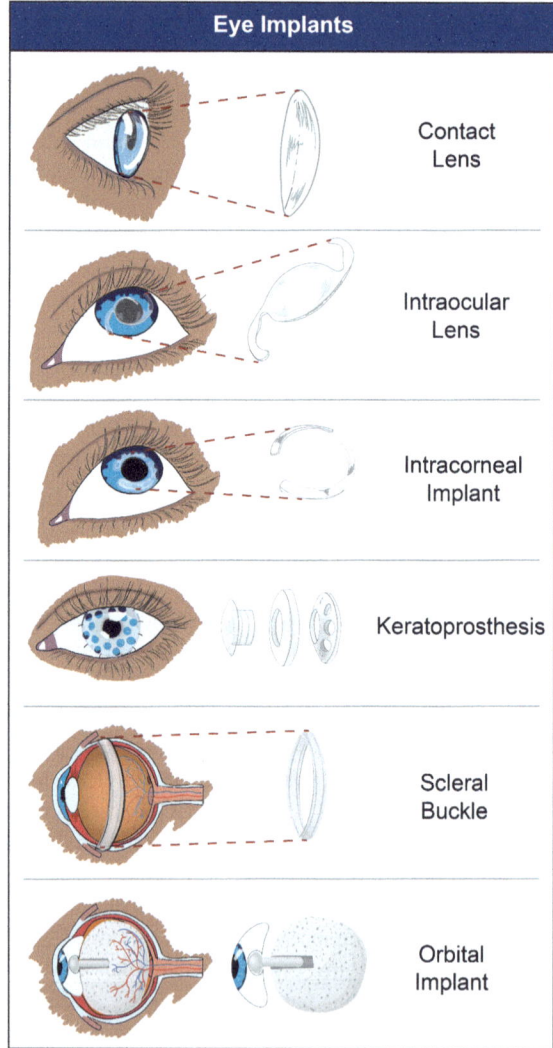

Fig. 5 Types of eye implants

3 Application of Biomaterials in Cosmetic Surgeries

Cosmetic surgeries are procedures mainly performed to increase the beauty quotient of the person. The term "plastic surgery" refers to cosmetic procedures performed to enhance the cosmetic appearance of oneself, was coined by Karl Ferdinand von Greffy in 1818. Although there have been cosmetic procedures for a long time, the use of safer biomaterials and new technologies has increased the demand for such surgeries. Biomaterials used in plastic or cosmetic surgeries are called "plastic cosmetic biomaterials" and are mainly characterized into two types, i.e., injectable biomaterials and prosthesis materials. Similar to most of the other applications discussed in this chapter, biomaterials used in cosmetic surgeries can also be classified as per their source, i.e., natural biomaterials and synthetic biomaterials.

Natural biomaterials have found wide use in cosmetic surgeries due to their biocompatible nature. Materials such as bio-protein glue, also called fibrin glue, are used in burn surgeries to reduce the amount of bleeding and fluid leakage from the body, thereby shortening the operation time and preventing infection. Decellularized tissue is mainly used in skin replacement after burn injury as it prevents host-versus graft immune response. The ECM materials prepared from decellularized tissue can be embedded in a hydrogel to act as an injectable biomaterial for filling irregular gaps. Collagen fibers are also used as facial soft tissue fillers, as they have the necessary mechanical strength and biocompatibility. They are also used in surgical sutures due to their excellent biodegradability which causes them to be absorbed in the body within 4–6 months. Artecoll is a type of medical cosmetic agent containing collagen with PMMA microspheres and is applied for the removal of wrinkles from the face. Hyaluronic acid is also used as a natural biomaterial in cosmetic surgeries. It is used as a temporary filler for induction of tissue repair. HA, along with gelatin, is mainly used in rhinoplasty (repair of the broken nose) due to its characteristic effectiveness and safety. It is also used to fill the wrinkles around the corners of the mouth, forehead, and eyebrows, and can even be used to repair the lips, chin, acne, and chickenpox scars.

Synthetic or organic materials have extensive applications in cosmetic surgeries wherein silicone, expanded polytetrafluoroethylene, and PMMA are widely used. Silicone is widely used as a preferred biomaterial for cosmetic implants since it is heat resistant, cold resistant, non-toxic, biological aging resistant, has inert physiological inertia, displays very little response to human tissues, and has good physical and mechanical properties. Silicone products have been used extensively to repair nasal deformities, skull cosmetic surgeries, and internal organs. It is also used as a biomaterial to repair maxillofacial and ear defects caused by cancer or trauma. Silicone gels and hydrogels are also used in breast enhancement surgeries. Although silicone is the most widely used synthetic material, it has some disadvantages such as strong hydrophobicity and susceptibility to bacterial infections. Expanded polytetrafluoroethylene is a commonly used tissue filler and is also used as a biomaterial scaffold due to its microporous structure and ability to adhere to tissue. PMMA is mainly used in contact lenses, to fill facial wrinkles, nipple depressions, acne scars, highlight the chin, and swell the cheeks. Other synthetic materials such as high-density polyethylene are used in the preparation of load-bearing implants for bone, and materials such as PLA and PGA are used in the preparation of biodegradable sutures known as Vicryl.

Ceramics are also used in cosmetic surgeries. Hydroxyapatite, which is the main constituent of bone, is mainly used as a substitute for large area bone defects like oral and maxillofacial regions and also a substitute for teeth. Although individual use of hydroxyapatite comes with high brittleness, low strength, and poor toughness, hydroxyapatite composite materials can lead to efficient repair of the defects. Bioglass materials can induce bone repair and regeneration and are mainly used in combination with other materials for the repair of oral cavities, orthopedic defects, and other cosmetic reconstructions.

Though biomaterials used in cosmetic surgeries have good biocompatibility, non-toxicity, and very little to no immune response, they still have certain shortcomings which need to be overcome. Recently, 3D printing technology has improved some conventional cosmetic surgeries and implants, including craniofacial reconstruction, ear and nose reconstruction, skin printing, and breast reconstruction. This emerging technology has led to the development of biocompatible scaffolds which have potential advantages in other areas of tissue regeneration. The development of biomaterials and technologies in cosmetic surgeries can lead to efficient and economic reconstruction of other organs soon [12].

4 Application of Biomaterials in Other Medical Devices

Apart from the above-mentioned applications of biomaterials, they are also used in various other medical devices when contact with biological material is necessary. Unlike most of the applications described in this chapter, the biomaterials used in other medical devices are mainly classified according to the type of material, i.e., metallic and polymer.

Metallic biomaterials possess high mechanical strength, elasticity, and thermal and electrical conductivity, and are resistant to wear and tear. Currently, the most commonly used metallic biomaterial in devices is titanium and its alloys, chromium-cobalt, and stainless steel. Titanium alloy such as Ti-6Al-4 V (titanium-aluminum-vanadium) alloy is commonly used in dental implants, bolts and column clips, pacemaker housing, artificial heart valves, and screws and staples used in bone fixation. The presence of aluminum and vanadium together can lead to toxicity problems; therefore, now titanium, aluminum, and niobium alloys are used since these materials exhibit better biocompatibility than vanadium. Cobalt-chromium alloy is also used in dental implants and artificial joints but is difficult to manufacture and mold due to its hard nature. Noble metals such as gold, silver, and platinum possess very high corrosion resistance and are generally observed in pacemaker wiring and dental crowns. Silver is mainly used to treat burn patients, in the stethoscope diaphragm, and germicidal coating fibers are used for wrapping wounds. Platinum is also observed in pacemaker electrodes and in hearing aids.

Polymers have now replaced metals in several applications due to their malleability and ease of formulation. Polyolefins, like polyethylene and its variants, are used in artificial joints, surgical cables, high-strength orthopedic sutures, catheters, stent-grafts, heart valves, and spinal disc substitutes. Teflon, a commonly used industrial polymer, is used as a biomaterial in the preparation of vascular grafts and as a soft tissue filler due to its porous nature and electronegative luminal surface which does not degrade. Polyvinylchloride has a pliable nature and does not react with any substance easily which is why it is used in the manufacturing of tubes, blood bags, and catheters. Polyesters such as polycaprolactone and polyethylene terephthalate are used in the preparation of membranes, filaments, nets, and vascular grafts. Polyamides, such as nylon, have high tensile strength, so are used in composites as balloons of the catheters used in angioplasty. Polyurethane is commonly used for coating breast implants, aortic and gastric balloons, male contraceptives, and also in surgical gloves. Natural polymers are generally used in combination with synthetic polymers mainly to increase biological compatibility. Natural rubber latex is a commonly used natural materials for medical needs and is commonly used in latex gloves, condoms, and dental equipment [13].

5 Conclusion and Future Perspective of Biomaterials

Biomaterial innovations and implant application is an active area of clinical practice and research to meet the market needs. Continued innovation helps to improve the performance of existing implants and find alternatives by combining them with novel materials and bioactive molecules to improve patient outcomes. The early concept of biomaterials being bio-inert is changed to be bioactive to promote tissue regeneration and host tissue integration to achieve better functional outcomes. Efforts are also focused on adding additional features to the traditional metallic implants with surface topography, ceramic coating, and porous architecture to mimic the trabecular bone. Overall, these efforts are focused on improving the performance of existing biomaterials and implants through better engineering designs. Additional efforts are also made to design semi-synthetic materials, and composites with nanomaterials to improve the strength and add a variety of other features including electrical conductivity. Overall significant progress has been made in creating implants biomaterials and implants to deliver physical and chemical stimuli to improve the biocompatibility and efficiency of the implant in the body. Efforts are also being made to develop bioactive factor eluting implant formulations to promote tissue healing. Additive manufacturing and bioprinting have the potential to fully replace the tissue or organ with an artificially grown organ that functions exactly like the natural.

Acknowledgments Prof. Kumbar acknowledges the funding support by the National Institutes of Biomedical Imaging and Bioengineering of the National Institutes of Health (#R56 NS122753 #R01EB020640 #R01EB030060 and #R01EB034202); the U.S. Army Medical Research Acquisition Activity (USAMRAA), through the CDMRP Peer Reviewed Medical Research Program under Award No. W81XWH2010321.

Questions and Answers with Explanation

Question 1: Collagen and hydroxyapatite in the bone are produced by

(a) Osteoclasts
(b) Osteoblasts
(c) Haversian canal
(d) Bone marrow

Explanation: Osteoblasts are the cells which are responsible for making new bone. The new is composed of collagen and hydroxyapatite which is produced by osteoblastic cells during proliferation.

Question 2: Which is considered as gold standard in bone and skin implants

(a) Allograft
(b) Autograft
(c) Xenograft
(d) Nanograft

Explanation: Autograft is considered as gold standard in bone and skin implants due to the fact that it is derived from patient's own body part hence they do not produce any immunogenic response which is commonly seen in allograft and xenograft.

Question 3: Disadvantage of synthetic biomaterial is

(a) Low biocompatibility
(b) Low cost
(c) High tensile strength
(d) Ease of manufacture

Explanation: Synthetic biomaterials are easy to manufacture, are low cost, and have high tensile strength although, due to their synthetic nature they are not able to mimic the human structures and thus have low biocompatibility as a disadvantage.

Question 4: Decellularized extracellular matrix is most commonly used in the manufacture of

(a) Skin implants
(b) Neuronal implants
(c) Cardiac implants
(d) Skeletal implants

Explanation: Decellularized extracellular matrix is used in all of the above implants although it is most commonly used in manufacturing Cardica implants as the cardiac tissue is most susceptible to any of the other biomaterials.

Biomaterials Application: Implants

Question 5: Dacron® a synthetic graft goes by the chemical name of

(a) Polyethylene
(b) Poly-lactide-co-glycolide
(c) Polycaprolactone
(d) Polytetrafluoroethylene

Explanation: Dacron® is a commercial name for synthetic graft made from polytetrafluoroethylene. It is a non-biodegradable biomaterial.

Question 6: Which was the first biomaterial to display promising result in neuronal growth

(a) Collagen
(b) Chitin
(c) Keratin
(d) Alginate

Explanation: Keratin was one of the first biomaterial which displayed promising results of neuronal growth. This paved the way for nerve regeneration.

Question 7: Which type of implants are used after evisceration or enucleation

(a) Orbital implants
(b) Keratoprostheses
(c) Contact lenses
(d) Scleral buckles

Explanation: Removal of some part of eyeball is called as evisceration whereas removal of complete eyeball is called as enucleation. In both the cases, orbital eye implants are commonly used to mimic or replace the eyeball.

Question 8: Which is the most commonly used material in preparation of hard contact lenses

(a) PGA
(b) PLA
(c) PVA
(d) PMMA

Explanation: Hard contact lenses are commonly manufactured using polymethyl methacrylate or PMMA due to its biocompatibility and ease of manufacturing.

Question 9: PLA and PGA biodegradable sutures are also known as

(a) Vicryl
(b) Apligraf
(c) Integra
(d) Osteofil

Explanation: Vicryl is a commonly used alternative name for suture having polylactic and polyglycolic acid (PLA and PGA). Apligraf and Integra are commercially available collagen scaffolds for skin regeneration whereas Osteofil is commercially available allograft bone paste.

Question 10: Which of the following metals is used in pacemaker wiring

(a) Copper
(b) Gold
(c) Chromium
(d) Titanium

Explanation: Pacemaker wiring is commonly made from gold as it is a non-reactive metal and does not cause any immunogenic response in the body after implantation.

References

1. Ratner BD, Hoffman AS, Schoen FJ, Lemons JE (2013) Biomaterials science. Elsevier, London
2. Dhandapani R, Krishnan PD, Zennifer A et al (2020) Additive manufacturing of biodegradable porous orthopaedic screw. Bioact Mater 5:458–467. https://doi.org/10.1016/J.BIOACTMAT.2020.03.009
3. Brydone AS, Meek D, MacLaine S (2010) Bone grafting, orthopaedic biomaterials, and the clinical need for bone engineering. Proc Inst Mech Eng H 224:1329–1343. https://doi.org/10.1243/09544119JEIM770
4. Qu H, Fu H, Han Z, Sun Y (2019) Biomaterials for bone tissue engineering scaffolds: a review. RSC Adv 9:26252–26262. https://doi.org/10.1039/C9RA05214C
5. Wei C, Feng Y, Che D et al (2021) Biomaterials in skin tissue engineering. https://doi.org/10.1080/00914037.2021.1933977
6. Kaur A, Midha S, Giri S, Mohanty S (2019) Functional skin grafts: where biomaterials meet stem cells. Stem Cells Int 2019. https://doi.org/10.1155/2019/1286054
7. Theus AS, Tomov ML, Cetnar A et al (2019) Biomaterial approaches for cardiovascular tissue engineering. Emergent Mater 2:193–207. https://doi.org/10.1007/S42247-019-00039-3/FIGURES/2
8. Esmaeili H, Patino-Guerrero A, Hasany M et al (2022) Electroconductive biomaterials for cardiac tissue engineering. Acta Biomater 139:118–140. https://doi.org/10.1016/J.ACTBIO.2021.08.031
9. Boni R, Ali A, Shavandi A, Clarkson AN (2018) Current and novel polymeric biomaterials for neural tissue engineering. J Biomed Sci 25:1–21. https://doi.org/10.1186/S12929-018-0491-8
10. Manoukian OS, Arul MR, Rudraiah S et al (2019) Aligned microchannel polymer-nanotube composites for peripheral nerve regeneration: small molecule drug delivery. J Control Release 296:54–67. https://doi.org/10.1016/J.JCONREL.2019.01.013
11. Lloyd AW, Faragher RGA, Denyer SP (2001) Ocular biomaterials and implants. Biomaterials 22:769–785. https://doi.org/10.1016/S0142-9612(00)00237-4
12. Peng W, Peng Z, Tang P et al (2020) Review of plastic surgery biomaterials and current progress in their 3D manufacturing technology. Materials 13:4108. https://doi.org/10.3390/MA13184108
13. Festas AJ, Ramos A, Davim JP (2019) Medical devices biomaterials – a review. 234:218–228. https://doi.org/10.1177/1464420719882458

Living Biomaterials

Caroline Hali, Adi Gross, and Boaz Mizrahi

Abstract Living biomaterials are the result of a combination of live organisms such as bacteria or cells with traditional biomaterials. This integration yields a unique system that enjoys the best of both worlds: a live manufacturer that can sense its environment, produce, and release biomolecules while exhibiting excellent stability in harsh physiological milieus. Such biomaterials can be natural, synthetic, or semi-synthetic as long as they retain their main pre-requisite characteristics of biodegradability and biocompatibility with body tissues. The integration of functional microorganisms into polymeric matrices imposes stringent requirements on material composition and engineering. The matrix must support growth and reproduction of the encapsulated organisms, while allowing the absorption and release of chemicals and biological molecules. The living component can be either natural (wild type) or genetically modified, each with its own advantages and disadvantages. For example, wild-type organisms are easier to produce and generally produce eco-friendly materials. On the other hand, thanks to remarkable advances in synthetic biology, genetically modified organisms can be reprogrammed to produce and secret desired biomolecules or to exhibit specific functionalities, but at the same time they elicit some inherent concerns regarding unintended ecological and health-related consequences. In this chapter, we introduce the concept of living biomaterials, explore various systems, and focus on the interactions and complexity of these multi-component systems. Then, we review some applications of living biomaterials with emphasis on the medical field. Finally, we present two case studies, in which we delve into the details of specific systems in an attempt to illustrate this fascinating topic.

C. Hali · A. Gross · B. Mizrahi (✉)
Laboratory for Bio-materials, Faculty of Biotechnology and Food Engineering, Technion-Israel Institute of Technology, Haifa, Israel
e-mail: bmizrahi@technion.ac.il

Graphical Abstract

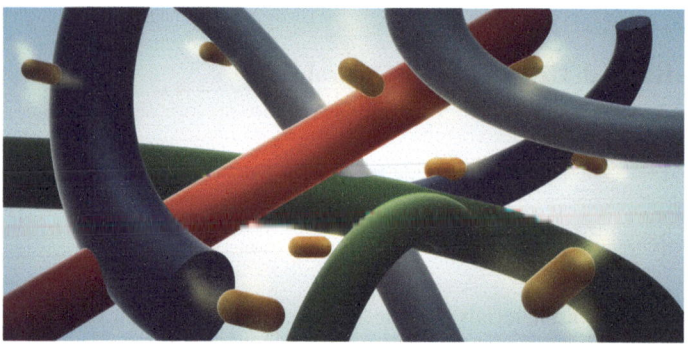

Keywords Living materials · Stimuli responsive · Smart materials · Sensing

1 Introduction to Living Materials

Living biomaterials are a new class of composite structures composed of living organisms, such as cell colonies or microorganisms, and a biomaterial. This integration of living and non-living components leads to the creation of a new, highly ordered system with exceptional properties such as self-healing, responsiveness to external stimuli, and biosensing. While these characteristics are provided by the living components, the polymeric matrices (i.e., the non-living component) provide three-dimensional (3D) structural integrity and self-assembly capabilities, as well as chemical and physical protection, communicating pathways, and bioactive cues for the living components to adhere or bind to the surface. Since the living cells can synthesize complex molecules from available precursors, the system can be seen as an "active device" or a "live manufacturer". Living biomaterials may integrate naive, wild-type cells or genetically engineered cells (the latter are also termed engineered living materials), both of which are used in health monitoring, disease treatment, and environmental remediation. Natural materials, such as bone, cartilage, collagen, and wood, are composed of living cells integrated within a bioscaffold. These materials can grow autonomously, sense their environment, secret essential biomolecules, and overall, regenerate. Moreover, these materials can quickly change their structural and mechanical properties when exposed to external stimuli, which can be endogenous, e.g., pH, redox, and enzymes, or exogenous such as temperature, magnetic or electrical signals, and light. Researchers have designed living biomaterials as an attempt to mimic natural materials without compromising simplicity and stability.

2 The Material Component

The integration of functional microorganisms into polymeric matrices imposes stringent requirements on materials engineering. The matrix is required to support growth and reproduction of the encapsulated organisms, but also to degrade into non-toxic byproducts within a desired period of time. In addition, it should also allow secretion of functional biomolecules produced by the living organism into the nearby environment, preferably in a controlled manner. The physical and structural properties, including interconnectivity, porosity, and surface morphology, also play a significant role in designing the appropriate matrix. These characteristics should therefore be studied in detail to select the most appropriate type of biomaterial. Some excellent examples of biomaterials used in living materials can be found in an excellent review by Rodrigo-Navarro et al. [1]. In this chapter, we review some of the most studied biomaterials used in living biomaterials based on their characteristics and chemical composition.

Agarose, for example, is a marine, water-soluble polysaccharide that represents reversible thermogelling behavior, excellent mechanical properties, high bioactivity, and switchable chemical reactivity for functionalization. Agarose gels also exhibit a porous structure that can be controlled by adjusting agarose concentrations and additives, making it possible to control the diffusion of various drugs through different mechanisms. As a result, agarose gels have received a great deal of attention in the fabrication of living biomaterials, in particular, as carriers for therapeutic agents and for 3D printing applications. For example, agarose hydrogel was used to engineer a living material system in which light-responsive protein-releasing *E. coli* bacteria were encapsulated within the agarose hydrogel simply by mixing at 37 °C [2]. The gels were formed by pouring the suspension into well plates and cooling to room temperature. These bacterial hydrogels allow spatially confined protein expression and dosed protein release over several weeks, thus showing great promise as a protein-based biopharmaceutical delivery system.

Pluronic F-127 is a synthetic copolymer made of amphiphilic copolymers consisting of units of ethylene oxide and polypropylene oxide (PPO). Pluronic F-127 hydrogel exhibits a reversible gelation mechanism around body temperature, which makes it ideal for dermal therapy (i.e., it is liquid at room temperature, and therefore injectable, but undergoes gelation inside the body). Pluronic F-127 is non-toxic, biocompatible, and biodegradable. Finally, Pluronic F-127 enhances cell attachment and collagen formation, making it favorable for augmentation of living cells.

Our group developed a unique hydrogel formulation based on Pluronic F-127, bacterial media, and live *Bacillus subtilis* (Fig. 1) [3]. This hydrogel hardens after administration on the skin, allowing the bacteria to continuously produce antifungal agents. We showed that this formula penetrates via the stratum corneum and accumulates in the epidermis, where Candida is usually confined, without penetrating the inner dermis layer. In vitro and in vivo results, all showed that Bacillus formulations completely inhibit Candida growth, demonstrating clinical

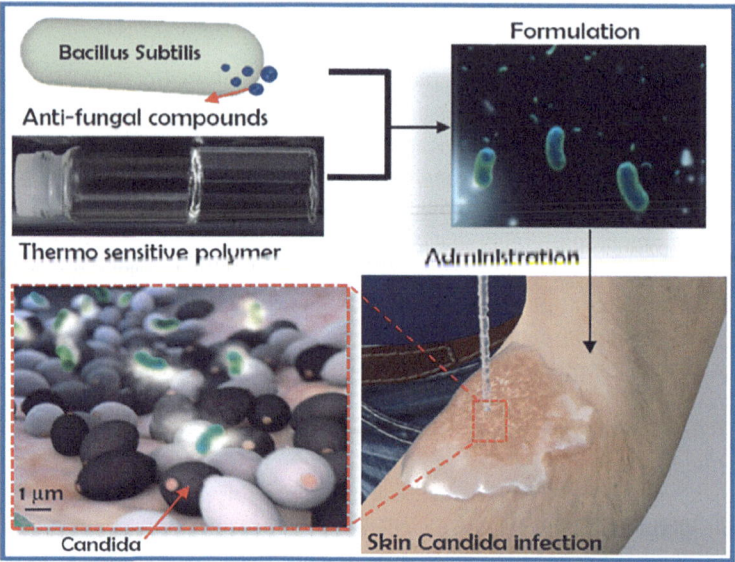

Fig. 1 Concept of the live system: *B. subtilis*, a food-grade bacterium, is encapsulated in a thermoresponsive Pluronic F-127-based formula that allows germination and proliferation. After administration, the transparent formula will harden and function as a unique "factory" that continuously produces and releases the natural antifungal agents locally. (Adapted from [3])

effects comparable to those achieved by the antifungal drug ketoconazole. Moreover, LC-MS/MS analysis of the bacterial formulation confirms the production and secretion of the broad antifungal biomolecule surfactin.

Other water-soluble, synthetic polymers have also been used as matrices in living materials, thanks to their hydrophilicity, biodegradability, and ability to penetrate organs such as the skin without toxicity to cells and tissues. Of these synthetic polymers, polyvinyl alcohol (PVA) is in exceptionally frequent use, since its excellent tensile, compressive, and adhesive properties render it ideal for the formation of fibers and particles.

There is a great need for a living biomaterial that effectively functions as an artificial pancreas and secrets insulin for the treatment of type 1 diabetes. Rat insulinoma cells (INS-1), which serve as a model of insulin-secreting cells, were incubated with or without gelatin hydrogel microspheres prepared using the w/o emulsion method [4]. Cell aggregates within the gelatin microspheres not only exhibited higher cell viability, higher reductase activity, and a larger number of live cells compared with naked cells, but also secreted significantly higher amounts of insulin. Similarly, a star-shaped PEO with 3,4-dihydroxy-phenylalanine (DOPA) end groups was suggested as a system for insulin-secreting cell immobilization [5]. When this polymer was oxidized in aqueous solutions with NaIO4, each DOPA end group attached covalently to a neighboring DOPA, forming a three-dimensional hydrogel structure. Donor islet cells were placed in the PEG-DOPA aqueous solution which was then oxidized. This adhesive material maintained an intact interface with the supporting tissue for up to 1 year, maintaining normoglycemia for over 100 days.

Biodegradable synthetic esters, such as poly-ε-caprolactone (PCL), polylactic acid (PLA), polyglycolic acid (PGA) and their copolymers, and poly(lactic-co-glycolic acid) (PLGA) copolymers are highly popular as cell-augmenting matrices, thanks to their defined and tailorable chemistry, easy processing, and modifications. These polymers are non-toxic and tissue compatible and are approved by the Food and Drug Administration for numerous applications, making them ideal as scaffolds for living biomaterials. Due to the presence of five methylene groups in its repeating unit moieties, PCL is the most elastic and degrades the slowest among these polyesters. They all, however, suffer from some inherent shortcomings related to their hydrophobic nature, alongside slower rates of degradation: it is hard to form hydrogels from them, so copolymerization with hydrophilic monomers is often required. In addition, they are generally not suitable for live-cell encapsulation techniques such as those applied during electrospinning.

Finally, natural polymers play an important role as matrices for living cells. Collagen, for example, the most abundant protein in mammals and the main component of the extracellular matrix, is considered ideal for this task, thanks to its abundance and easy processing, flexibility, water solubility, biocompatibility, and biodegradability. Indeed, collagen supports cell attachment while promoting a chemotactic response. Hyaluronic acid, a glycosaminoglycan that is also found in the extracellular tissue in many parts of the body, has also been used to encapsulate living cells. Hyaluronic acid offers numerous advantageous properties, including a range of crosslinking techniques for enhanced mechanical properties, bioresorbability, and biodegradability. Its functional groups, namely carboxylic acids and alcohols, can be used to introduce cell adhesion ligands and growth factors for an enhanced rate of tissue regeneration.

3 The Living Component

Bacteria, fungi, algae, and animal cells have traditionally been incorporated into polymeric matrices to engineer living materials. These organisms, being "alive", can play a fundamental role in the synthesis of functional biomolecules, sensing, reacting, and self-organizing in response to small changes in their environments. In addition, like all other "live" organisms, the presence of a living entity allows self-repair and healing, which is beyond the capabilities of regular synthetic materials. As these natural systems grow, self-repair, and adapt to the environment, they exhibit distinctive "living" attributes that are beyond the reach of most existing synthetic materials. *Bacillus subtilis*, for example, is a non-pathogenic gram-positive bacterium naturally found on the human skin, in epithelial wounds, and in soil. Being ubiquitous, *B. subtilis* has developed several adaptive strategies to kill and limit the growth of competing organisms (or to modulate their metabolism). For maximum advantage, *B. subtilis* relies mainly on the production and secretion of a wide array of potent antibacterial agents, including subtilin, subtilosin, bacilysin, fengycin, iturin, and surfactin. The latter, for example, is a cyclic lipopeptide and metabolism-altering molecule and among the most powerful biosurfactant in nature. The strong antibacterial activity of surfactin has been attributed to the disruption of bacterial cell wall and to secretion of growth and virulence factor that are toxic to other microbes. Moreover, *B. subtilis* has an excellent, adaptable metabolism, which makes it easy to cultivate rapidly and effectively on many natural and synthetic biomaterials.

The living component, with an emphasis on bacteria, can be composed of natural or genetically modified organisms. While the former is the wild-type, non-modified version, the latter contains DNA that has been altered using genetic engineering techniques. Wild-type organisms have some advantages over genetically engineered organisms: they generally produce ecofriendly materials and are easier and therefore cheaper to produce. That said, wild-type organisms often give low yields and have low titers of their natural products, properties that ultimately depend on the organism type, the environment, and the specific conditions.

The introduction of synthetic biology and the ability to reprogram organisms with complex functionalities have revolutionized the field of living materials. Instead of looking for the most appropriate organism, scientists can now pinpoint an individual gene (e.g., one that encodes the synthesis of a particular protein), separate it from the original organism, and transfer it into the most suitable cell, bacterium, or virus. Thus, synthetic biology has enabled precise genetic manipulation of organisms with enhanced functionalities for biotechnological applications, including cell programming, biosynthesis of functional molecules, and gene regulation. However, the mere insertion of new genes into an organism raises some inherent concerns about unintended ecological and health-related consequences. For example, when using antibiotic-resistance bacteria and when crossbreeding between gene-modified organisms and wild species, the potential health and environmental impact must be carefully assessed by the authorities, while considering the organisms and the genes involved.

4 Applications of Living Biomaterials

Humans coexist with a diverse microbial community that lives within and upon us – our microbiota. The skin and gut, for example, each function as a mechanical and biological protective barrier between the outer environment and the sterile interior of our body. The naturally occurring microbiota within the human body opens up numerous opportunities for therapies based on living materials. For example, a living formulation that continuously produces and delivers potent biomolecules and therapeutics is potentially more effective than a traditional therapy, as it may provide continuous treatment for a chronic illness. Since the industrial production of some biomolecules is extremely expensive, the living materials concept may also reduce high production costs. In the broader context, by targeting lab-culturable members of each population, it may be possible to use engineered living components so as to enable them to continuously secrete a range of proteins and drugs with health benefits. Being administrated in situ, living materials may be used to administer anticancer drugs in situ, thus decreasing systemic adverse effects or to combat the rising number of antibiotic-resistant diseases.

Traditional wound care formulations have some inherent shortcomings. Antiseptics, for example, may be toxic to human keratinocytes and fibroblasts and may therefore increase the intensity and duration of skin inflammation. In many cases, the antibacterial activity of topical therapy lasts less than a few hours and wound dressing must be replaced regularly. Topical formulations containing an oily phase such as ointments and emulsions tend to dry out, and the dressing may stick to the surface of the wound, causing pain, cell damage, and impaired wound healing. Live bacterial formulations that continuously produce antibacterial molecules and deliver them to the surface of the wound surface may be a successful alternative therapy. For example, a dissolvable polymer such as polyvinyl alcohol (PVA) can be used as a vehicle for live bacteria that continuously produce and secrete antimicrobial molecules. Specifically, PVA microparticles containing *B. subtilis* were prepared [6] using a spray dryer apparatus (Fig. 2). PVA/*B. subtilis* solutions were prepared by suspending harvested bacteria pellets in sterile aqueous PVA solution and incubating overnight. Then, the bacterial suspension was connected to the designated spray drier pump, with inlet and outlet temperatures set to 110 °C and 45 °C, respectively, and the dried microparticles were collected from the cyclone. Numerous fabrication approaches can be taken into consideration for designing

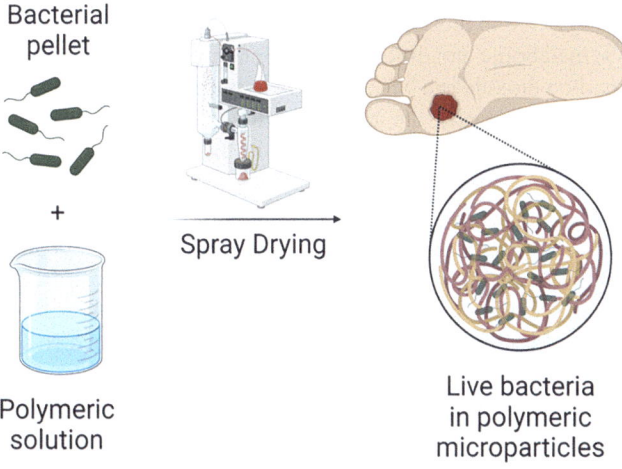

Fig. 2 Preparation and administration of live *B. subtilis* microparticles. Microparticles, formed by spray drying, were administered directly onto open ulcers and lesions

wound care formulations based on living materials, including hydrogels, emulsions, fibers, microparticles, and 3D printed devices.

The living material approach may also be used in cancer treatment where local delivery of the therapeutic agent is critical for successful treatment. Since some bacteria demonstrate an amazing ability to colonize tumors selectively, they may be seen as programmable vehicles for administrating anticancer treatment. *Salmonella, for example,* is suitable for tumor therapy, thanks to its ability to reside within tumors and suppress their growth. With proper metabolic engineering, *Salmonella* can be designed to be a viable therapeutic system that continues to grow inside the tissues of a tumor, releasing genes in situ. Similarly, bacteria can be designed to release anti-inflammatory or antibacterial agents, with the ability to control the release kinetics using external stimuli such as light or essential metabolites.

Under inflammatory conditions such as inflammatory bowel disorders, in which mucosal healing plays a critical role, live materials with targeted administration and activity in the gut can be a promising alternative treatment. For example, bacteria that instead of secreting therapeutic proteins into the inflamed bowel, not only have an anti-inflammatory capability but can also produce a scaffold that allows it to grow in the bowel and also fuse into the inflamed, damaged mucosal layer and seal it.

Living materials can also be utilized for biodetection, for example, by encapsulating "sensing" biological cells within hydrogels or particles [7]. Hydrogel particles that encapsulate engineered *E. coli* capable of detecting chemical and biological elements, from nitro compounds and lactam species to heavy metals and report by fluorescent signals. This detection functionality is often attributed to an enzymatic or biochemical activity of the sensing cells. To elegantly convert metabolic signals into easily detectable optical signals or color changes, luminescent O_2-sensing nanoparticles, for example, can be integrated into the hydrogel in proximity to the living cells. In this manner, oxygen and carbon dioxide can be detected, and the cell's metabolite condition can be sensed accurately and dynamically. Glucose and carbon dioxide can be detected in a similar manner for various biosensing applications. Alternatively, the actual sensing capabilities can be utilized using fast-growing cells, such as yeast cells, thanks to their ability to induce significant shape/size changes. According to this method, the input detected by the system can be either chemical or optical.

Living biofilm-forming materials have been applied in a diverse range of biomedical applications. During biofilm formation, bacteria can be adhered to a surface and begin secreting exopolysaccharides, DNA, and proteins that function as a biofilm scaffold. Proteins secreted by the cells or bacteria can also self-assemble into more complex structures to form the basis of the extracellular matrix biofilm. The biofilm that is formed can replace damaged natural biofilm or add to an existing biofilm a new functionality such as resilience or self-regeneration. For example, a living biofilm glue can be designed as a tissue adhesive or sealant that degrades in the tissue in a controlled fashion, thus supporting the natural healing process. In order to increase the adhesiveness of these systems to the tissue, additional natural or genetically encoded adhesive components can be integrated, including enzymes, sticky end groups (e.g., 3,4-dihydroxyphenylalanine and aldehydes), and mucoadhesive polymers.

Other applications of bacterial biofilms include the support of probiotic formulations. Live bacterial systems can be used as oral formulations for treating intestinal disorders and infections. The polymeric carrier can be used to support and deliver beneficial bacteria, protect them from the harsh conditions of the gastrointestinal tract, and release them in the desired location. For example, probiotic biofilm-forming bacteria, such as *Lactobacillus fermentum* and *Lactobacillus acidophilus,* may be entrapped in self-assembled, mucoadhesive fibers and particles. Moreover, food products, such as yogurts, ice creams, and cheese, can contain beneficial bacteria without compromising their organoleptic characteristics. In a similar fashion, the same technology can be used to prevent the formation of biofilms such as those created in medical device-associated infections. Probiotics can be added to a pre-formed biofilm with the intention that the "new" bacteria will overcome the existing bacteria, thus hindering biofilm growth.

5 Case Studies

5.1 Case Study I: Engineered Bacillus Subtilis *Biofilms as Living Glues [8]*

Main goal: To develop functional cellular medical glues with engineered *Bacillus subtilis* biofilm (Fig. 3).

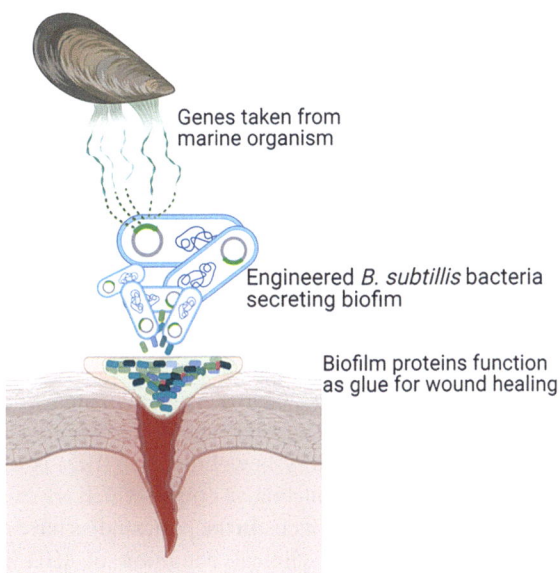

Fig. 3 Genetically engineered *B. subtillis* bacteria with enhanced capabilities to synthesize and secret biofilm proteins for wound healing

The material component: Amyloid and mussel proteins and fibers formed by the bacteria. This is not a "classic" system in which the living component is injected next to or into a biopolymer.

The living component: Genetically engineered *B. subtillis* bacteria with enhanced capabilities to synthesize and secret biofilm proteins. Several factors were inserted into the *B. subtills* genome including TasA, which encodes amyloid fibers, BslA, for hydrophobic surface layer protein, and EPS (exopolysaccharide).

Biocompatible adhesives are highly required in the field of biomedicine for diverse tissue adhesion, as a skin filler, and for bonding of injured skin. Marine adhesive systems demonstrate highly adhesive abilities in hydrophilic environments, rendering them a great inspiration for biomaterial glues. Barnacles, mussels, and sandcastle worms exhibit complex protein systems for underwater adhesion. In addition, development of adhesive coatings was inspired by the amyloid fibrous structure of barnacles.

Bacillus subtilis was chosen since it is "generally regarded as safe" (GRAS), and since it has a single outer membrane, making it ideal for the secretion of large quantity of enzymes and proteins. *B. subtilis* forms clusters, known as biofilms, by integrating bacteria into an extracellular matrix. Such protective biofilms are formed in response to environmental threats such as antimicrobial agents, liquids, and gases. In this study, adhesive components and performance, environmental tolerance, and other features of the live glues were tested. The main step of the study included testing the integrated "living glue" components in various experiments and quantifying the effect of each factor on the adhesion properties. For example, the effects of various environmental conditions (pH, humidity, presence of detergents) on the curing time of the living glue were investigated. It was concluded that this living material biofilm glue holds great potential to further improve its adhesive strength; however, additional experiments and development are required before the system can be introduced to the clinic.

5.2 Case Study II: Layer-by-Layer Microencapsulation of Ligilactobacillus Salivarius *for Bowel Inflammation Relief [9]*

Main goal: To determine whether layer-by-layer (LbL) encapsulation of *Ligilactobacillus salivarius Li01* (Li01) can improve the beneficial functionality of the bacteria as a probiotic treatment.

The material components: Chitosan and alginate.

The living component: The wild-type bacteria *Ligilactobacillus salivarius Li01*.

Background: Inflammatory bowel disease (IBD) is a collective name for diseases such as Crohn's and ulcerative colitis, which are caused and enhanced by genetic and environmental effects. In the past, studies have demonstrated the effect of the microbiome on the clinical condition of IBD patients. Thus, using orally administrated probiotics may play a key role in the treatment and relief of IBD symptoms.

The lactobacillus bacterium Li01 is known for its anti-inflammatory effect, which probably results from an increase in the levels of anti-inflammatory cytokines alongside a reduction in the levels of inflammatory cytokines (i.e., TNF-α and IL-6). This bacterium also has the ability to protect the intestinal barrier and increase the abundance of the "beneficial" gut bacteria. In order to develop a successful probiotic treatment, several issues need to be considered: (1) the bacteria must remain viable until they reach the intestine, including storage time, (2) the bacteria must adhere to the intestine's mucus layer in order to colonize there and prevent the adhesion of pathogens. Since Li01 is very sensitive to environmental factors and since on their journey to the small bowel, the bacteria go through some extreme conditions, oral administration is far from ideal.

Polysaccharides are used by the gut microbiota for fermentation and their prebiotic effects can benefit human health. These natural polymers are capable of protecting biomolecules from the extreme conditions of the gastrointestinal tract. Alginate is a negatively charged polysaccharide composed of β-D-mannuronate and α-L-guluronate. It is mucosal biocompatible with limited mucoadhesivness. Chitosan, on the other hand, is a positively charged polysaccharide (pKa = 6.3) composed of β-D-glucosamine and N-acetyl-D-glucosamine. In vivo studies have shown the positive effect of chitosan on microbiome changes and ulcerative colitis, as was demonstrated in dextran sodium sulfate (DSS)-induced mice. The researchers hypothesized that LbL encapsulation of Li01 in alginate/chitosan matrix can protect the cells from the environment and keep them viable for longer periods of time.

Moreover, the mucoadhesive properties of the system, attributed to the polymers, enhanced the adhesion of the bacteria to the intestine, thus improving its colonization. The results revealed that in both simulated gastric and simulated intestinal (SIF) fluids, the encapsulated bacteria had higher survival rates compared (SGF) with the free cells, a finding that was also validated through confocal microscopy. The effect of the carrier on the adhesion was inconclusive according to two different methods. To understand whether LbL-Li01 systems are better at alleviating colonic inflammation compared with free Li01, an in vivo model using colitis DSS-induced mice was used. Mice treated with the encapsulated Li01 formula showed a 100% survival rate compared with 70% for the free bacteria group.

6 Future Prospective

In this chapter, we focused on living materials in the broad medical context. Living material strategies combine the desired properties of both biomaterials and living biological organisms. They are stable, biocompatible, and biodegradable, yet carry biological cells that impart superior abilities/properties such as sensing, self-assembly, and adaptation. In addition, cell or bacteria can be designed to synthesize and secrete desired biodrugs and release then locally in the treated area. Thus, the living formulation can be seen as a small "factory" that continuously produces and releases its products at the relevant site. More importantly, in order to allow the living component to flourish, prosper, and release its agents for longer periods of time, either wild-type or genetically engineered bacteria can be chosen.

Next-generation advanced materials should include "live" functional properties that surpass existing capabilities, such as adaptation to environmental cues, the ability to dynamically switch between different material states, and self-healing. Therapeutic biofilms and numerous delivering implants have been explored, including skin patches for wound healing, and adhesive bacterial matrices for the treatment of chronic inflammation in the intestine or sealing of blood leakage in vascular tissues. Most of these systems, however, are still based on pathogenic strains of *E. coli*, so they are irrelevant for in vivo uses. We anticipate that as research continues, new organisms and techniques for fabricating living materials as well as novel applications, still unimaginable, will be developed.

Quiz

Question 1: Which of the following is NOT a characteristic of living component?

(a) Sensing its environment
(b) Self-assembling
(c) Batch-to batch similarity
(d) Secreting biomolecules

The living component can sense its environment, produce, and release biomolecules while exhibiting excellent stability in the harsh physiological milieus. Batch-to-batch similarity has nothing to do with the living components and in fact can be in contradiction to it.

Question 2: Which of the following is expected from the material component of living biomaterial?

(a) Support the reproduction of the encapsulated organisms
(b) Support the growth of the encapsulated organisms
(c) Support the metabolic exchange of the encapsulated organisms
(d) Secrete biomolecules

The matrix is required to support growth and reproduction of the encapsulated organisms and allows secretion of functional biomolecules produced by it. The metabolic exchange is assisted by the surroundings.

Question 3: Which of the following cannot be a characteristic of the living component?

(a) Be composed of synthetic cells
(b) Be a self-healed version of a cell

Living Biomaterials 195

(c) Be in its wild-type form
(d) Be genetically modified

The living component should be alive.

Question 4: Which of the following system *does not* describe a living materials concept?

(a) Bacteria releasing hyaluronic acid in dermal microneedles
(b) *Bacillus subtilis* microparticles for wound infections healing
(c) Probiotic bacteria capsules
(d) None of the above

All of the mentioned describe a living component in a scaffold forming a living material.

Question 5: Referring to Case Study 1, What makes this system extraordinary compared to "classic" living materials?

(a) The live component is synthetic
(b) The living component is not injected next to or in to a biopolymer
(c) The bacteria do not release any beneficial molecule
(d) This system presents a self-healing of itself

In a classic living material system, the live component is integrated into the polymeric matrices. Here, the material component, amyloid, mussel proteins, and fibers are formed by the bacteria.

Question 6: Referring to Case Study 1, which of the following is the real reason for using *B. subtilis*?

(a) Its small size
(b) It has a single outer membrane
(c) It is a hydrophobic bacterium
(d) It is presented in the human body

Bacillus subtilis was chosen since it is "generally regarded as safe" (GRAS), and since it has a single outer membrane making it ideal for the secretion of large quantity of enzymes and proteins.

Question 7: Referring to Case Study 2, which of the following does not take part in the living material system?

(a) Chitosan
(b) Alginate
(c) Dextran sodium sulfate (DSS)
(d) Li01

Chitosan and alginate form the LbL encapsulation for the Li01 bacteria. Dextran sodium sulfate (DSS)-induced mice were used as animal model.

Question 8: Referring to Case Study 2, what is the charge on chitosan in the stomach and in the intestine, respectively?

(a) Positive; No charge

(b) Negative; No charge
(c) No charge; Positive
(d) No charge; Negative

Chitosan, pKa ~ 6.4, has *amino groups* from the deacetylated units at C-2. At pH lower than pKa such as the stomach, chitosan molecules are protonated ($-NH_3^+$) to the quaternary form, giving a positive charge to the polymer.

Question 9: Referring to Case Study 2, polyglutamic acid is a negatively charged polymer with pKa-4.45. Which of the following polymers in the LbL system can potentially replace polyglutamic?

(a) Chitosan
(b) Alginate
(c) Both
(d) None

Alginate, like polyglutamic acid, is negatively charged and can form electrostatic interactions with the positively charged chitosan.

Question 10: What can be true regarding next-generation living materials?

(a) Will include "live" functional properties that surpass existing capabilities
(b) Will include new organisms and techniques
(c) Will be made of non-hazardous organisms that are non-toxic to the human body
(d) All of the above

Next-generation advanced living materials will include "live" functional properties that surpass existing capabilities by using novel organisms and materials with improved properties

References

1. Rodrigo-Navarro A et al (2021) Engineered living biomaterials. Nat Rev Mater 6(12):1175–1190
2. Sankaran S, del Campo A (2019) Optoregulated protein release from an engineered living material. Adv Biosyst 3(2):e1800312
3. Lufton M et al (2018) living bacteria in thermoresponsive gel for treating fungal infections. adv funct mater 28(40):1801581
4. inoo k, bando h, tabata y (2018) enhanced survival and insulin secretion of insulinoma cell aggregates by incorporating gelatin hydrogel microspheres. Regen Ther 8:29–37
5. Brubaker CE et al (2010) Biological performance of mussel-inspired adhesive in extrahepatic islet transplantation. Biomaterials 31(3):420–427
6. Ben David N et al (2021) Bacillus subtilis in PVA microparticles for treating open wounds. Acs Omega 6(21):13647–13653
7. Liu S, Xu W (2020) Engineered living materials-based sensing and actuation. Front Sensors 1:1
8. Zhang C et al (2019) Engineered Bacillus subtilis biofilms as living glues. Mater Today 28:40–48
9. Yao M et al (2021) Improved functionality of Ligilactobacillus salivarius Li01 in alleviating colonic inflammation by layer-by-layer microencapsulation. NPJ Biofilms Microbiomes 7(1):58

Therapeutic Polymer Conjugates and Their Characterization

Victor M. Quiroz, Joshua Devier, and Joshua C. Doloff

Abstract Drugs, proteins, and nanoparticles delivered into the body are exposed to numerous mechanisms for premature elimination and deterioration within the body. Polymers can be covalently bound to this cargo as either bioactive or inert carriers of such therapeutics to impart protective characteristics. These include improved biodistribution, first-pass clearance reduction, specific targeting, and immune evasion. These polymers can be either synthetic or naturally derived and degradable or inert. Both the polymer and the cargo of interest may be joined together using a series of controlled and specific conjugation chemistries that target functional groups on either species to form covalent bonds. In addition, selectively cleavable linkers may be utilized to engineer cargo release at specific locations under unique physiological conditions such as low pH, oxygen, or biomolecule-laden environments. We will discuss not only different conjugation chemistry strategies and how they should be considered for targeted applications but also how one might reconcile them with different desired downstream characterization as well as in vivo uses.

V. M. Quiroz · J. Devier
Department of Biomedical Engineering, Translational Tissue Engineering Center, Wilmer Eye Institute/Smith Bldg., Johns Hopkins University School of Medicine, Baltimore, MD, USA

J. C. Doloff (✉)
Department of Biomedical Engineering, Translational Tissue Engineering Center, Wilmer Eye Institute/Smith Bldg., Johns Hopkins University School of Medicine, Baltimore, MD, USA

Department of Materials Science and Engineering, Institute of NanoBioTechnology, Johns Hopkins University, Baltimore, MD, USA

Department of Oncology, Sidney-Kimmel Comprehensive Cancer Center, Johns Hopkins University School of Medicine, Baltimore, MD, USA

Bloomberg-Kimmel Institute for Cancer Immunotherapy, Johns Hopkins University School of Medicine, Baltimore, MD, USA
e-mail: jcdoloff@jhu.edu

Graphical Abstract

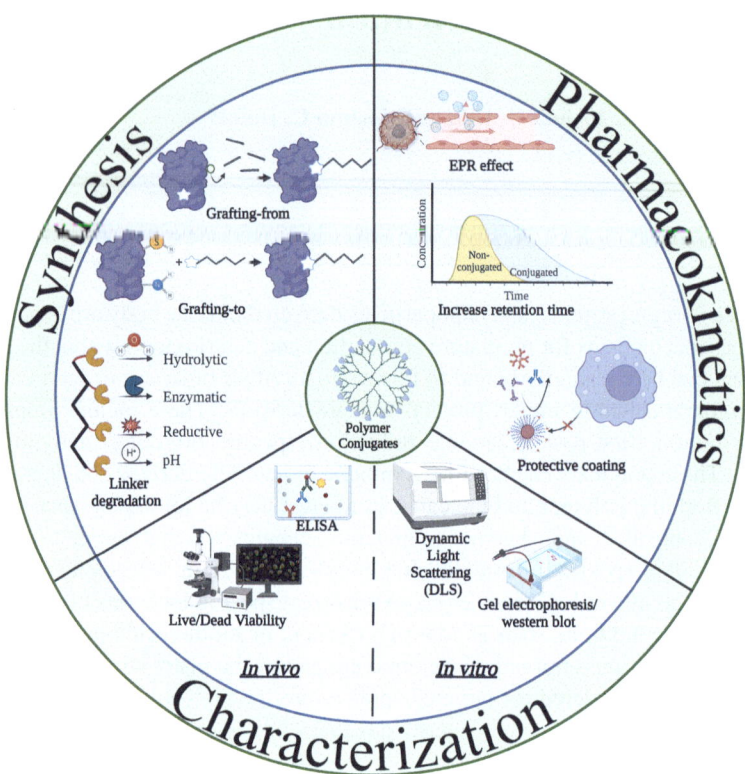

Keywords Polymer conjugates · Pharmacokinetics · Polymer characterization · Protein conjugation · Therapeutic polymers

1 Overview

Up until now, this text has discussed a wide array of approaches and polymers, such as hydrogels and nanoparticles, that are being used to release therapeutic interventions in the body. These techniques use a framework of separating and protecting delicate cargo from the harsh and turbulent environment within the body. This framework generally involves completely enveloping cargo within a polymer structure through non-covalent interactions such as steric entrapment, ionic attraction, or hydrophobic separation until the cargo is released. Nanoencapsulated therapeutics can have issues of low drug loading (w/w) necessitating higher amounts of delivered particles. Additionally, care must be taken to prevent burst release of encapsulated drug/cargo from polymers and hydrogels. We will now explore another widely used approach of binding polymeric molecules directly onto a "cargo" of interest to impart protective and functional qualities. These new therapeutics are known as *polymer conjugates* and are principally used in engineering at molecular and nanometer length scales for drug delivery applications. Such conjugates feature a four-component system including: (1) the polymer or carrier, (2) a linker or bond that joins the two together, (3) a therapeutic molecule/cargo of interest that primarily consists of small molecule drugs or proteins/biologics, and (4) an optional targeting or labeling moiety.

2 Therapeutic Polymers

Polymer selection is crucial in determining the ultimate fate of the conjugate within the body. Characteristics such as charge, length, molecular weight, steric hindrance, and hydrophilicity greatly influence accumulation, blood circulation time, stability, and removal/excretion from the body. The polymers used within the area of conjugates can be broken into either synthetic or natural and further categorized into degradable and non-degradable.

3 Non-Degradable Synthetic Polymers

Inert therapeutic polymer carriers are referred to as *excipients* or inactive substances that serve as the vehicle for a drug. Many polymers used that are biologically inert are synthetically derived. Of these, no polymer is as widely used and well characterized for conjugation than polyethylene glycol (PEG). It's considered as the gold standard for excipients. In fact, the first protein conjugate approved by the FDA in 1990, Adagen, was a PEGylated (or coated with linear PEG chains) adenosine deaminase (ADA) conjugate used to treat severe combined immunodeficiency syndrome (SCID) caused by deficiency of ADA. PEG is used in most conjugates approved by the FDA because of the many qualities that make it desirable for drug delivery. PEG is exceptional when used as a coating for insoluble drugs as it's an electrostatically neutral polymer that is hydrophilic. It is also commercially available with a wide variety of functional groups featuring low polydispersity. While these qualities make PEG highly versatile, there are deficiencies associated with it. In general, PEG is non-biodegradable but is considered bio-eliminable, meaning there is the potential that it may be broken down in the body via reactive oxygen species but is well regarded as being inert towards traditional degradation mechanics. Otherwise, PEG may also accumulate in tissues if its unable to be excreted by the kidneys. This can lead to acidosis and hypercalcemia [1]. Also, studies have shown that some patients have developed anti-PEG antibodies, which was perhaps bound to happen given its widespread use as a polymer in medical applications that is also foreign to the body.

Another widely used synthetic polymer is poly N-(2- hydroxypropyl) methacrylamide (pHPMA), a hydrophilic copolymer. First generations of pHPMA were created using non-degradable backbones and relied on degradable linkers to release the drug. Second-generation pHPMA also has additional incorporated degradable spacers within its backbone, often in the form of enzymatically degradable peptide sequences or hydrolytic bonds [2]. This allowed for higher molecular weight pHPMA to be used, as it could be cleaved into smaller fractions. Second-generation pHPMA also had longer circulation times than their lower molecular weight first-generation counterparts. Thus, despite being biologically inert and not metabolized by the body, pHPMA and PEG can be functionalized using combinations of other polymers and spacers to impart bioactive properties.

4 Degradable Synthetic Polymers

One alternative to having to functionalize PEG and HPMA is to use synthetic polymers that are degradable by the body. Polylactic-co-glycolic acid (PLGA) is one such synthetic polymer that is cleavable through hydrolysis because of its ester bond. Once cleaved, it is broken into glycolic acid and lactic acid which are recycled by the body. In addition, such metabolized byproducts may also be cleared and excreted from the body through the urine.

5 Non-degradable Naturally Derived Polysaccharide

Alginate is a polysaccharide often derived from algae that is non-degradable by mammals because of an inability to produce the enzyme that breaks it down, alginase. It is comprised of two monosaccharides arranged in block-repeating groups: L-guluronic acid (G) and D-mannuronic acid (M). They form a polysaccharide structure with a negative charge due to G- and M-group carboxylic acid epitopes, which can be utilized in conjugation chemistries. Alginate can also be crosslinked to form larger structures like encapsulating particles using cationic affinity-based interactions. Cations such as Ca^{2+} crosslink alginate molecules by being surrounded by the negatively charged monosaccharides forming what is referred to as an "egg-box" structure. The structure that alginate takes is dependent on the ratio between the G and M groups. Repeat M groups make flexible formations while G-group epitopes form more rigid zig-zag formations due to the orientation of the cyclic groups (equatorial and polar, respectively). Lastly, alginate can be combined with other naturally derived molecules such as chitosan and peptides to add positive domains for the formation of micelles.

6 Degradable Naturally Derived Polysaccharides and Polypeptides

Other polysaccharides often used for drug delivery include negatively charged hyaluronic acid (H.A) and positively charged chitosan. Both have the benefit of being degradable by the body. H.A is a large hydrophilic macromolecule with a molecular weight (MW) > 1000 kDa made from D-glucuronic acid and N-acetyl-D-glucosamine [1]. It can be metabolized by hyaluronidases and glucosidases in the liver creating water, urea, and CO_2 [1]. H.A has also been widely studied for its wound-healing properties as well as its capability to shift immune phenotypes from inflammatory towards immunoregulatory. Chitosan, on the other hand, is a hydrophobic polysaccharide made up of N-acetyl-D-glucosamine and D-glucosamine [1]. Chitosan predominately accumulates in the liver in addition to kidney and spleen similar to H.A. With alternating charge and hydrophilicity, these two polysaccharides offer many avenues of conjugation chemistries.

Finally, peptides are naturally derived, biodegradable, non-immunogenic (when derived from the host to be treated), and produce non-toxic metabolites. Structures made from peptides can be low or high MW and can be combined with other synthetic and natural polymers to impart qualities such as degradability. While there are many amino acids, only 6 residues are primarily used for conjugates. These include Arg (Arginine), Lys (Lysine, positive), Glu (Glutamic acid) and Asp (Aspartic acid, negative), Ser (Serine, uncharged), and Tyr (Tyrosine, uncharged hydrophobic) [3]. The most used are Glu, Asp, and Lys because of their functionalization potential. Besides enzymatic degradation, peptide conjugates have broad functions. One such example is using proline for intracellular mitochondrial targeting [3]. However, the main drawback is that there is low manufacturing potential compared to PEG and pHPMA. Additionally, because of their positive charge, peptides have issues in circulation but can be shielded with techniques such as amidation of amine side groups to remove some positive charge or even PEGylation [3].

6.1 Classification

It is the cargo that ultimately defines the classification of conjugates. While there are many subtypes or classifications within the literature, we will discuss them as broad categories: proteins, drugs, and nanoparticles. Typically, researchers begin the process of creating conjugates with an already determined drug or protein of interest with appropriate polymer and conjugation strategies needing to be selected and validated. However, if the polymer or the linker is being tested, then often well-studied drugs like doxorubicin (DOX) and FDA-approved proteins like insulin [4] are typically utilized. Additionally, cargo that has high immunogenicity, high clearance, or poor solubility can be used to see if polymer or linker aids in its delivery.

6.1.1 Protein Conjugates

Proteins make up the most diverse and thus largest subset of polymer conjugates. In fact, all but a few conjugates available in the clinic are proteins. *Protein conjugates* include any macromolecule comprised of mostly amino acids. This includes cytokines, antibodies, and their fragments, peptides, enzymes, interleukins, and growth hormones. Proteins, as we will see in synthesis (Sect. 2), can be bound to polymers with and without the use of linkers by conjugation chemistries that directly use amino acid residues as functional groups.

6.1.2 Drug Conjugates

Following proteins, much of the work within the field of drug delivery is focused on *drug conjugates,* primarily for cancer therapy. The word drug is a broad descriptor for any molecule with a therapeutic effect but otherwise isn't comprised predominantly of amino acids. This would include both organic and inorganic molecules. Drug conjugates necessitate a different approach compared with proteins because the pre-functionalized groups of amino acid residues aren't available. This necessitates the use of linkers to be able to bind to carrying polymer(s) and release the drug at a desired time and tissue, often requiring functionalization of the drug. One example of a drug conjugate that is FDA approved is Movantik, a PEG-naloxegol conjugate where the drug is an opioid antagonist to prevent overdose and the PEG moiety functions to reduce constipation often seen as a result of opioid use [5].

6.1.3 Nanoparticle and Nucleic Conjugates

Another major type of conjugate involves nanoparticles. This often comes in the form of PEGylation or adsorption of polymers to the surface of a synthetic/natural nanoparticle structure. Nanoparticles have been reviewed at length in previous chapters and will not be expanded at length, though, two structures considered as nanoparticles will be highlighted: micelles and dendrimers. Lastly, nucleic acids are often used in conjunction with nanoparticles but will not be focused on within this chapter.

6.1.4 Linkers

Proteins such as enzymes and antibody fragments can still function while conjugated if they're not internalized or otherwise irreversibly bound. In general, a drug conjugate is inactive while it is traveling through the body [6]. This is beneficial in some ways as it can reduce off-target dosing of distal tissue. It is only once the polymer is separated from the drug that it can exert its therapeutic effect, typically through internalization into a cell. To accomplish this, *linkers* are used that undergo cleavage under varying conditions. Typically, there are four variants of linkers used that are sensitive to pH, hydrolysis, reduction, and enzymatic degradation [2, 7]. pH-sensitive linkers are useful, as both inflamed and tumor tissues have a pH of around 6.5 whereas normal tissue and plasma has a pH of 7.4 [7]. The most used pH-sensitive functional group is that of hydrazone found on ketones and aldehydes. Hydrazone is generally more stable and more widely used than other pH-sensitive groups like Imines. These bonds are cleaved at a pH of 5–6 but are stable at 7.4 [7]. Enzymatic degradation is another widely used method to release drugs. Often, oligopeptide sequences that are enzymatically cleavable are used such as GGFG (Gly-Gly-Phe-Gly) or GFLG (Gly-Phe-Leu-Gly) [2]. Peptide sequences are degraded by proteases such as cathepsins and papain (cysteine protease) while others by matrix metalloproteases (MMPs) and even cell-specific enzymes. Hydrolytically degradable linkers are degraded by water and release more as a function of time than location and concentration like the other linker types. Hydrolytically degradable bonds such as esters are often used. Linkers can be made sensitive to reducing intracellular environments through the inclusion of a disulfide bond. Additionally, glutathione is an antioxidant that assists in release of redox-sensitive bonds and is at a higher concentration intracellularly (1–10 mM) versus extracellularly (5μM) [3].

Besides the actual bond that is cleaved, cargo release kinetics are also dependent on linker sequence, length, steric hindrance, and charge to name a few. Additionally, the structure and loading amount of both the polymer and cargo are important. For example, steric hindrance of the polymer can keep an enzyme-peptide complex from forming or a drug from being internalized into a cell. These considerations need to be optimized for an effective therapeutic. If the linker is broken, too soon doses can spread throughout the body causing side effects. However, if the linker bond is too strong and release is delayed, therapeutic efficacy may be lost.

6.1.5 Stimuli-Responsive Polymers

Polymers can have more properties than just being degradable or acting as excipients. They can also be *stimuli responsive* such that the structure, charge, and other properties change under defined conditions. Some polymers are partially micelleable where below a lower critical solution temperature (LCST), they exhibit a phase change due to a decrease in solubility and the polymers collapse reversibly into micelle structures [7]. For example, elastin-like polymers are insoluble above the LCST. This can be used in hypothermia-induced cancer treatment to increase retention of conjugates near a tumor [7]. Stimuli-responsive polymers also involve selective cleavage and degradation by the same types of bonds used for linkers. These cleavable bonds are placed between different polymer units within the backbone of a copolymer system. This gives the added benefit of degradation preventing prolonged accumulation and longer serum retention. Of note, the key is to make a conjugate that can be degradable into non-toxic byproducts and where degradation rate is slow enough to provide increased plasma half-life but fast enough that it allows eventual clearance.

6.2 Purposes for Conjugation

We have now seen that polymer conjugates can take many forms but are categorized into broad classes including protein and drug conjugates, among others. With such a diverse set of cargo that can be conjugated to polymers, so too are the mechanisms by which polymers improve the therapeutic potential of the cargo.

6.2.1 Changes in Pharmacokinetics

The primary reason for polymer conjugation is to improve the pharmacokinetic profile of a therapeutic. As a brief review, *pharmacokinetics (PK)* is a field of study that quantifies the transfer of a particle through the body while pharmacodynamics is how the body acts on this particle. Delivery of a therapeutic polymer conjugate into the body follows a four-step life cycle of: absorption, distribution, metabolism, and elimination *(ADME)*. *Absorption* is how the conjugate enters the body and the amount that crosses a barrier to do so [1]. For example, injection strategies result in different plasma concentration profiles of drug going from fastest to slowest: intravenous (IV) > intraperitoneal (IP) > intramuscular (IM) > subcutaneous. *Distribution* is all of the route(s) the conjugate takes within the body once it has been administered. These compartments include systemic blood and lymphatic circulation, organs, tissues, and capillaries. This is often the key stage for which the polymer is chosen and functionalized because of the well-studied bottlenecks that reduce biodistribution, such as adsorption of proteins in the blood-like complement that can lead to mononuclear clearance. *Metabolism* is how the conjugate is broken down by the body into constituents. Usually, polymers are first degraded by hydrolysis or enzymatic degradation, both systemically in the blood or in organs such as the liver. This is often more of a consideration of safety because metabolites can accumulate within the body leading to potentially undesirable effects. For instance, PLGA (poly-lactic and glycolic acid) polymers are favored because the ester bond can be hydrolyzed to produce lactic acid and glycolic acid, both of which are naturally found in the body and are recycled but can lead to acidosis and inflammation if accumulated in a particular tissue [1]. Finally, *excretion* is how the polymer conjugate as well as all metabolites/constituents leave the body. A common method of excretion includes renal filtration by passage through the glomerulus going into the urine and feces.

Distribution and Clearance Mediated by the Size of Particles The first and most sought-after property of polymer conjugation is the increase in the *stokes radius (R_s)* or *hydrodynamic radius* (R_H) of the cargo. The two terms are used interchangeably for polymers and are experimentally derived values given to macromolecules such as proteins and polymers to estimate their size by assuming they are hard spherical particles in a given solution. This follows the Stokes-Einstein equation:

$$R_H = \frac{K_b T}{6\pi\mu D} \quad (1)$$

where D is the diffusion coefficient for a particular solvent, K_b is the Boltzmann constant, T is the temperature in Kelvin (typically body temperature is 310 K), and μ is the viscosity of the solution that the molecule is in. Higher molecular weight polymers generally have a higher hydrodynamic radius. It is important to note that properties such as charge, hydrophobicity, and branching structures of the polymer can aggregate water molecules differently which will increase or decrease the hydrodynamic radius. For example, a higher molecular weight polymer may have a smaller hydrodynamic radius compared to a lower molecular weight polymer if it has a lower affinity to water.

Conjugates may not be globular, or spherical, and may be more ellipsoid in structure which can potentially increase aggregation and circulation time [1]. To account for this, another experimentally derived measurement called *the radius of gyration* (R_g^2) can be used where the root-mean-squared positional average of each monomer of the macromolecule is summed:

$$R_g^2 = \frac{1}{2N^2} \sum_{k=1} (r_k - r_m)^2 \quad (2)$$

where the monomers on the polymer chain are denoted as r_k = 1, 2, …, N, and r_m is the mean distance from the core or center of the polymer. Both values are found using light scattering assays such as dynamic light scattering (DLS) for hydrodynamic radius and small angle x-ray light scattering (among others) for radius of gyration and are outlined in greater detail in Sect. 3. Each approximation can be used in conjunction with each other to measure the extent to which the conjugate is spherical by using the relationship [8]:

$$R_g^2 = \left(\frac{3}{5}\right) \times R_s^2 \quad \text{or} \quad R_g = .775 \times R_s \quad (3)$$

For instance, a conjugate or macromolecule is more globular if Rg is lesser (Rg/Rs < 0.775) and non-spherical if Rg is greater (Rg/Rs > 0.775).

Both the size and shape of the conjugate are what primarily determines the extent of distribution. In order for a therapeutic to go from systemic circulation (plasma) into a target tissue, it would have to be able to diffuse from the pores of blood capillaries. Non-fenestrated capillaries (found in muscles, bone, adipose tissue, and mesentery) have a pore size of 5 nm while fenestrated capillaries (e.g., found in connective tissue, glands, and intestinal mucosa) have pores around 6–12 nm [6, 9]]. Lymphatic vessels, which are more permeable than non-fenestrated capillaries, have a pore size of around 6 nm [6]. Figure 1 summarizes some pore size variations found throughout the body in addition to clearance sizes of key organs. Thus, conjugation can help skew where accumulation takes place.

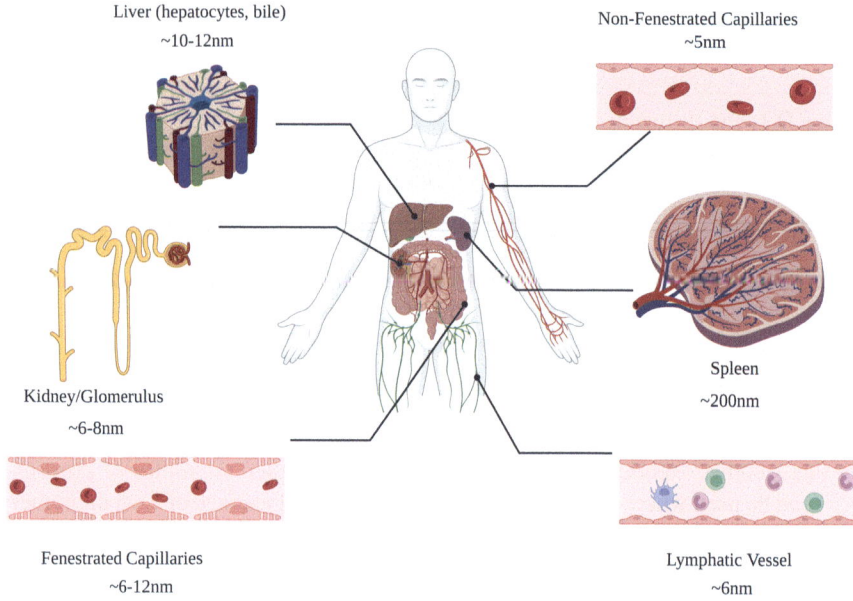

Fig. 1 Size cutoff ranges for organs and tissues within the body. (Image created with BioRender)

Particles smaller than these pore sizes can rapidly diffuse out of capillaries and the respective tissues while larger particles will be retained in the plasma for longer periods before crossing. This increases the likelihood of *clearance* which is a function of metabolism and excretion. Clearance is primarily mediated by three organs: the liver (metabolism/excretion), the kidneys (excretion) and the spleen (metabolism), and they do so in a size- and shape-dependent manner (Fig. 1). Kidneys, through glomerular filtration, remove particles 6 nm or less from the blood and into urine. In general, particles over 8 nm do not get cleared by the kidneys. Particles in-between 6 and 8 nm are cleared via a charge-dependent mechanism in order of Positive > Neutral > Negative. If a particle is not removed by the kidneys and falls under a size between 10 and 20 nm then it will most likely be cleared by the liver through two mechanisms [10]. (1) Either the particles are internalized and enzymatically degraded by hepatocytes and Kupffer cells (liver resident macrophages) or (2) hepatocytes transfer the particles into the bile (biliary excretion) where it gets sent into the gut [6]. Particles that are larger than 200 nm are cleared by the spleen through phagocytosis mediated by mononuclear cells.

Polymer conjugation ultimately helps prevent first-pass clearance where the delivered particles are immediately cleared by filtering organs. The most common approach is to PEGylate a drug or protein to increase *serum half-life*, or the amount of time it takes for half of the amount of a therapeutic to be cleared by the body. Increased half-life in turn allows for more conjugated drug to reach the target destination and be retained for longer. This is important for maintaining concentration within the *therapeutic window*, or the concentration needed for a therapeutic effect without going too high (overdose) or too low (no effect). However, it is important to note that clearance of a drug and its metabolites is necessary as prolonged accumulation of drugs can be toxic. A secondary consideration of conjugation is choosing a preferred clearance mechanism. For instance, renal clearance may be preferred if either the polymer or cargo has metabolites that can cause side effects or toxicity within the liver.

EPR Effect We have discussed endothelial and capillary permeability size cutoffs for penetrating tissues with polymer conjugates. Interestingly, tumors have been shown to have erratic capillary beds with leakier endothelial walls thought to be the result of rapid vascular endothelial growth factor (VEGF)-mediated angiogenesis. This results in a phenomenon known as the *enhanced permeability and retention (EPR)* effect where particles that are between 10 and 200 nm in diameter accumulate within tumors preferentially over other healthy tissues. Thus, a conjugated chemotherapeutic drug can have diameters above the 5–12 nm size cutoff for normal blood capillaries while systemic circulation time can be increased as less is being immediately cleared by the liver and kidneys. Of note, however, the EPR effect has had some controversy as the phenomenon varies between tumor types.

6.3 *Immune Evasion*

Once a therapeutic is delivered into the blood, they are susceptible to *opsonization* whereby immunogenic proteins, or opsins, adhere to the surface of a substrate or particle. These proteins include clotting factors, compliment, and other humoral defenses such as antibodies. This effect is mediated mostly by the charge of the particle in addition to its hydrophilicity and size. Opsonization increases the hydrodynamic radius of the delivered cargo and can change its clearance. Once something is tagged with opsins, it signals to passing mononuclear cells (e.g., monocytes/macrophages) to phagocytose the particles where they undergo oxidation and enzymatic degradation. This mechanism of clearance predominately occurs in the liver and spleen since these organs have high concentrations of mononuclear cells. These phagocytic cells are collectively known as the *reticuloendothelial system (RES)*. Protein adsorption protection through polymer conjugation is the primary mechanism of preventing this. Conjugation also offers protection from the immune system by also blocking immunogenic recognition epitopes on proteins. Polymers such as PEG can offer "steric stabilization" whereby hydrophilicity of PEG can repel circulating proteins. However, even with PEGylation, conjugates that are larger/higher molecular weight or have a higher density of PEG are susceptible to increased protein adsorption. Zwitterionic polymers, those that have alternating positive and negative charge, also counteract opsonization [6].

Shielding therapeutics thus relies on being able to sterically block detection and adsorption of molecules. The density of polymer coatings also contributes to the extent of hydrophilicity that allows for immune evasion. For instance, the MW of an ethylene glycol monomer has a length of 3.5 Å and weight of 44 Da [10]. If linear PEGs are conjugated to the surface of a protein, would the length of PEG simply be 3.5 Å times the number of the monomers repeat units? Not quite so. Polymer coatings take on different confirmations depending on the density of the molecules adsorbed/conjugated to the surface of a particle. The confirmation of linear polymer coatings can be described as *brush*, tightly packed and fully extended, or *mushroom*, collapsed and partially covering the surface [10]. Each chain has what is known as a *Flory radius (R_F)* or radius of possible confirmations and is given by:

$$R_F = a \times N^\nu \tag{4}$$

where N is the number of monomers in the chain, a is the length of monomer, and ν is the Flory constant where 3/5 is given for a good solvent [11]. A specific confirmation depends on R_F and the length of extension of the chain (L) and distance of the chains (D). For a particle such as a protein or drug, D is given by:

$$D = \frac{\sqrt{4\pi r^2}}{N} \tag{5}$$

where r is radius of the particle. As the density of PEG increases, there is more steric hindrance (D ~ R_F) and the chains begin to extend out until L > 2R_F and a fully formed brush confirmation is formed. Below this, chains stack on themselves in a collapsed mushroom manner. Thus, grafting density is low when D is greater than R_F and high when R_F is less than D. Figure 2 outlines the different confirmations of polymer coatings. PEG-brush confirmations generally reduce opsonization and have increased biodistribution more so than mushroom confirmations (Fig. 2). While these properties increase drug life in vivo, they may negatively impact the activity of the therapeutic. For example, PEGylation of enzymes doesn't impact efficiency of low MW targets but it can reduce cleavage of high-MW substrates [12]. While epitopes may be shielded from immune detection, they can compromise binding epitopes for intracellular transport.

6.4 Specific Targeting/Delivery

Up to now, we have discussed using polymer conjugation to increase the circulation time or decrease clearance so the probability of delivery to a tissue of interest is increased. This technique is known as *passive targeting*. Passive targeting can also be made more efficient by utilizing selectively degradable bonds to release cargo. For example, during inflammation, endothelial cells loosen their gap junctions, increasing the porosity for particles to travel through. Immune cells also release cathepsins during an inflammatory response. Therefore, high molecular weight conjugates with linkers and backbones that have oligopeptide sequences that are

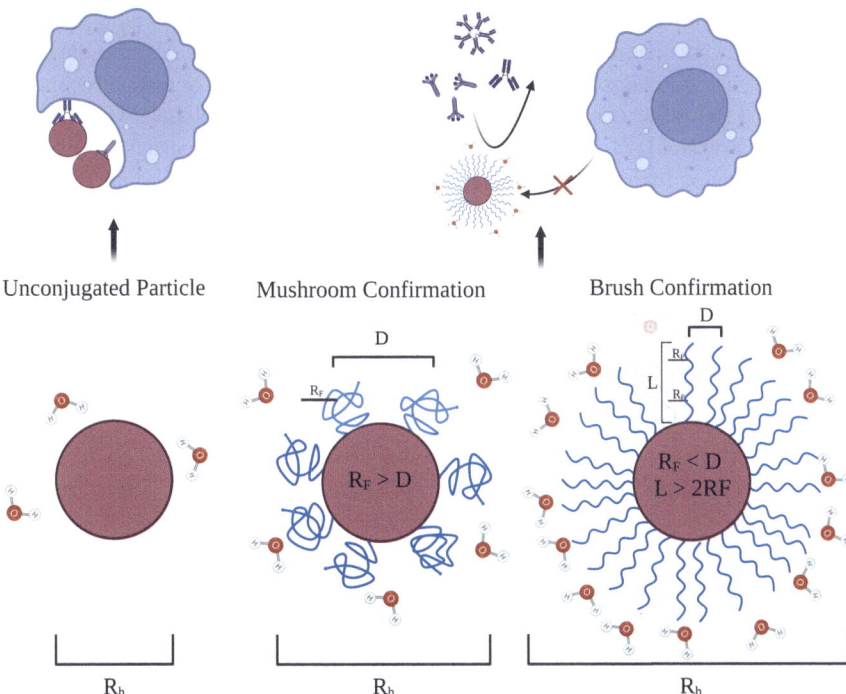

Fig. 2 Adsorption configuration of linear polymer chains on the surface of a particle, and how it allows vs. inhibits interactions with cells and various circulating biomolecules. (Image created with BioRender)

enzymatically degradable by cathepsins could preferentially accumulate and release within inflamed tissue. Alternatively, polymer conjugates can include targeting moieties that allow for *specific targeting*. Ideally, the target is differentially expressed only on cells of interest over all others (e.g., diseased vs. healthy or liver vs. non-liver) cells. This targeting moiety should also lead to internalization of the cargo into the cell to have its therapeutic effect. For example, folic acid preferentially accumulates on certain cancer cells because they overexpress the folate receptor. Additionally, conjugates can be functionalized with targeting moieties including ligands, antibody fragments (Fab), or single-chain variable fragments (scFv). Antibody fragments are often used instead of full-sized IgG antibodies because they are relatively large (150 kDa) and can affect the PK of conjugates. Other strategies include incorporation of peptide sequences such as RGD (Arg-Gly-Asp) to selectively bind integrin receptors. Peptides can also aid in accumulation in tumor sites as MMPs are increased within the high growth and remodeling tumor microenvironment. *Intracellular targeting* involves using conjugates to traffic cargo into specific organelles within the cell. Therapeutics are internalized into cells using endosomes and released upon their merging with lysosomes (organelles containing extreme pH and degradative enzymes) to form endolysosomes, which lead to vesicle membrane destabilization and cargo release. Amine or protonating groups can be added onto conjugates to provide endosomal escape thus destabilizing the negatively charged endosome membrane. Furthermore, steroid hormone receptors can be used to traffic conjugates into cell nuclei by functionalizing with steroid ligands like cortisol [12]. Endosomes can also function to release drugs from polymer carriers using hydrazone bonds as endosomal and lysosomal pH dips to as low as 5 [3]. Lastly, *negative targeting* is a technique where conjugate properties result in pharmacokinetic exclusion of certain organs like kidney, liver, and spleen. This can be done by conjugation of large polymers such that the R_h is larger than 10 nm. This approach can decrease the chance of negative side effects of the drug if metabolites are toxic in specific organs/tissue, as such modifications will prevent them from entering those tissues in the first place.

6.5 Increasing Solubility

Many cancer therapeutics and chemotherapies are hydrophobic and have poor solubility within the body. This limits their therapeutic effect because of poor distribution, and often necessitates the use of higher doses to achieve therapeutic efficacy. Unfortunately, these drugs may also suffer from dose-dependent toxicity and/or metabolic instability. Therefore, conjugation with hydrophilic polymers such as PEG or HPMA may increase their distribution and reduce the doses required.

6.6 Regulatory Considerations for Polymer Conjugates

First, we discuss the area of biomedicine, and compounds generally regulated by the US Food and Drug Administration (FDA). Of note, there are different therapeutic categories of biologics, such as nucleotides (e.g., DNA, mRNA, siRNA) and proteins (e.g., monoclonal antibodies, cytokines), small molecules (e.g., chemical compound drugs-like receptor tyrosine kinase inhibitors), as well as biomaterial (e.g., PLGA, PEG, etc.) (Fig. 3, ***upper left***). When entities across different categories are conjugated together, they must meet standards and gain approval by all relevant subcommittees in order to be used in the clinic as a drug-conjugate system. It may seem like a minor distinction, but one major technicality to note for FDA approval consideration is that short peptides ≤40 amino acids are not considered biologics and are instead regulated as conventional small molecule drugs [5].

Other than biomedical applications in humans, polymers are also used as stand-alone as well as conjugate systems in agriculture (e.g., as fertilizer components for plants, antimicrobials in livestock, or as insecticidal pesticides) as well as in industry (Fig. 3, ***lower left and right***), including packaging/coatings (also those that are color-changing), actuators, electronics, photovoltaics, sensors, as well as solar cells. Generally, such polymer conjugate systems are advantageous due to being low cost, stable, and light weight, as well as having useful optical and electrical properties.

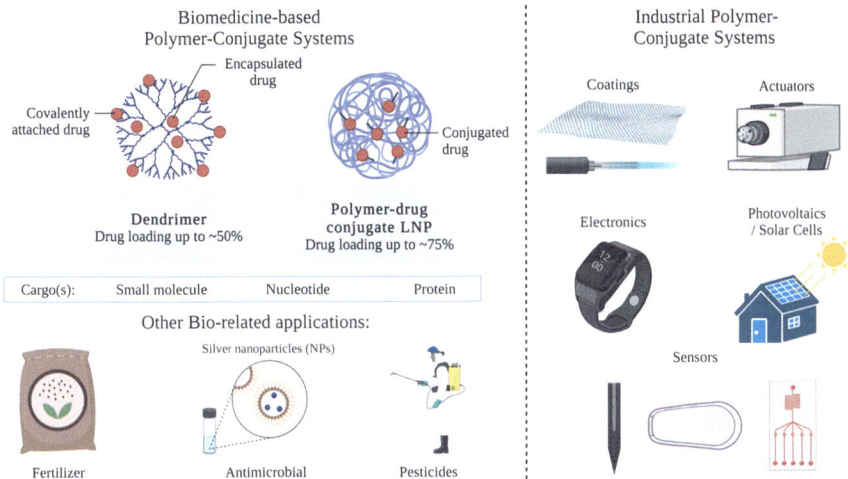

Fig. 3 There are different therapeutic categories of drug—biologics, proteins, and small molecules—as well as biomaterial. Other than biomedical applications in humans, polymers are also used as standalone as well as conjugate systems in agriculture as well as in industry. (Image created with BioRender)

7 Chemistry and Synthesis

7.1 Therapeutic Polymer Structure

The varying forms of polymer synthesis allow for a breadth of different confirmations and structures. There are predominantly five types of structures used for conjugating polymers and include linear, grafted, branched, dendrimer, and micellar (Fig. 4). Each of these structures uses covalent bonds to bind the polymer carrier to the therapeutic though micelles may also involve non-covalent interactions. Observed structures are not necessarily unique to any one polymer or macromolecule. Synthesis techniques exist to fabricate such conformations using a variety of polymer backbones (e.g., copolymers, block polymers, etc.) with natural and/or synthetic formulations being used.

7.1.1 Linear

Perhaps the most straightforward of the structures comes in the form of linear polymer chains. The polymer backbone can be comprised of the same monomers (homopolymer) or differing polymers (copolymer). These structures are easily synthesized and have a greater conjugation efficiency than the other structures. While conjugation is straightforward, there are less-functional sites than with other more-complex configurations [5]. Conjugation sites can be at the ends of the linear chains or at side

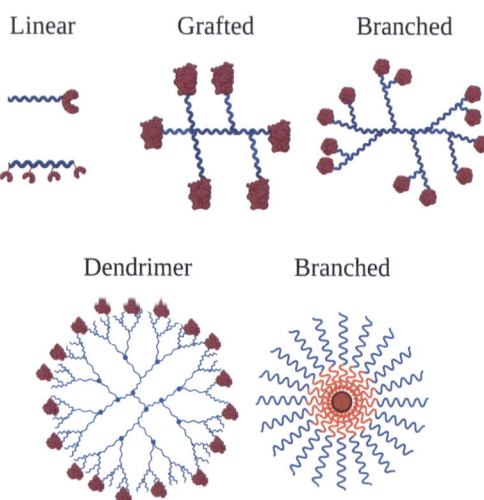

Fig. 4 The major structures of polymer conjugates. (Image created with BioRender)

groups along the backbone. Linear polymers generally have a higher degree of freedom allowing them to enter pores that can be smaller than the radius of similar molecular weight constructs. One drawback is that complex structures increase the half-life of proteins and drugs while linear conjugates are not retained in circulation as long.

7.1.2 Grafted

Grafted polymer chain structure consists of a main polymer backbone with secondary polymer chains covalently bound as side chains. These secondary side chains can give more functional groups for conjugation sites. Additionally, the multiple polymer chains increase the hydrodynamic radius of the structure.

7.1.3 Branched

Branched structures are comprised of homopolymers, or copolymers, covalently bound by using a crosslinker to create higher molecular weight multifunctional group structures. Linear grafted and branched structures don't differ in drug release much if bound by hydrolytically cleavable bonds [12]. The difference between the structures, however, does impact drug release if enzymatic spacers are used. Complex architecture such as branched can suffer from lower drug release efficiency because of steric hindrance. Generally, enzyme-substrate complexes are easily accessible at the end of a linear chain than on side chains and crosslinked branches. This can result in decreased drug activity [12].

7.1.4 Dendrimer

Dendrimers are a form of branched polymers that extend out from a center molecule to form a star-shaped macromolecule. A popular polymer used within the field is poly(amido amine) PAMAM because of its ease of synthesis and is an excellent example of star-like conjugates [7]. Generally, ethylene diamine is used as the core molecule and is polymerized with methyl acrylate to form double the number of branches for each functional group. These two reactions can be repeated in cycles called generations. Importantly, there is a linear increase in radius and an exponential increase in functional groups with each subsequent generation leading to very monodisperse dendrimers. Dendrimer structures feature many functional groups at the end of each branch (128 amino groups after 5 generations) allowing for conjugation of drugs and other moieties on their surface. These globular structures have improved biodistribution than linear and branched conformations because of their larger hydrodynamic radius. Their structure and protective qualities for cargo are attractive for applications in crossing barriers such as the gastrointestinal tract. Besides covalent binding at the surface, dendrimer branching structure can also be used to encapsulate drugs. Early in their development, the drawback to dendrimers was that they were formulated with non-degradable polymers like HPMA-PAMAM block polymers. Newer formulations, however, feature peptide and disulfide spacers for degradation into excretable fractions.

7.1.5 Micellar

Micelles are what are known as *colloids* or a macro mixture consisting of insoluble particles suspended within a solution. This structure differs from the others mentioned above in that it generally does not involve covalent binding of the cargo to the polymer. Micelles are comprised of self-aggregating *amphiphilic polymers* or polymers that have both a lipophilic (hydrophobic) and hydrophilic domain along the polymer backbone [7]. Amphiphilic polymers are unimers in low concentrations, but at high concentrations they spontaneously form aggregates with a hydrophobic core and a hydrophilic outer surface. This is mediated through a spontaneous thermodynamic and entropic process. The concentration where micelles are formed is known as the *critical micelle concentration (CMC)* [7]. The CMC is most impacted by the hydrophobic domain structure and length. Generally, lower CMC is associated with longer or stronger phobic chains. Many formulations use block polymers to combine properties in addition to adding further functional groups on the surface. This is a useful characteristic to have, as many drugs are hydrophobic and have poor solubility. The hydrophobic polymer domains surround the drug while the hydrophilic domains increase solubility leading to increased biodistribution and longer retention times in circulation. Similar to dendrimers, polymer backbones can be made pH sensitive by adding groups that are protonating/deprotonating such as carboxylic acid or sulfonic acid. While micelles are stable at high concentrations, stability drops as they are diluted within the body leading to premature release of the drug. Thus, micelles offer an option of non-covalent delivery using hydrophobic interactions, but their in vivo stability is limited.

7.2 Grafting-to Vs Grafting-From Conjugation

Despite an array of structures, there are two popular methods of conjugation [4]. Either a preformed polymer is functionalized with a reactive group and is bonded to another functional group on its cargo, known as the *grafting-to* method. Or the cargo is functionalized using the same kind of reactive group on the polymer as graft-to but with an additional electron radical initiator and the monomers are polymerized from the cargo, known as the *grafting-from* method. Grafting-to is the more traditional way of polymer conjugation because it can use a wide variety of polymers and the use of click-based chemistry is very versatile. However, it also has lower efficiency because steric repulsion on proteins or drugs may limit access to certain functional groups. Grafting-from is a more novel approach and has certain benefits over graft-to. First, because polymerization occurs at the drug or protein, the technique is better at avoiding steric issues. Additionally, it is easier to purify than grafting-to. This is because grafting-to may have unreacted protein and/or unreacted polymer, and the conjugate produced by grafting-from is robust in having high-efficiency conversion to conjugates [13]. Grafting-from can also form more brush confirmations than grafting-to, which can be beneficial for certain applications. One drawback to grafting-from is that it suffers from high polydispersity compared to grafting-to.

7.2.1 Protein Conjugation

To conjugate a polymer to a protein, two components are used to bind the two: a reactive group on the polymer/initiator and a functional group of interest on the protein. The protein reactive groups as well as the functional groups on the proteins are essentially the same for both types of grafting methods [4]. Despite similarities, protein conjugation may differ between grafting-from and grafting-to depending on the monomer to be polymerized. This may necessitate different conjugation chemistries. The classical strategy is to use native amino acids and other functional groups on the protein for conjugation. Conjugation chemistries modify the reaction conditions such as pH as well as careful selection of protein reactive groups on the polymer/initiator to selectively target native functional groups over other reactive sites on the protein surface. The most widely used groups are N-terminus, C-Terminus, Lysine, Cystine, and Tyrosine [4]. One-step chemistries targeting each of these groups have been developed and will be explored below.

Besides native conjugation, newer methods have also been developed to increase the selectivity of conjugation. This includes genetic engineering (recombination) to introduce either canonical or non-canonical amino acids (ncAA) into the protein of interest [4]. Once highly reactive AA groups are added, conjugation chemistries can be made more selective in spatial location of interest. Meanwhile, ncAA insertion allows for chemical handles that otherwise are not normally found on natural proteins allowing for a wide breadth of biochemical reactions for conjugation [4].

7.3 Other Considerations

Stability throughout manufacturing processes (synthesis, purification, storage, delivery, and in vivo testing) is a key attribute for a chemistry of choice. For instance, certain groups like esters are hydrolyzable so purification techniques like dialysis can denature the conjugates. Another consideration is that the reactive sites should be sufficiently away from the bioactive epitope of proteins so as to not introduce steric blocking. There are some important differences between conjugation to small molecules and other drugs compared to proteins. First, proteins being large macromolecules suffer from steric repulsion more so than drugs. Depending on the tertiary structure, certain amino acids and other functional groups can be tucked away within the hydrophobic core of proteins or otherwise not on the surface of the protein. Additionally, proximity of other amino acids can cause undesirable side reactions depending on the conjugation chemistry (as we will see with N-terminus conjugation chemistries). Polymer conjugates may also be directly bound to the protein without the use of a linker though linker use is feasible with proteins as well.

7.4 Residue Targeting

7.4.1 Lysine

Lysine is one of the most common amino acids (AA) on the outer layer of proteins, making up to about 6% [4]. As a result, chemistries targeting lysine are temporally non-specific in that conjugates will be speckled along the surface of the protein [4]. When choosing lysine, an important consideration is whether heterogenous lysine conjugation is fine or if precise chemistries are required. This may differ based on the application. For instance, if a protein is too immunogenic or if longer blood circulation times are needed, then lysine conjugation may be sufficient for such applications. The first FDA-approved conjugate, Adagen, utilized non-selective lysine conjugation in order to attach PEG to the protein. The functional/reactive group on lysine is the ε-amine group, where the amino group is located on the carbon atom at the position ε to the carboxy group. This amine group has a pKa of 10.5 at a neutral to basic pH making it *nucleophilic* [4]. Nucleophilic molecules or atoms tend to donate electrons or react at electron-poor sites. Two of the most widely reported chemistries for conjugation of polymers to the amine of lysine are N-hydroxysuccinimide (NHS) ester amidation and reductive amination using an aldehyde group (Fig. 5a) [4, 13]. NHS esters react with amines on lysine to create an amide (NH=O-O-R) at neutral to slightly basic pH (7.2–9). NHS esters, however, undergo hydrolysis on the scale of hours at neutral pH so reactions need to be fast enough to not be limited by hydrolyzable esters. Thus, if lysines are on the interior of the tertiary structure of the protein, the reaction may not be efficient. This can also affect grafting-to approaches as the ester can hydrolyze during dialysis for purification. Besides esters, reductive amination is another common method. Reductive amination involves binding an aldehyde (CH=O) to a primary amine (NH_2) resulting in a (NH-R) formation. This is done in a reducing environment using a reducer such as $NaBH_3CN$ or $NaBH_4$ [4].

7.4.2 Cysteine

Cysteine (Cys) is typically the first AA of interest used in site-specific conjugation of proteins. This is because it is less abundant in proteins in addition to being more nucleophilic than lysine with a pKa of 8 vs 10 [4]. Cysteines contain a thiol functional group (-SH) that is excellent for conjugation chemistry. Cysteine is often located in the interior of a protein and forms disulfide bonds essential to the structure of proteins, making it hard to access for chemistries. Thus, chemistries involving Cys predominantly use two approaches (Fig. 5b): disulfide re-bridging using a dibromo-maleimide, or disulfide exchange with pyridyl disulfide (PDS) [4, 13]. In disulfide re-bridging, the disulfide bonds of two cysteines are reduced to free the thiol groups for conjugation. Then, maleimide/dibromo-maleimide undergoes a fast addition reaction with one of thiols followed by the second addition reaction of the other thiol [4]. This fast kinetic allows for "re-bridging" the disulfide bond without losing the tertiary structure of the protein. Disulfide re-bridging is best used for conjugation in smaller molecules. One drawback is that the maleimide-thiol bond is reversible in vivo via retro-Michael addition because of other thiol groups that are

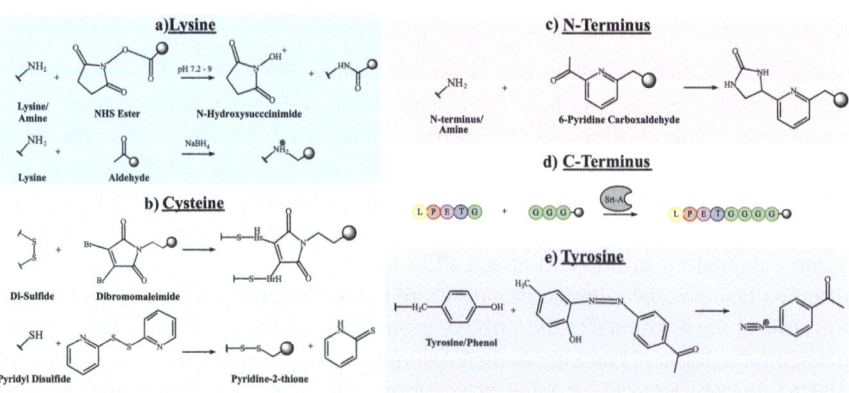

Fig. 5 Conjugation chemistries for varying amino acid residues on proteins. (Image created with BioRender)

abundant in plasma. Additionally, the need for a reducing agent may destabilize sensitive proteins. Disulfide exchange is the process where a thiol (on the protein) reacts with a disulfide bond-containing functional group (such as PDS) resulting in a disulfide bond connecting the protein to the polymer. This can be performed using physiological pH under mild conditions. This bond is stable but is susceptible to cleavage in low pH environments. This can be used advantageously to release proteins inside of compartments near tumors or intracellular spaces. Free cysteines are rarely found on the outer surface of a protein for direct modification. However, recombinant proteins can be made by inserting cysteines at specific epitopes of the protein for downstream conjugation. In fact, cysteine is often the amino acid of choice for recombinant proteins for site-selective thiol reactions. The classical method of conjugating to a free Cys is to use the Michael acceptor maleimide for alkylation of the carboxy group.

7.4.3 N-Terminus

Each amino acid in nature has an amine (NH_2) and a carboxylic acid (COOH) that bind to each other in a chain to form a peptide or protein leaving an amine or *N-terminus* that caps the start of a protein and a carboxy group or *C-terminus* at the end of a protein. The N-terminus amine has a pKa of ~8, which allows for conjugation over Lys if the pH of the reaction is controlled [4]. Thus, techniques for conjugation of amines on lysine also conjugate amines of N-termini. One of the most widely used chemistries for conjugation is reductive amination using 2-pyridine carboxaldehyde (2-PCA) [4]. This is a one-step method where the aldehyde forms an intermediary imine with the N-terminus and forms a cyclic group with the amide in a cyclic imidazolidinone (Fig. 5c). However, if a Cys residue (pKa = 8) is located near the N-terminus, then undesired reductive alkylation side reactions can occur [4]. Another drawback is that up to 20–30% protein can be denatured after the reaction [4]. Again, reductive amination is best used with smaller proteins as the larger the protein, Lys can lead to off-target conjugation.

7.4.4 C-Terminus

While there are more expansive methods to conjugate N-termini, strategies to conjugate polymers and drugs to proteins on the *C-terminus* have been developed. The classical method for C-terminus conjugation is to use enzymatic ligation (Fig. 5d). This is accomplished using recombinant proteins to insert peptide sequences such as LPXTG/A at the C-terminus of a protein, where X denotes any canonical amino acid [4]. Then, a polymer that is functionalized with a Gly-Gly-Gly- (GGG) epitope chain is conjugated to the peptide tag sortase-A (Srt-A) enzyme. This creates a protein-LPXT-GGG-polymer complex. While this method is used for N- and C-terminal conjugation, it requires the use of recombinant proteins. A more novel strategy of chemical ligation has been developed using single electron transfer

(SET) where a visible light catalyzer, lumiflavin, along with a Michael acceptor leads to alkylation of the carboxy group [4]. The reaction selectively modifies C-termini over Glu and Asp residues, which can be undesired side reactions in other conjugation strategies.

7.4.5 Tyrosine

Lastly, tyrosine (Tyr, Y) is an amino acid with a hydrophobic phenol functional group. Though not as utilized as lysine or cysteine, direct modification strategies have been developed to use tyrosine as an alternative residue. One widely reported chemistry is azo coupling using diazonium salts with a ketone on the para position of the phenol [4]. The diazonium molecule forms an azo (R-N=N-R) bond to the phenol group of tyrosine leaving a ketone group for further functionalization (Fig. 5e). Then, conjugation with a polymer functionalized with an aminooxy group ($O-NH_2$) forms an oxime-linked conjugate.

7.4.6 Recombinant Techniques

Direct modification and conjugation to native amino acids on a protein or peptide has been the classical approach to creating novel therapeutics. However, newer approaches using engineered recombinant proteins have been developed whereby non-native amino acids or peptide chains/tags can be inserted to add functional groups at specific epitopes for conjugation or enzymatic ligation. There are two strategies when using protein engineering. The first is to introduce *canonical amino acids* into the protein of interest, or the 21 amino acids found within humans. Typically, cysteine is frequently added because of the established site-selective chemistries. Other residues that are inserted include tryptophan, tyrosine, methionine, histidine, and selenocysteine [4]. Generally, these other amino acids are less reactive than lysine and cysteine. Yet, considerations such as reaction conditions and polymer selection can favor certain conjugation strategies over others. *Non-canonical amino acids* (ncAA) are not a part of the proteome and are used to add unique chemical handles normally not found on natural proteins such as azides (R-N-3). This allows for the use of *biorthogonal reactions* or reactions on molecules with functional moieties without effecting their biological/biochemical functions [14]. These usually can be performed at physiological temperature, pH, and are generally non-toxic. A prime example of this is copper-catalyzed azide-alkyne cycloaddition (CuAAP), a click reaction that is widely used for reacting alkynes and azides [14]. Using novel copper catalyst ligands, reactive oxygen species mediated by copper are mitigated, allowing for non-destructive modification of alkynes and azides on proteins and cells [14]. The main drawback to this is that the use of transition metals can lead to toxicity from unreacted copper that leeches into surrounding tissue. This limitation sparked the creation of strain-promoted azide-alkyne cycloaddition (SPAAC) which uses cyclic strain to drive reactions rather than transition metals. For these biorthogonal chemistries, reaction rate is important to optimize.

The faster the reaction, the less stable the bond. However, because proteins such as antibodies are in low quantities in reactions involving conjugation, slow reaction rates are typically not ideal.

7.4.7 Small Molecule Conjugation

A major facet of a therapeutic polymer's promise lies in its conjugation to small molecule drugs, allowing for targeted delivery of tracers and pharmaceuticals. The delivery of small molecule drugs follows a few schemes. One is controlled release via polymer degradation, where covalent polymer bonds degrade at pre-determined kinetic rates. Considering the kinetics of bond breakage following a Poisson distribution, the degradation of bonds can be modeled as follows:

$$p(t) = ke^{-kt} \qquad (6)$$

Probability that time between bonds breaking at rate k is t.

$$t_{break} = \frac{1}{k} \qquad (7)$$

Time to break n bonds

$$t_{reaction} = t_n = \sum_{n}^{i=1} \Delta t_i = \sum_{n}^{i=1} \frac{1}{k_i} = \frac{1}{k} \sum_{n}^{i=1} \frac{1}{i} \approx \frac{1}{k} \ln(n) \qquad (8)$$

Reaction time constant

$$t_{diffusion} \sim \frac{L^2}{D} = \frac{\pi L^2}{4D} \qquad (9)$$

Reaction time constant for diffusion constant D and length scale L.

$$\varepsilon = \frac{\tau_{diffusion}}{\tau_{reaction}} = \frac{L^2 \pi k}{4 * D_{H_2O} \left[\ln(L) + \frac{1}{3} \ln\left(\frac{bN_{AV}\rho}{M_o}\right) \right]} \qquad (10)$$

Erosion number for b (# degradable bonds/monomer), polymer density r, and monomer MW M_0.

Utilizing polymers or copolymers of a specific erosion number, e, allows for specified degradation and engineered release. Whether small molecules are directly conjugated to polymer chains or shielded within a polymer matrix, controlled release disperses drugs into a circulatory/buccal/mucosal space to maximize therapeutic effect over a set time frame without spiking drug concentration into a toxic/undesired range.

The first variants of small molecule-polymer conjugates tested therapeutically utilized cleavable peptidyl linkers bonding HPMA copolymers to doxorubicin. This drug, PK1, was more effective than unlinked doxorubicin for cancer treatment with a 17x-77x improved tumor accumulation that led to a 250x concentration difference in target/non-target tissues. Synthesis proceeds from the attachment of enzyme-degradable peptide linkers to HPMA or PEG monomers. This can be accomplished by monomer incubation with the chain transfer agent, containing the peptide chain-of-interest and an initiator. The peptide CTA may be produced via solid-phase peptide synthesis (SPPS) using a fluorenylmethyloxycarbonyl (FMOC) protection/deprotection scheme, followed by a terminal functionalization to the monomer (e.g., the N-terminus of oligopeptide GFLG linked to 4-cyanopentanoic acid dithiobenzoate via DIC, HOBt attachment). A facilitating initiator (e.g., 2,2′-azobisisobutyronitrile [AIBN]) then allows for RAFT (reversible addition-fragment chain-transfer) polymerization of conjugated and un-conjugated HPMA/PEG monomers to produce multiblock therapeutic polymers.

Alterations to polymer-drug conjugates gave greater depth to targetability via stimuli-responsive release mechanisms. pH or ion-sensitive polymers are useful in their ability to differentiate ion gradients between the stomach, small intestine, tumors, or intracellular compartments. Amino-linked cationic polymers have found usage in targeting the low pH of the stomach. The FDA-approved drug Eudragit E (aminoalkyl methacrylate copolymer) is one such compound having high solubility at pH $\leq$ 5, with drug-loaded microspheres synthesized through a spray-drying technique. A water-in-oil-in-water technique is also often used for microsphere formulation using the cationic polymer polyvinylacetal diethylaminoacetate. Anionic polymers better suit the high pH of the intestine, allowing protection of ion-sensitive drugs from stomach acid. FDA-approved polymers like Eudragit S/F/L [poly(methacrylic acid-co-methyl methacrylate)] or hydroxypropylmethylcellulose (e.g., acetate succinate or phthalate) derivatives allow for tunable pH-sensitive release. Eudragit S/F/L varies based on the ratio of carbonyl groups to ester groups, with more ester groups increasing pH-solubility thresholds. Similarly, increasing the number of phthalates, succinoyl, and acetyl groups in a hydroxypropylmethyl-cellulose polymers alters pH-release parameters. By incorporating an internal oligosaccharide shell, colon-specific delivery can be achieved due to the combination of the location's low pH and microbiota.

The Warburg effect, the shift of cancer metabolism to glycolytic ATP production, is well documented for its ability to significantly lower tumoral pH (pH ~ 6.5–7.0) compared to that of homeostatic tissue (pH ~ 7.4). Combined with the EPR effect, the low pH of tumors allows for highly specific accumulation and release of small drug-polymer conjugates. Polymeric targeting can be achieved through the incorporation of imidazole groups, predominantly through linked histidine that takes on a positive charge at pH ~ 6.

PLGA [poly(lactic-co-glycolic acid)] is a poly(a-hydroxy-ester) utilized in sustained release systems and is useful due to its first-order kinetic hydrolysis into lactic and glycolic acid monomers that are further metabolized by the body into CO_2 and H_2O. Being more so hydrophobic, the PLA ([poly(lactic) acid) component reduces copolymer hydrolysis of the connecting ester bonds, while PGA ([poly(glycolic) acid) enhances the product's degradation rate.

8 Intrinsic Characterization

Once polymers have been conjugated to cargo of interest, unreacted reagents can add noise in subsequent characterization assays. Typically, this includes unreacted/partially reacted polymer, cargo, reagents, vs. the fully formed conjugate of interest. Protein conjugation is unique in that there is much more steric repulsion/interference of polymer-binding chemistries than with drug and nanoparticle conjugation. This poses challenges in purification because polymers and macromolecular proteins may have similar molecular weights which would make separation non-trivial. Comparatively, small molecule conjugation would be easier to separate because the molecular weight disparity is much larger. Besides the molecular weights of constituents, the polymerization chemistry can also impact purification. Choosing a grafting-from approach allows for an easier purification because of the decrease in unconjugated components and the increase in disparity between formed conjugates and monomers. Common methods to purify conjugates include dialysis and centrifugation while others such as electrophoresis and size-exclusion chromatography both purify and characterize. Dialysis is a simple way to remove unreacted reagents and fragments by submerging products of conjugation in water within a semi-permeable membrane. The pores of the membrane should have a permeability/molecular weight cutoff that is smaller than the conjugate of interest. Dialysis should not be performed on hydrolyzable polymer conjugates and conjugation chemistries with degradable protein-binding epitopes such as maleimide. Centrifugation is another simple method whereby density gradients are used to selectively sediment undesired product. While both are simple and inexpensive, they don't provide additional information of the separated product.

The therapeutic polymer's intrinsic properties are comprised by its molecular weight, its propensity for aggregation, propensity for denaturation, as well as its drug-loading capacity and its ability to release ions. *Gel electrophoresis* is a characterization technique factoring in both the size and charge of a given polymer (Table. 1a). This method applies an electric field onto a particle solution placed into a gel matrix between a cathode and an anode. After laying the gel horizontally and parallel to the electronic current in a conductive buffer (traditionally Tris-EDTA (TE) buffer), a solution of interest is pipetted into precast wells in the gel (traditionally agarose gel). A voltage is then applied, pulling particles to an electric pole of opposite polarity at a strength dictated by the particle charge. Furthermore, particle movement is impeded on the basis of size due to the gel matrix. Whereas *size exclusion chromatography* (SEC) matrices slow smaller particles due to the extra pathlength introduced by small-diameter pores, gel electrophoresis agarose matrices slow larger particles. Overall, these factors result in a banding pattern where highly charged, smaller-sized particles reach their respective electric poles faster than a

Table 1 Strengths and limitations of common assays used to characterize the material properties of polymer conjugates

Technique/Instrument	Strengths	Limitations
(a) Electrophoreses/Western blot	Bands can be removed from the gel (agarose) for further purification or analysis. Migration is charge, size, and shape dependent. Gives MW estimation during separations by using a standard	Polydisperse polymer chains lead to band drift
(b) SDS-PAGE	SDS degrades hydrogen backbone binding and separates non-covalently bound protein subunits. Allows for rapid determination of molecular weight without shape-based effects	SDS denaturation limits downstream analysis to linearized proteins (as opposed to western blot methods). SDS does not denature disulfide bridges in protein tertiary structure
(c) Chromatography SEC Ion exchange Hydrophobic Affinity HPLC	Better suited for grafting-from strategies. Small MW drug conjugates readily separated. Conjugates better separated in ion exchange and hydrophobic interaction LC. HPLC good for small and stable proteins. SEC can purify while characterizing the size of conjugates	Protein-polymer conjugates similar in MW may not be separated efficiently. HPLC may not have enough solid phase interaction for separation. HPLC can denature protein structure in medium. SEC hard to do if unreacted protein, polymer, and conjugate are the same MW (less than two-fold)
(d) Mass spec	Mass spec can determine conjugation site	Difficult if conjugates not easily ionizable. Necessitates the use of acids but can lead to fractionation
(e) MALDI-MS	Rapid analysis of sample composition. Multiplexable sample analysis. Where necessary, MALDI systems have a spatial resolution of 10μm	Decreases sensitivity for analytes >20 kDa
(f) Dynamic light scattering (DLS)	Acquire concentration of particles. Size of particle's (RH)	10 nm–1μm size. Only gives bulk measurements, so large dispersity is average thus, not good for heterogenous/polydisperse conjugates

lower-charged and larger-sized particle of the same polarity. Due to the standardized charge distribution of DNA backbones and lesser-secondary structure effects, gel electrophoresis is readily used for DNA fingerprinting and molecular weight (MW) correlation. Though, bands can smear or drift depending on the polydispersity of the conjugation chemistry or polymerization. The method requires adaptation for polymers where secondary structure and shape will influence migration rate, however. Protein/peptide polymers, for example, are often denatured in sodium dodecyl sulfate (SDS) to break protein backbone hydrogen bonds and linearize the sequences prior to gel electrophoresis. Binding at a ratio of 3 molecules to 1 amino acid, SDS also imparts a negative charge across the polymer, greatly reducing the effect of the chain's intrinsic charged groups. This technique, *SDS-PAGE (polyacrylamide gel electrophoresis)*, allows for standardized assessment of polymer chain MW that hinders the movement of high-MW polymers (Table 1b).

PAGE is further refined into a *western blot* protocol that allows for epitope-based detection/confirmation of a polymer composition (Table 1a). Here, the substrate is loaded and migrated vertically in a polyacrylamide gel. After completion, an applied voltage is used to transfer the substrate horizontally onto a transfer membrane. The membrane is embedded with primary antibodies binding epitopes-of-interest and then secondary antibodies that bind to the Fc region of the primary antibodies. Functional tags linked to the secondary antibodies allow for recognition of the polymer-of-interest and are visualized via a variety of methods (colorimetric, chemiluminescent, or fluorescent outputs from enzymes such as horseradish peroxidase (HRP) or alkaline phosphatase (AP)).

Isoelectric focusing is a variation of PAGE that characterizes the isoelectric point (pI) of a polymer solution, the pH at which the charge of a molecule is in equilibrium. This is performed by loading a polymer sample through a pH gradient gel in an ampholyte solution. After applying an electric field, charged polymers will migrate until they reach a region in the pH-gradient gel that neutralizes their charge, halting field-induced movement. The final stopping location of the polymer indicates pI and can be visualized via fluorescent methods.

Column-based techniques (Table 1c) for particle size analysis are also widespread. SEC characterizes size by passing the polymer solution through a matrix (traditionally silica- or agarose-based) with a variety of larger and smaller selective pore sizes. Due to the selective nature of pore diameter, polymers of a larger size will be excluded from longer smaller-diameter channels that a small-sized polymer will enter. The resultant effect of size-exclusive channels is an increase in the retention time (also referred to as elution volume) for small particles relative to larger particles. Measuring retention time via UV/Visible (UV/Vis) light absorbance readings between samples or compared to a standard, a chromatogram can be generated that gives an indication of particle diameter, as well as the opportunity for sample extraction/isolation. Other useful column chromatography types are ion-exchange chromatography, hydrophobic-interaction chromatography, and affinity chromatography. *Ion-exchange chromatography* refers to passing a sample through a column that retains an ion of particular polarity and is split into binding, elution, and regeneration. Here, a cation exchanger column (e.g., sulfobenzyl, sulfoethyl/carboxylate) is negatively charged and would immobilize cations, while an anion exchanger column (e.g., diaminoethylamino) is positively charged and would immobilize cations. These bound particles are then released from the column using a charge-screening salt pumped at increasing concentrations to elute increasingly higher-charged particles. *Hydrophobic interaction chromatography* works very similarly but utilizes a hydrophobic column that binds hydrophobic regions of polymer as they are pumped through in water. Following binding, a hydrophobic organic solvent (e.g., acetonitrile, hexane, methanol) is pumped through the system at increasing concentrations to elute increasingly more hydrophobic materials. *Affinity chromatography* is a fourth very versatile method for substrate characterization or isolation. Here, polymer solution is pumped through a specialty column that binds a ligand of interest. The uniqueness of ligand-receptor interactions allows for effective characterization/sequestration of antibodies, lectin-bound molecules, or a variety of enzyme/ligand pairs. This technique has significant usage in poly-peptide isolation via the usage of nickel-nitrilotriacetate (Ni-NTA) columns that bind poly-histidine sequences in polymers. Here, the resin's central Ni^{+2} ion (internally bound to 3 oxygens and nitrogen) coordinates to the nitrogen of 2 adjacent histidine residues.

Mass spectrometry (often abbreviated Mass Spec or MS) similarly factors size and charge through determination of the mass/charge ratio (m/z) of particles in solution (Table 1d). The underlying mechanism of MS first involves vaporization and then electron-beam ionization of a sample. This ionized material is then accelerated down a vacuum tube and deflected using an electromagnet. Following deflection of the particle's path by the magnetic field, the ion hits and is registered by an ionic detector. The degree of magnetic deflection and time-of-flight (TOF), measured by the detector, determine the m/z ratio intrinsic to the polymer solution. One major type of mass spectrometry in therapeutic polymer research is *MALDI: Matrix-Assisted Laser Desorption/Ionization* (Table 1e). This technique uses a laser to hit a solid substrate material, composed of the analyte embedded in a matrix, suitable for solid materials where "burst" reads are more appropriate than constant vaporization. MS is also often coupled with chromatography techniques, such as *gas chromatography (GC-MS)* or *high-performance liquid chromatography (HPLC-MS)*, to elucidate mass spectrometric data alongside retention time. GC features a liquid stationary phase and a gaseous mobile phase, while for HPLC the stationary phase is solid and mobile phase is liquid. Unlike previously mentioned techniques, GC/HPLC-MS feeds eluted particles into a mass spectrometer rather than a UV/Vis detector. The need for heating and vaporization of material in GC limits usefulness to volatile materials. Additionally, charged samples (such as samples with ionic interactions) are not compatible with GC analysis and are run in a liquid-phase systems instead. HPLC uses high-pressure liquid injection and is better suited to assessment of structured polymers (whole/partial polymer assemblies, proteins), while GC-MS may better characterize subunits (certain polymer monomers, amino acids), liquid chromatography techniques are well suited for understanding small molecules bound to polymers, such as glycan moieties on proteins. Following polymer incubation in a glycan-cleaving enzyme (such as the N-glycan-cleaving PNGase F for proteins), the products may be run in HPLC-MS to understand a given polymer's glycosylation pattern. A primary drawback of HPLC systems is the propensity for carrier solvent to alter polymer and protein structure during analysis, leading to altered characterization.

One of the most fundamental intrinsic properties is polymer size. *Dynamic light scattering (DLS)* is one method for understanding the hydrodynamic size distribution of a therapeutic polymer (Table 1f). DLS works on the principle that particles in a given solution will scatter light at a specific scattering angle θ. Brownian motion of particles will lead to changes in measurements of scattered light intensity, with smaller particles leading to rapid intensity changes and larger polymer particles having slower intensity fluctuations. Ultimately, the intensity fluctuation rate of a given polymer solution can be correlated to the hydrodynamic size distribution of polymers present, relaying average size and uniformity of polymers in solution. This information is useful to understand the intended outcome of polymer synthesis but can also indicate the desired or undesired aggregation propensity of polymer units. The specific range of DLS measurements varies by instrument but is often from approx. 0.1 nm to 10μm.

8.1 *Functionality and Biological Evaluation*

Once the material products are purified and characterized to meet design criteria, the pharmacokinetic and pharmacodynamic properties of the conjugate should be tested. These experiments can be broken down into in vitro and in vivo methods. In vitro assays test for viability of cells after exposure as well as protein adsorption to the conjugate. In vivo assays offer a more robust and extensive modality to test the efficacy of therapeutics. Such tests include PK studies using live imaging within animal models such as radio labeling, *positron emission tomography (PET)*, and fluorescent imaging. Metabolism of conjugate and polymer fractions can also be monitored using *nuclear magnetic resonance (NMR)*. Other studies for general biocompatibility include immune response, peripheral mononucleated cell count, as well as hemolysis tests.

8.1.1 In Vitro

In vitro testing affords the engineer the ability to quickly test parameters needed for the efficacy of the therapeutic. One such assay that monitors initial safety is a *live/dead viability assay* where conjugates and their fractions/metabolites are exposed to cells in culture and the percentage of viable cells is measured over time (Table 2a). There are many forms of viability testing such as the MTT [i.e., (3-(4,5-dimethylthiazol-2-yl)-2,5-diphenyltetrazolium bromide) tetrazolium] colorimetric assay whereby the metabolic activity of cellular mitochondria is measured and used to estimate the number of functioning cells left in culture. Though, the most used assay involves fluorescence microscopy using the fluorophores ethidium homodimer (red, stains dead cells) and calcein AM (green, stains live cells) where the fluorescence may be imaged, or even measured on a plate reader for the number of cells. When performing such viability tests, it's important to experiment on multiple cell lines, especially those that are targeted (such as cancer cells), and cells where metabolism and excretion takes place (such as hepatocytes). Additionally, a wide range of doses

Table 2 Strengths and limitations of common assays used to characterize the pharmacokinetics and biocompatibility of polymer conjugates in vitro and in vivo

Technique/instrument	Strengths	Limitations
In vitro	–	–
(a) Viability (live/dead fluorescence) MTT/MTS	Estimate toxicity of chemistries or materials on a variety of living cells Multiple cell types can be readily tested	Terminal assay (cannot use the cell lines after viability tests) Mechanism of toxicity won't be known
(b) Fluorescence correlation spectroscopy	Protein adsorption can be quantified via R_H Particles can be sized in blood or plasma (no separation needed) Stability in serum can also be estimated	Conjugate needs to have a fluorophore
(c) PBMC proliferation and cytokine release	Immunogenicity of conjugates can be determined before injection CD4+ helper T cell proliferation can be quantified	PBMC, specific responding immune cell won't be known
(d) Hemolysis assay	Best used on pH-responsive polymers and linkers to test membrane disruption for intracellular delivery	Variation in hematocrit for patients may affect absorbance measurements. RBC membranes are not the same composition of endosomes of all cell types
In vivo	–	–
(e) ELISA (also for liver toxicology)	Serum concentration can be measured over time in intervals ALS and ATS concentrations can determine liver toxicity Sensitive and easy to perform Cellulate concentration excreted	Not real time
(f) Positron emission tomography (PET)	High signal-to-noise ratio Real-time imaging Needs expensive equipment	Radioactive isotope can destabilize or fraction the conjugate leading to potential excretory routes Only one radioactive isotope can be imaged at one time
(g) Live fluorescent labeling	Affordable Can choose targets Numerous stains exist	Requires bound fluorophore Degradation on fluorophore containing epitope can mislead accumulation Fluorophore emission may not penetrate thick tissue
(h) Mass spec imaging (MSI)	Mass spectra overlaid on histology tissue slice gives spatiotemporal information	Does not give metabolic species Low signal-to-background ratio
(i) NMR	Determine metabolic species Non-destructive	Low sensitivity, necessitating higher concentrations or more sample Solvents can denature proteins

should also be used to study the potential for toxic effects up to 5–10 times higher a dose. Lastly, special care should be taken to test for the effects of unreacted chemistry components such as metals and organic solvents.

In addition to toxicity, other tests for biocompatibility include measuring for the adsorption of proteins onto the surface of the polymers. We have already discussed methodologies to reduce such problems by using hydrophilic and high-density polymer conjugation. However, it is difficult to a priori predict whether combinations of chemistries and conjugate components will be sufficient to keep adsorption within acceptable ranges. To determine adsorption several strategies can be employed. The first is to use *fluorescence correlation microscopy (FCM)*. The benefit to using this over other previously mentioned light scattering techniques such as DLS is that you can measure it in medium containing whole blood or plasma (Table 2b) [3]. FCM works by measuring the diffusion time and the intensity fluctuations of a particle conjugated to a fluorophore within a confocal microscope to determine the diffusivity of the particle within the medium. Then, using the Stokes-Einstein equation (Eq. 1), the hydrodynamic radius of the particle can be deduced. This allows for the characterization of protein absorption by measuring the R_H before and after incubation in whole blood, serum, BSA (bovine serum albumin), or HSA (human serum albumin). The drawback to using this technique is that it requires the conjugation of a fluorophore to the conjugate which can impact its PK or at the least require additional testing to ensure unchanged therapeutic effect. A second method to test for protein adsorption is MS. The conjugates can be analyzed for the weights of each ionized fraction before and after incubation with proteins to identify proteins adsorbed to the surface.

In addition to protein binding, assays can be run using donated blood to gauge potential immunogenic responses. Whole blood can be collected and separated using gradient centrifugation into red blood cells (RBCs) and peripheral blood mononuclear cells (PBMCs) which include neutrophils, lymphocytes, monocytes, dendritic cells, and NK cells. The PBMCs can be used to determine immune response via incubation with the conjugates and measuring proliferation markers on CD4+ T cells using flow cytometry. These activation and proliferation markers include CD40, CD80, CD69, and CD25. Additionally, ELISAs or multiplexed antibody-bead assays in conjunction with flow cytometry can measure changes in inflammatory cytokines released from the PBMCs (Table 2c). Immunogenic cytokines include TNF-α, INF-γ, IL-1β, IL-2, IL-4, and IL-5 to name a few. Differentially expressed markers and released inflammatory cytokines suggest that the conjugates are susceptible to phagocytosis and other immunogenic clearing mechanism. Besides using the PBMCs to study immunogenicity, the RBCs separated from whole blood can be used in a *hemolysis assay* (Table 2d). This is especially useful in pH-responsive conjugates carrying drugs meant for cytosolic release. RBC lipid bilayers mimic endosomes such that the amount of hemoglobin released is an approximation of efficiency/potential of endosomal escape within cells. A hemolysis assay involves incubation of the conjugates along with RBCs in buffers consisting of pH correlating to extracellular (7.4), early endosomal (6.8), and late endosomal (>6.8) conditions [7]. Then, the amount of hemoglobin detected in a spectrophotometer compared to a full lysis control determines the % lysis of the RBCs due to the phospholipid bilayer disruption mediated by the conjugate. The ideal scenario is that there is no statistical difference in lysis compared to a no treatment control at pH of 7.4 while significantly increased lysis of phospholipid bilayers of the RBCs occurs at a lower pH.

8.1.2 In Vivo

Assessments that are done within animal models give valuable information regarding the resulting pharmacokinetics (PK) mediated by the polymer in addition to any determination of pharmacodynamics (PD) of the cargo on the disease or model of interest. PK can be determined primarily through a few methods. The first is through using *an enzyme-linked immunosorbent assay (ELISA)* that specifically detects the cargo or a tag on the conjugate whereby the concentration over time can be measured, as conjugation can increase the time that the conjugate is detectable in serum or tissues of interest. In addition to detecting serum concentration of particles, ELISAs for liver toxicity can be used that detect aspartate transaminase (AST) and alkaline phosphatase (ALT) which are released within the body when there is liver damage and cell death (Table 2e). While ELISAs are very sensitive, they lack real-time monitoring of therapeutic distribution.

The primary method to monitor PK of conjugates is through using some form of real-time imaging. Radiolabeling is a method where radioactive isotopes (such as carbon-14 and technetium-99) replace certain atoms within the molecular structure of the conjugate to impart radioactivity. Then, the distribution of the particles can be tracked in real-time throughout the body by using PET [4]. PET features a high signal-to-noise ratio and thus has little background in imaging (Table 2f). Some drawbacks to using radiolabeling are that only one isotope can be detected at a time because PET cannot discriminate between radioactive signatures. Additionally, the radioactivity of the labeled atom can destabilize the conjugate.

Fluorescence labeling and imaging is another strategy to track the biodistribution of particles in vivo. This requires attachment of a fluorophore preferably by covalent binding though other methods such as streptavidin-biotin binding, enzymatic ligation and recombinant techniques have been used to attach fluorophores (Table 2g). Fluorescently labeled particles and imaging allow for non-invasive tracking to locate aggregation in vivo. Some of the drawbacks that are associated with live fluorescent imaging are that the light emitted from the fluorophore may not be able to penetrate through thick tissue. One important aspect to consider is that degradation of the conjugate or polymer will result in fractioned species where some carry the fluorophore and others don't. This can lead to misperceived accumulation of the non-degraded conjugate in an area of interest when really it is the degraded fractions that are being tracked. Lastly, addition of another covalently attached moiety can alter PK and possibly functionality.

Another technique that can be used to identify particles within tissues is *mass spectrometry imaging (MSI)*. This technique uses the principle of MALDI-MS to ionize a grid pattern along a tissue surface to analyze particles by weight (Table 2h). Then using software, images can be generated based on the mass spectra. Drawbacks to this analysis are background noise and an inability to detect metabolites.

While the above-mentioned assays are excellent in tracking the biodistribution of conjugates, there are still the aspects of metabolism and excretion to consider. While imaging can offer some insight into accumulation in excretory organs such as the kidneys, bladder, and liver, information on the metabolites would not be possible. One method to study this is to use NMR (Table 2i). Often H^1 or C^{13} atoms have their intrinsic spin measured under a magnetic field to obtain information of neighboring atoms allowing for the elucidation of molecular structure. This is key to studying degradation products and metabolism of therapeutics. One advantage to NMR is that it leaves the samples intact during analysis though solvents may be needed that can potentially denature conjugates. One major disadvantage is relatively low sensitivity necessitating higher concentrations of sample.

9 Summary

Polymer conjugates are a type of drug/biomaterial particle whereby the cargo is covalently bound to a macromolecule. The carriers used in conjugates are highly diverse in charge, hydrophobicity, and structure. This makes them suitable for a diverse set of applications. They can be either synthetic or naturally derived and biodegradable or non-biodegradable. The cargo of interest defines the classification of the conjugate. The most widely applied therapeutics and the focus of this chapter are proteins and drugs. Linkers are utilized for controlled release using bonds cleavable by different chemical environments via pH, reduction, enzymatic degradation, and hydrolyses, each of which is ideal for different applications. In addition to linkers, stimulus-responsive polymers have also been utilized to alter PK profiles under different temperatures.

Conjugation can increase the retention time in systemic circulation thereby increasing its therapeutic potential. This is mediated by an increased hydrodynamic radius of the particles which considers the apparent size of the polymer and cargo. The shape of these conjugates can also be determined using the radius of gyration and can be used to determine if the conjugate is globular or ellipsoid. The size and shape of the particles are primary determinants of how long the conjugates stay within tissue of interest due to the varying size limits of pores in capillaries. Size also dictates the clearance mechanism used by the body further altering the retention time while also introducing new considerations such as altered metabolic activity at the organ of elimination.

Covalently bonded coatings of polymers are also used to protect therapeutics from being coated in other proteins within blood that can increase the hydrodynamic radius. Furthermore, adsorbed opsins can trigger phagocytes to clear the therapeutics through the RES system. Polymer conjugation sterically covers epitopes on the cargo from binding to receptors which can offer immune protection.

This coverage of polymer on cargo can take on different shapes based on the density of polymer chains and the size of allowable confirmations called the Flory radius.

Conjugates can take on many covalently bound structures including linear, grafted, branched, and dendrimers. Others use non-covalent attachment like micelles. Preformed polymers are chemically bound to the cargo known as grafting-to or are polymerized from an initiator on the cargo called grafting-from. For protein conjugation, selective reactions are used to target amino acids on the proteins resulting in epitope-specific conjugation. Small molecules are often attached to polymers using degradable linkers for targeted controlled release. Engineered release of such drug-conjugates via linkers can be modeled using the erosion number which is a ratio of rate of diffusion over rate of reaction/degradation.

Once manufactured, polymer conjugates are first purified using a compatible separation technique that often relies on molecular weight differences between final product and unreacted components. Once purified, conjugates must be analyzed for their intrinsic properties such as molecular weight, charge, size and shape, and conjugation site. Then, biological functionality must also be assessed with the new conjugate to ensure safety and efficacy.

Within the world of medicine, there exists stakeholders that are affected by the products biomedical engineers create. Stakeholders can be the companies that create or manufacture the product, the doctors administering the dosage, patients receiving treatment, and even their families. As we have seen, conjugation of optimized therapeutic polymers to administered drugs offers protection from premature clearance and increases accumulation within a tissue of interest. This can reduce the dosage of a particular drug required for a therapeutic effect. This translates to cost savings for both patients and the company manufacturing it. Lowering dosages requirement is particularly enticing for drug conjugation as the field is pushing for FDA approval of chemotherapy-polymer conjugates. And chemotherapy treatment often brings an onset of debilitating side effects such as nausea, pain, and hair loss. Therefore, the potential for conjugation chemistries to aid in eliminating such negative effects in patients is great. In the world of nanoparticles and nucleic acids, approaches using liposomal encapsulation of mRNA have made a paradigm shift in the speed at which vaccines can be developed. The obvious example being the mRNA vaccines developed and approved under emergency use in 2020 for immunization against SARS-COV-2. Regarding long-term market outlook, drug conjugates are generally considered advantageous, as they may enable reinvigorated interest in a compound that is no longer under patent protection. By creating a new drug form with altered efficacy, safety, or properties in vivo, one is able to file patents for new intellectual property, essential for companies to ensure profitability following heavy upfront expenditures needed for multi-stage clinical trial testing. Combined with the expanding library of naturally derived as well as synthetic polymers, there are many possible drug-conjugate forms that may provide efficacy and/or safety improvements as part of future potential drug systems.

Quiz Questions (Multiple Choice)

Question 1: A positively charged hydrophilic PEG-small molecule dendrimer is found to have a hydrodynamic radius of 3 nm. Which of the following organs would play the biggest role in clearing the conjugate from a patient's systemic circulation?

A. Liver.
B. Spleen.
C. Kidney.
D. Lungs.

Answer: C. Kidneys have a size cutoff range of 6-8 nm. Factors such as charge can influence glomerular filtration as well.

Question 2: A protein-peptide conjugate is created to sterically block an immunogenic epitope on the protein. Thus, its diffusivity is similar to a protein of about 10^{-6} cm^2/s. What would be the expected hydrodynamic radius of such a particle in water?

A. 2.27×10^{-9} m.
B. 2.88×10^{-8} m.
C. 2.13×10^{-13} m.
D. 2.16×10^{-6} m.

Answer: A. First, the diffusivity of a protein is converted to m^2/s. Using the Boltzmann constant of 1.38×10^{-23} m^2*kg*s^{-2}*K^{-1}, an average body temperature of 310 K, the viscosity of water (10^{-3} Pa) can all be plugged into the Stokes-Einstein equation. After converting the units, the denominator yields a value of 1.884×10^{-12} leaving all units cancelled while the numerator gives 4.278×10^{-21} m. Dividing these two values gives the final answer of 2.27×10^{-9} m or 2.27 nm radius.

Question 3: Your company is testing a new experimental branched pHPMA-drug conjugate that has hydrolytically degradable linkers along its backbone for controlled release of the drug. Which of the following purification modalities would likely NOT be suitable for your prototype therapeutic?

A. Dialysis.
B. Centrifugation.
C. Size exclusion chromatography.
D. Gel electrophoresis.

Answer: A. Dialysis uses a semipermeable membrane to filter out unreacted components of a reaction using water. A hydrolytically degradable bond has the potential to be denatured or cleaved during this process.

Question 4: Your research lab hopes to create an amphiphilic micelle structure to increase the solubility of a hydrophobic drug. Which of the following assays would be most useful to characterize the shape of the conjugates. Additionally, which values indicate a spherical particle?

Dynamic Light scattering, $R_g/R_H > 0.775$.

   A. Small angle x-ray diffraction and dynamic light scattering, $R_g/R_H < 0.775$.
   B. Small angle x-ray diffraction, $R_g/R_H > 0.775$.
   C. Small angle x-ray diffraction and dynamic light scattering, $R_g/R_H > 0.775$.

Answer: B. DLS gives hydrodynamic radius R_H while small angle x-ray diffraction gives R_g. Thus, both would be needed. The ratio of R_g/R_H for a globular (circular) particle is less than 0.775.

Question 5: A 100 Da PEG chain has a monomer length of 3.5 Å. If a nanoparticle has a radius of 10 nm. The distance between PEG monomers is 100 Å. which of following confirmations would the pegylated nanoparticle take?

   A. Mushroom.
   B. Brush.
   C. The particle will have both mushroom and brush confirmations.
   D. Not enough information.

Answer: A. Using the equation for the Flory Radius yields 55.47 Å. Since the R_F is lower than 100 Å, the confirmation of the coating will likely be mushroom.

Question 6: You are designing a drug that disrupts the metabolic pathway of a cancer cell and have decided to use a degradable linker to facilitate intracellular release. Which of the following is the least suitable for this task?

   A. Hydrazide (pH).
   B. Oligo-peptide chain (enzyme).
   C. Ester bonds (hydrolytic).
   D. Disulfide bond (reduction).

Answer: C. Engineered intracellular release can be facilitated by using bonds sensitive to cleavage of enzymes, lower pH, and a reducing environment. Hydrolysis is a less ideal method due to potential degradation in systemic circulation.

Question 7: A dendrimer contains encapsulated Doxorubicin (DOX) and you're tasked with studying the release of the drug over time. Doxorubicin is fluorescent when it leaves the nanoparticle. The polymer has the composition such that it is comprised of hydrolytically cleavable ester bonds. You want to prevent burst release of DOX from these nanoparticles. Which of the following rates would best represent such as release?

   A. $\tau_{reaction} > \tau_{diffusion}$.
   B. $\tau_{reaction} < \tau_{diffusion}$.
   C. $\tau_{reaction} = \tau_{diffusion}$.
   D. there is no difference in outcomes.

Answer: B. Burst release can be characterized by a rapid cleavage of the hydrolytically degradable bond such that the rate ($\tau_{reaction}$) is larger than diffusion ($\tau_{diffusion}$). Thus, for slower release of the drug, $\tau_{reaction} < \tau_{diffusion}$.

Question 8: You're studying different molecular weight PEG chains to increase the hydrodynamic radius of a chemotherapeutic. However, with increased retention time, your tasked to test possible toxicity relating to these new formulations. Which of the following assays would NOT be suitable?

A. Live/dead viability in vitro.
B. MTT assay.
C. ATS/ALS ELISA.
D. Live fluorescent imaging.

Answer: D. Live fluorescent imaging is useful for tracking a particle in vivo in real time. It would not be ideal to investigate toxicity.

Question 9: Which of the following are NOT considerations when conjugating proteins?

A. Reduction of functionality due to steric hindrance.
B. Similar molecular weight of polymer chains leading to difficulty characterizing.
C. Linkers must be used to bind the initiator to the functional group.
D. Neighboring amino acid residues may lead to off targeting chemistries.

Answer: C. Linkers do not necessarily have to be used in protein conjugation strategies. Polymers may be covalently bound to proteins using a functional amino acid residue and a functional group on the polymer and or initiator.

Question 10: You are researching possible amino acid residues to target on your protein for conjugation to a biomaterial carrier. You determine that you need a site-specific reaction for this. Which of the following would NOT be appropriate?

A. Reductive amination of a lysine group.
B. Disulfide exchange of a cysteine.
C. Recombination of the protein to introduce a non-canonical amino acid.
D. Enzymatic ligation of a C-terminus.

Answer: A. Lysine is generally the most abundant amino acid on the surfaces of proteins. Thus, conjugation strategies targeting lysines are usually performed for the purpose of speckled coatings and not for epitope or site-specific reactions.

References

1. Su C, Liu Y, Li R et al (2019) Absorption, distribution, metabolism and excretion of the biomaterials used in Nanocarrier drug delivery systems. Adv Drug Deliv Rev 143:97–114. https://doi.org/10.1016/j.addr.2019.06.008
2. Kostka L, Etrych T (2016) High-molecular-weight HPMA-based polymer drug carriers for delivery to tumor. Physiol Res 65:S179–S190. https://doi.org/10.33549/physiolres.933420
3. Melnyk T, Đorđević S, Conejos-Sánchez I, Vicent MJ (2020) Therapeutic potential of polypeptide-based conjugates: rational design and analytical tools that can boost clinical translation. Adv Drug Deliv Rev 160:136–169. https://doi.org/10.1016/j.addr.2020.10.007
4. Ko JH, Maynard HD, Angeles L, Angeles L (2019) Polymer Conjugates By Rational Design 47:8998–9014. https://doi.org/10.1039/c8cs00606g.A

5. Ekladious I, Colson YL, Grinstaff MW (2019) Polymer–drug conjugate therapeutics: advances, insights and prospects. Nat Rev Drug Discov 18:273–294. https://doi.org/10.1038/s41573-018-0005-0
6. Longmire M, Choyke PL, Kobayashi H (2012) Clearance properties of Nano-sized particles and molecules as imaging agents. Consideration and Caveats. 3:703–717. https://doi.org/10.2217/17435889.3.5.703.Clearance
7. Larson N, Ghandehari H (2012) Polymeric conjugates for drug delivery. Chem Mater 24:840–853. https://doi.org/10.1021/cm2031569
8. Smilgies DM, Folta-Stogniew E (2015) Molecular weight-gyration radius relation of globular proteins: a comparison of light scattering, small-angle X-ray scattering and structure-based data. J Appl Crystallogr 48:1604–1606. https://doi.org/10.1107/S1600576715015551
9. Sarin H (2010) Physiologic upper limits of pore size of different blood capillary types and another perspective on the dual pore theory of microvascular permeability. J Angiogenes Res 2:1–19. https://doi.org/10.1186/2040-2384-2-14
10. D'souza AA, Shegokar R (2016) Polyethylene glycol (PEG): a versatile polymer for pharmaceutical applications. Expert Opin Drug Deliv 13:1257–1275
11. Cruje C, Chithrani DB (2014) Polyethylene glycol functionalized nanoparticles for improved cancer treatment. Reviews in Nanoscience and Nanotechnology 3:20–30. https://doi.org/10.1166/rnn.2014.1042
12. Kopeček J, Yang J (2020) Polymer nanomedicines. Adv Drug Deliv Rev 156:40–64. https://doi.org/10.1016/j.addr.2020.07.020
13. Messina MS, Messina KMM, Bhattacharya A et al (2020) Preparation of biomolecule-polymer conjugates by grafting-from using ATRP, RAFT, or ROMP. Prog Polym Sci 100:101186. https://doi.org/10.1016/j.progpolymsci.2019.101186
14. Winssinger N (2018) Bioorthogonal chemistry Chimia (Aarau) 72:A755. https://doi.org/10.1038/s43586-021-00028-z.Bioorthogonal

Biocompatibility of Polymers

Ruba Ibrahim , Abraham Nyska , and Yuval Ramot

Abstract The use of biomaterials is growing in our modern healthcare systems, and there is special increase in the use and development of biodegradable materials made of different polymers. However, biocompatibility of biomaterials remains a great challenge for the manufacturers during their development. In this chapter, we define biocompatibility and outline the material-host interactions that are expected with the use of biomaterials. In addition, we review the major guidelines that define the standards and regulations for evaluating biocompatibility of medical devices. The chapter also describes the factors that need to be considered when evaluating biocompatibility of materials and the numerous in vitro and in vivo tests that have been developed for assessing biocompatibility. We finish by highlighting the central role of the toxicologic pathologist in the evaluation of medical devices, and provide a look for the future of biocompatible medical devices.Graphical Abstract

R. Ibrahim · Y. Ramot
The Department of Dermatology, Hadassah Medical Center, Jerusalem, Israel

The Faculty of Medicine, The Hebrew University of Jerusalem, Jerusalem, Israel

A. Nyska (✉)
Consultant in Toxicologic Pathology, Tel Aviv, Israel

Tel Aviv University, Tel Aviv, Israel
e-mail: anyska@nyska.net

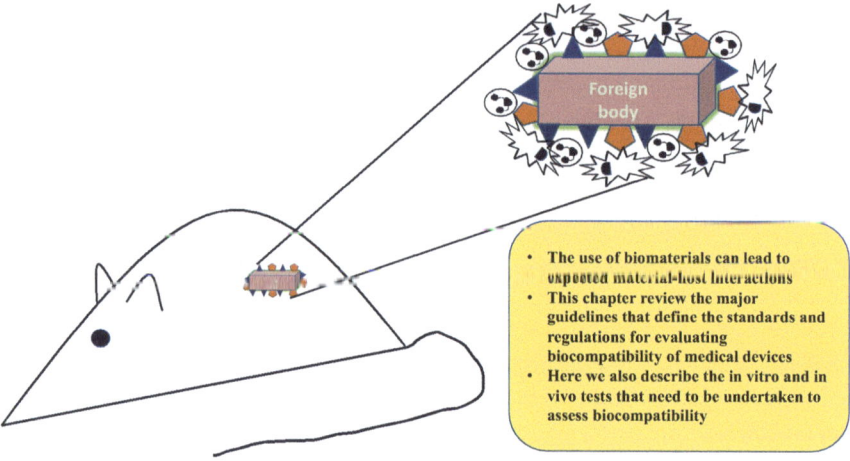

Keywords Biocompatibility · Polymers · Safety · Toxicology · Foreign body reaction

1 What Is Biocompatibility?

The term biocompatibility is defined as the capability of a material to execute specific tasks in medicine with an appropriate host response. Certain factors are referred to when considering biocompatibility. These factors focus on the interaction between the biomaterial and the host tissue and its surrounding environment. In order to achieve favorable long-term outcomes, such parameters should be taken into consideration when designing biocompatible materials.

In vitro and in vivo assays can be used to assess biocompatibility. Polymers can be used to release products, and residual monomers from the biodegradation process might react with the cells (in vitro) and the surrounding tissues (in vivo) or affect the organism. Various chemicals and physical interactions play an important role in in vitro assessment, but leached substances (secreted substances) play the major role in biocompatibility assays. The response of the cells exposed to the leachables helps us measure whether the reaction is causing negative or positive outcomes. To ensure that these leachable components do not produce local or systemic toxicity, many standardized methods have been introduced to assess the suitability of materials. Defined tests for toxicity associated with leachable substances are provided by the ISO 10993 standards.

An acute inflammatory response due to infection can sometimes mimic an inflammatory response secondary to foreign-body response (FBR) due to poor biocompatibility. Bacteria and their cell wall components as well as fungi, such as *Candida*, are organisms that are capable of causing an intense inflammatory reaction when colonizing implants. These organisms may replicate and trigger an acute inflammation. Such reaction is characterized by attracting polymorphonuclear leukocytes and macrophages (phagocytosis) as first responders of inflammatory cells to defend from such invading pathogens. This reaction can lead to local tissue destruction and cause signs of infection such as heat and redness as well as thick, dense foreign-body capsules. This acute inflammatory reaction induced by infection is caused by ineffective sterilization. Thus, this response is not described as poor biocompatibility, even though it shares similar characteristics with poor biocompatibility. Another factor which is important when considering biocompatibility is the mechanical effects of the implant. The implant should be designed in certain shapes and sizes that cause minimal irritation or damage to the implanted site.

Tissue-biomaterial interactions are influenced by properties of the tissue as well as of the material, and by the fluid transportation around the implant. These properties affect the cellular FBR, which is initiated by the tissue to protect itself from the foreign implant and produce a biocompatible environment. Protein adsorption is the first process occurring after a biomaterial is implanted and comes in contact with biological fluid. Proteins begin to adhere to the surface of the tissue thus infiltrating the site and creating a monolayer of proteins. The adsorbed proteins interact with receptor proteins that are expressed by the tissue cells, dominated by neutrophils and macrophages at this stage. The macrophages release cytokines that can act as pro-inflammatory or anti-inflammatory cytokines. Biocompatibility can be determined by the cytokine release response. The biomaterial is considered biocompatible when a response is described as a mild inflammatory reaction (between 3 and 6 weeks), resolving spontaneously and leading to thin fibrous encapsulation that isolates the implant from the tissue. Such inflammatory stimuli can be persistent and thus turning from acute to quiescent, chronic inflammation. The chronic response may be mild and ongoing and with no other local or systemic adverse responses. In some situations, when the macrophages are incapable of degrading the foreign material, they fuse and form multinucleated foreign-body giant cells (FBGCs). The FBGCs may release more contents and cause a more aggressive response, therefore indicating that the biomaterial is less biocompatible (Fig. 1) [8, 12].

2 Clinical Significance of Biocompatibility

Although most implants have good outcomes and function long-term with mild inflammatory response only, complications of biomaterials may result from biomaterial-tissue interactions. These outcomes can effect negatively on both the patient and the healthcare system.

Certain implants can cause delayed-type hypersensitivity reaction (Type IV). Such allergic reactions may cause significant clinical effects. Cutaneous reactions may be observed both on the skin adjacent to the implant site or generalized. Contact

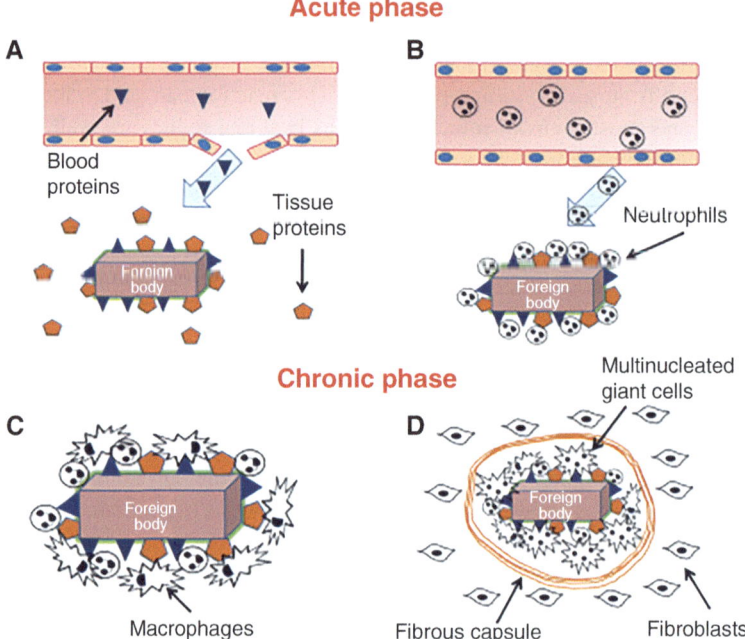

Fig. 1 Schematic representation of the acute and chronic phases of the tissue foreign-body reaction. (Reproduced with permission from Ref. [8])

dermatitis may present as erythema with scaly plaques or papules remote to the implanted site. When the hypersensitivity reaction is systemic, generalized dermatitis is observed. Other cutaneous reactions that may be induced include urticaria (hives) and vasculitis (inflammation of the blood vessels).

Additional adverse reactions include chronic inflammation and excessive fibrosis surrounding the implant, necrosis, aseptic loosening (osteolysis) of the implant, and formation of pseudotumors, which may result in significant morbidity for the patient [12].

3 Standards and Regulations Aimed at Evaluation of Medical Devices

The multi-step approach of evaluating the biocompatibility of medical devices and implantable drug delivery systems begins with initial material screening, safety evaluation, product testing, and finally product analysis. In order for the product to meet international standards, certain biocompatibility requirements are addressed. The International Organization for Standardization (ISO) standard ISO-10993 guidelines prepared by Technical Committee ISO/TC 194 combines existing data evaluation from all sources, which helps manufacturers to meet the standards required for device biocompatibility. The guidelines include 20 parts. Part 1 of the standard is the Guidance on Selection of Tests, Part 2 covers animal welfare requirements, and Parts 3 through 20 are guidelines for specific test procedures or other testing-related issues, including genotoxicity and reproductive toxicity, cytotoxicity, degradation, and local tissue and systemic response. Europe and most Asian countries accept biological studies that comply with ISO 10993 recommendations. The FDA accepts ISO guidelines, although it demands more strict requirements in certain areas. The Blue Book Memorandum G95-1 "Required Biocompatibility Training and Toxicology Profiles for Evaluation of Medical Devices" was issued by the FDA in 1995 and was updated in 2016. This document describes a wide range of tests required to prove the safety of the biological device under development. Such guidelines serve as general framework only and not as a systemic checklist. Thus, it is up to the manufacturer to select the proper tests when developing a biological product. According to the European Medical Devices Directives, safety of medical devices is an essential requirement. Generally, all biocompatibility testing should comply with Good Laboratory Practice (GLP) regulations which apply to the studies regarding safety. The GLP assures that the studies are conducted in accordance with regulatory submissions [2].

Biocompatibility evaluation is also required for drug delivery systems and combined medical devices such as contraceptive implant devices, implantable infusion pumps, catheters, and chambers as well as patches for transdermal drug delivery and nano-based drug delivery systems.

The effect of the drug on the device and its interaction within the biological environment should be investigated for cytotoxicity, irritation, and hemocompatibility. Cytotoxicity is investigated by exposing the studied sample to extraction media and the resulting extract is then exposed to certain cells in which the effect is then determined. Polymer degradation is evaluated by placing samples in either an oxidative or hydrolytic solution. Later, the samples are placed in certain sites for a period of time. Their local effects on tissues are observed and examined for inflammation, necrosis, or fibrosis. Such tests do not necessarily predict the clinical performance of medical devices.

Another factor that needs to be considered is chemical characterization. Chemical assessment is discussed in Part 18 of ISO 10993 in which identification and quantification of the material and its chemical constituents along with assessment of the overall safety of the device is described. Toxicological data from peer-reviewed scientific literature also provides valuable information.

The aim is to provide a functional and biocompatible device that results in only mild effects on the host. Thus, it is necessary to monitor the whole process carefully and determine the host response. It is important to follow the standards and regulations to fulfill the requirements needed to achieve a biocompatible medical device. In vivo and in vitro studies should be undertaken. Studies on animal model, such as ovine, canine, or primates, have shown to provide important data [16].

4 Points for Consideration When Choosing Biocompatible Materials

Choosing suitable biocompatible materials when manufacturing a medical device is an essential step. The chemical and physical properties should be carefully selected to minimize toxicological risks that might be induced by the medical device. The material and its leachables should be assessed on their reaction with the cells and the surrounding tissue. Part 18 of ISO 10993 as well as other scientific literature provide data on material characterizations that should be determined. Therefore, the manufacturer should carefully study the material and fully understand the manufacturing process in detail with the aid of the available data on similar devices in order to conduct an efficient biocompatibility evaluation. This is an important part of the evaluation since certain materials may seem suitable initially but may change throughout the processing steps. Data obtained from the literature may help reduce the extent of biocompatibility testing. Proof of an already approved device with equivalent physical and chemical properties should be provided.

Chemical properties are important since they affect the cell response when in contact with the material. Such reaction is influenced by multiple surface properties, including the chemical composition and function, roughness, smoothness, wettability, surface mobility, electrostatic force effects, crystallinity, and heterogeneity of biological reaction. Polymeric surface properties may be modified in a controlled manner to improve the biocompatibility and clinical performance of the device. Wettability of the materials surface may influence the cell response significantly. Whether the surface has hydrophilic or hydrophobic characteristics is determined by the measurement of the contact angle. It is considered hydrophilic when the contact angle is below 90°. Protein adsorption, cell adhesion, and hemo-compatibility are influenced by the hydrophilicity of the surface. For example, if a surface is hydrophilic, there is decreased interfacial free energy, resulting in lower protein adsorption and cell adhesion and better hemo-compatibility. This may be achieved by varied techniques that may help the device become more suitable. Surface energy and wettability play a critical role in thrombosis. Adhesion and activation of platelets to the surface is regulated by the surface energy and wettability when material and blood are in contact.

Methods widely used in surface characterization include contact angle measurements, IR Spectroscopy Atomic Force Microscopy (AFM), Scanning Electron Microscopy (SEM), X-ray Photoelectron Spectroscopy (XPS), and Energy-Dispersive X-ray Analysis (EDX).

Polymer composites determine the degradation and stability of the material. Understanding the chemical characterization, including the final product and the leachables and extractables are important in assessing biological risks. Standards that may be used in the evaluation of biological risks are mentioned in ISO 10993-17, 18, 19, and 35. Extractables and leachables of polymers can be composed of organic and inorganic substances which include lubricants, additives, accelerators, monomers and high molecular weight oligomers, and residual solvents. Medical devices often need to be radiopaque for observation of the device in vivo by radiography. Devices constructed of materials such as organic polymers usually lack sufficient radiopacity. Therefore, they are filled with heavy metal compounds in the form of crystalline inorganic salts such as barium sulfate. However, these compounds may be released and may cause cytotoxicity. Surface chemistry may also control the device performance. Additives in the polymer, for instance, may change with time resulting in the blooming phenomenon that can result in modification of the material properties and decreased degradation resistance as well as enhancement of bacterial adhesion.

Physical changes, such as color, surface appearance, strength, flexibility, surface dimensions, or weight of the polymer, are also affected by the chemicals. The reaction of the chemicals affects solvent permeability, adsorption, and also induces stress cracking.

Investigating the manufacturing process is critical and should be done carefully. Biocompatibility factors may change throughout the process even in the final steps. Thus, it is important to keep in mind that it is not guaranteed that the final product is biocompatible when using biocompatible raw materials. Every step of the manufacturing process (including molding, polymerization, fiber forming, etc.) needs to be documented. Sterilization of polymer-based implantable devices is crucial. Some methods may have few adverse effects while others may leave toxic residues. Established processes for sterilizing polymer-based medical devices include steam, ethylene oxide dry heat, or irradiation-based methods. However, these methods have the potential to modify the material properties [2].

5 In Vitro and In Vivo Tests for Assessing Biocompatibility

Safety evaluation of all devices that will be in contact with the human body is an important process to ensure patient safety. The ISO 10993-1 provides a review of existing data including test selection and additional experiments on safety. ISO 14971 "Application of risk management to medical devices" provides a thorough explanation on the assessment and risk control related to medical device use. The manufacturer must document all the materials which have been used in the device production, contact route with the tissue as well as the time duration. Data on similar devices used and production processes must also be documented for potential biological hazards. Potential hazards may be short-term (thrombosis, hemolysis, irritation, acute toxicity) or long-term (genotoxicity or carcinogenicity) effects. The tests are chosen according to the population of the patients, duration of the product (limited – less than 24 h; prolonged – up to 30 days; long-term – above 30 days) as well as the level of contact with the tissue. Devices are divided according to exposure routes, including surface devices with external contact (such as skin, mucosa, or breached/compromised surfaces), external communicating devices (which include blood, circulating blood, bone, tissue, dentin), and implant devices (including bone and tissue) [2].

Documenting quantitative information on the ingredients level, residue level, degradation products, and amount of leached material is important in estimating the potential adverse effects that might arise. Therefore, evaluating the medical devices in their final product form is preferred.

Any change throughout the process must result in re-evaluation of the biomaterial for risks, and even adding or changing evaluation tests, if necessary, to ensure that the biological performance is not changed. Both in vivo and in vitro investigations are performed in the evaluation of the final product safety [15].

Reference materials are usually used to serve as experimental controls in most biocompatibility tests. Negative controls are reference materials which may be used mostly in the form of blanks or include extraction vehicles. Blanks may be used when comparing the effects of test material extracts. Positive controls help demonstrate the suitability of the test system. For example, a high-density polyethylene may be used as negative control and organotin-stabilized polyurethane as a positive control (zinc diethyldithiocarbamate and zinc dibutyldithiocarbamate polyurethans). Polyvinyl chloride-containing organotin additives can also be used as a positive control [2].

Surface devices that are in contact with the skin include different kinds of bandages, tapes, electrodes, monitors, patches, and external prostheses. The devices should be tested for cytotoxicity, irritation as well as sensitization. This allows investigation of the risk for allergic reaction to the devices or leachables. These tests may be performed on Guinea pigs or rabbits [8].

Surface devices in contact with mucosal membranes include contact lenses, dental prostheses, orthodontic devices, bronchoscopes, endotracheal tubes, urinary catheters, intra-intestinal or intra-vaginal devices, and other drug-administrating mucoadhesive devices. These devices should be assessed for toxicity and genotoxicity as well as pyrogenic response.

Devices that are in contact with blood or circulate in blood need to undergo assessment for hemocompatibility risks, such as hemolysis (red blood cell breakdown) and thrombosis (clot formation) that may be caused by the medical devices. In addition, factors such as flow dynamics and the material, device, and blood interaction should be analyzed. The ISO 10933-4 "Selection of tests for interactions with blood" describes the required tests.

Devices that are implanted and are in contact with the tissue, bone, and fluids such as the blood, including pacemakers, endovascular stents, heart valves, and hemodialysis membranes, need to be assessed for implantation reactions (implantation-induced or system toxicity and hemocompatibility) and in some cases for carcinogenicity. The evaluation begins from the initial phase of implantation and the direct blood contact. Evaluation of the device after it is explanted is also required to investigate the presence of tissue response such as inflammation, fibrosis, necrosis, or degradation products. It is noteworthy to mention that the blood reactivity of the animal species differs from that of human blood. Therefore, data need to be carefully interpreted. Studies on permanent contact devices or devices containing a source of energy should also be tested for reproductive toxicity to evaluate teratogenicity and prenatal and postnatal development.

Once the device is implanted, the first process that occurs (usually within minutes to hours) is adsorption of proteins such as albumin, fibrinogen, fibronectin, immunoglobulin G, and von Willebrand factor. Following this step, there is cellular interaction with the protein layer, in which a layer of proteins is created. This interaction is called the Vroman effect. The interaction usually depends on both the physical and chemical properties of the materials as well as the proteins that compose the layer. Afterward, the neutrophils and macrophages interact with the device proteins, leading to an inflammatory response, which involves destructive enzymes, hydrogen peroxide, and superoxide anions. Colonization of bacteria is also determined by protein adsorption. Adsorption profile is a major factor in determining the bioactivity and the cell response. Therefore, evaluation of the composition and amount of the adsorbed protein and the degree of surface changes is an important step and may be challenging. This may be used as a rapid screening test to compare polymers in biomaterials development. Usually, an increase or decrease in surface protein adsorption depends on the clinical application. In medical devices that have contact with the circulation, for example, decreased protein adsorption is preferred. Polymer-coated devices reduce the protein adsorption. Poly (ethylene oxide) or poly (ethylene glycol) and phosphorylcholine are examples of modified polymer surfaces. On the other hand, increased protein adsorption on the surface enhances osseous integration as well as osteoblast attachment and proliferation.

Biochemical reaction induced by plasma protein adsorption on polymer surfaces may lead to thrombosis and therefore remains a challenging issue. As factor XII interacts with negatively charged surfaces it is auto-activated, thus initiating the intrinsic pathway, resulting in blood coagulation. Fibrinogen and von Willebrand factor also play an important role in platelet adhesion. To detect, identify and quantify the adsorbed proteins, different methods are available. ELISA, ultraviolet-visible spectroscopy, and conventional colorimetric methods, such as BCA, Bradford, or Lowry-based assays, may be used to quantify proteins in solution. Dynamics and kinetics of adsorbed proteins may be detected with the aid of atomic force microscopy (AFM), surface plasmon resonance (SPR), Quartz Crystal Microbalance with Dissipation (QCM-D), and Ellipsometry.

Cell culture techniques have provided a suitable evaluation index for cellular biological response. Cell lines must be chosen carefully to assess cytocompatibility. Immortalized cell lines (continuously proliferating cells) which may be derived from viruses, mutations, or neoplasms, are used in the initial cytotoxicity testing. In the next step, the cells are chosen for cytocompatibility assessment according to the interaction and according to the medical application. For example, devices that are in contact with the circulatory system need to be evaluated for interaction between endothelial cells and blood platelets. Fibroblasts, L929 cells derived from mouse fibroblast, are used in assessment of cytotoxicity of materials with skin contact. Odontoblasts, gingival fibroblasts, and periodontal ligament cells are used in dentistry. The use of human conjunctival or corneal epithelial cells is suitable in evaluating contact lenses and their care solutions.

The ISO 10993012 "Sample Preparation and Reference Materials" includes recommendations on the contact methods and sample preparations. There are 3 methods of contact tests that are suggested by the standards, including direct contact, indirect contact, and extract methods. The direct contact is the most sensitive and therefore recommended for low-density materials. On the other hand, extract methods are used more commonly for higher density devices. The indirect method and the extract dilution methods are applied to detect leachables under normal and exaggerated conditions. Different extraction mediums, temperatures, and times are chosen according to the purpose of the test, physical and chemical properties of the material, and the final product and its leachables [2].

There are various methods to measure cytotoxicity. Cell counting (viable/nonviable cells) technique using a cytometer, or an automatic cell counter, is a basic initial step in cytotoxicity screening but might be time-consuming. For example, in the clonogenic assay, the number of colonies that are growing in contact with the test sample is counted and compared with those of the control. Another method is using dyes, such as trypan blue dye or neutral red. On the other hand, biochemical-based assays may be more reliable and therefore are most widely used. Formazan-based methods may be used to assess the cell viability and proliferation, but they do not detect the mechanism of cell death. Another method used in cell viability measurement is lactate dehydrogenase (LDH) release assay. ELISA BrdU (bromodeoxyuridine) colorimetric immunoassay may be used in measuring cell proliferation based on the BrdU incorporation during DNA replication. Cell density may be determined using the sulforhodamine B (SRB) assay. Fluorescence-based assays, using dyes such as 7-aminoactinomycin D or propedium iodide, can penetrate non-viable cells and help determine the healthy and damaged cell ratio. Radioactive elements such as chromium 51 (51Cr) may also be used as a method for the quantification of cell damage. Another method uses the radioactive nucleoside, 3 H-thymidine, which is incorporated during DNA mitosis to determine the level of cell division that occurred with the test sample and then compared with controls.

Cell death is an important process that may serve as a defensive response or may be due to an unwanted reaction. Apoptosis is a programmed process that occurs as a homeostatic mechanism to maintain the cell population and may also occur as a defensive process. Characteristics of an apoptotic cell include caspase activation, nuclear fragmentation, and apoptotic body formation. Necrosis is due to cellular insult secondary to an injury or blood supply insufficiency. Different methods are available for detecting necrosis and apoptosis.

Apoptosis with high caspase activation may be detected using fluorescent probes. A common method is detection of phosphatidylserine (PS) on the cell surface by using PS-binding protein annexin V. Its detection serves as a hallmark for the early stage of apoptosis. Certain dyes such as JC-1, JC-10, or TMRE may be used to monitor the mitochondrial membrane potential to determine the cells fate. Chromatin condensation of apoptotic cells may be observed with the aid of nuclear stains and microscopy or flow cytometry. DNA fragmentation may be observed using terminal deoxynucleotidyl transferase (TUNEL) assays and DNA ladders may be investigated by electrophoresis. Other methods, such as protease biomarkers, are also used to determine cell viability.

Genotoxicity tests are performed to detect genetic damage that might be caused by the biomaterial. This is a mandatory step for devices with a contact surface more than 30 days and for devices that will be implanted for more than 24 h. OECD test guidelines provide all the recommended tests needed for chemical safety. However, the American and the European guidelines recommend the performance of three in vitro assays in order to detect DNA damage, mutations, and chromosomal abnormalities.

Oxidative stress, an imbalance between reactive oxygen species (ROS) production and a cell's antioxidant mechanism, may lead to cell damage and cell death by necrosis or apoptosis. Therefore, investigations of oxidative stress include go/no go tests in which the total amount of ROS is measured. ROS can also be detected using electron spin resonance or using methods based on transformation of radicals into stable molecules. 2,7-dichlorodihydrofluorescein diacetate (DCFH-DA) is commonly used. Various cellular enzymes play a role in the antioxidative mechanism to scavenge free radicals, and their failure to do so may lead to apoptosis. Thus, measurement of the enzyme levels may be performed using Ellman's Reagent (5,5'-dithio-bis-(2-nitrobenzoic acid) (DTNB) and glutathione reductase (GR), as well as the thioreactive fluorescent dye 5-chloromethylfluorescein diacetate (CMFDA). Further investigations on gene and protein regulations such as NFkB, COX-2, Egr-1, JNK, iNOS, c-jun, c-fos, and c-myc in oxidative stress may also be performed.

The host inflammatory response is an adaptive response triggered by a physiologic or pathological mechanism. This inflammatory reaction is referred to as FBR in implanted biomaterials. It is a crucial process in determining the implantation response and in distinguishing whether the inflammatory response is an adverse event or an adaptive response. Initial response begins with an acute inflammation in which leukocytes are activated and are attracted to the perivascular tissue and the implantation site. During the chronic inflammatory phase, the cell types that predominate are macrophages and lymphocytes. As part of the wound healing process, development of a granulation tissue may be observed. In prolonged inflammation, serious adverse effects such as aseptic loosening or osteolysis may develop and thus it is important to evaluate the biological response to implanted devices. The inflammatory activity and relevant inflammatory cytokines and factors which are secreted, including TNF-α, IL-1β, IL-6, MCP-1, MIP-1α, IL-2, VEGF, IL-4, and IL-10, should be assessed. ELISA (Enzyme-Linked Immunosorbent Assay) is a method used for quantification of proteins but may measure one cytokine per assay. More improved methods such as cytometric bead array (CBA), a flow cytometry-based method, can quantify several cytokines. Immunohistochemistry is used to detect inflammatory cells and the protein distribution around the device.

A surface with minimal platelet adhesion and activation with reduced thrombogenic potential as well as inflammation is needed to provide an ideal biocompatible environment with a thrombosis-resistant material device. To fulfill these expectations, ISO 10993-4 provides standards for hemocompatibility testing which are classified into 5 different categories, including thrombosis, coagulation, platelets, hematology, and immunology. For devices that contact blood, evaluation of hemocompatibility using in vitro models, such as human blood, may be adequate. Surface charge, energy, and topography of a biomedical device influence thrombogenicity. Most of the methods are performed in vivo to evaluate the flow reduction, occlusion percentage, and mass of the thrombus.

Scanning electron microscopy (SEM) is used to measure platelet adhesion and aggregation. Fibrin and platelet activation may be determined using specific antibodies. Measuring and interpreting platelet-leukocyte aggregates with the aid of flow cytometry is an additional method. Coagulation can also be assessed by measuring the hemoglobin within the erythrocytes. The prothrombin time (PT) assay and the partial thromboplastin time (PTT) assay are additional tests used in measuring the time it takes for blood to coagulate. Measuring fibrin and fibrinogen degradation product concentrations may also be evaluated since increased levels imply increased fibrinolysis which can be seen in thrombotic states. Production of kallikrein and factor XII once it is in contact with blood is another parameter that may be assessed in blood coagulation. Additional tests include ELISA as well as radioimmune detection of specific coagulation factors.

P-selectin, which is a membrane glycoprotein, is present in the α-granules of the platelets. Detection of membrane P-selectin represents activated platelets on the surface of a device. P-selectin detection may be performed with the aid of ELISA, flow cytometry, immune fluorescence, or by scanning electronic microscope. Atomic force microscopy (AFM) is another recently suggested method for the imaging of activated platelets. An additional method uses agents that can stimulate aggregation (such as thrombin and collagen) from platelet-rich plasma to detect prolonged aggregation time resulting from contact with the tested sample. Leukocytes are also activated once there is an inflammation, thus their activated state may be determined by detecting expression of molecules, such as L-selectin, on the leucocyte surface. Another significant factor for screening is hemolysis caused by the interaction of erythrocytes with the material. Hemolysis index (HI) is a technique used to predict hemolysis risk by calculating hemoglobin release from erythrocytes.

The complement system is made up of plasma proteins that interact with pathogens. Complement activation evaluation may be used to determine the extent of the activation induced by the device material. The 50% hemolytic complement activity of serum (CH50) test may be performed by using lysed sheep red blood cells (SRBC) pre-coated with rabbit anti-sheep red blood cell antibody (hemolysin) and then incubated with test serum resulting in complement activation and hemolysis. Endothelial cells also produce various molecules during the coagulation process and therefore may also be studied. Nitric oxide, which prevents platelet aggregation and vascular smooth muscle cell proliferation, may be evaluated. Tissue factor (TF) and thrombomodulin (TM), also expressed by the endothelial cells, may be analyzed using ELISA. High TM and low TF may indicate a non-thrombogenic state. von Willebrand Factor (vWF), prostacylin (PGI2), tPA/PAI-1, and cell adhesion molecules, such as VCAM-1, ICAM-1, PECAM, and E-selectin, which are also involved in the coagulation cascade, may be evaluated.

Immunogenicity risk assessment is critical, since unwanted immune response may be induced by the biomaterial. Guidelines on methods for testing immunotoxicity are found in ISO 10993-20156. In vitro assessment of potential skin irritation may be analyzed using models such as SkinEthic™ RHE, EpiSkin®, and modified EpiDerm SIT® according to ISO 10993-10157. Bovine Corneal Opacity and Permeability (BCOP), Isolate Chicken Eye (ICE), Cytosensor Microphysiometer, and Fluorescein Leakage are methods to determine ocular irritation. Chemical immunotoxicity may be assessed through various useful in vitro assays, such as predicting contact allergens by measuring keratinocyte or dendritic cells inflammatory response or assessment of chemical-induced specific gene expressions with the help of chip technology [2, 8].

In vivo tests are performed to assess the biocompatibility of the implanted materials using animal models. Immunohistochemistry is used to detect inflammatory cells and the protein distribution around the device. Macrophages are the key cells in an immune response. Activated macrophages are divided into M1 and M2 macrophages, among which M1 macrophages are mainly involved in the pro-inflammatory response and M2 macrophages are associated with anti-inflammatory response. There is a switch from M1 to M2 macrophages, in which transforming growth factor-β1 is released and stimulate fibroblasts and myofibroblasts leading to FBGCs and fibrosis formation as part of the wound-healing process [16].

ISO10993 includes guidelines on the evaluation of biocompatibility in animal models. ISO 10993-6 (2016) also includes a scoring system that may define the rate of reactivity and irritability of the inflammatory response based on cell prevalence, density, neovascularization, fibrosis, and few other features. Interpretation of the irritancy/reactivity score may be challenging and may affect the ranking of the true biological effect of the product.

The mouse-derived J774A.1 and RAW264.7 macrophage cell lines are mostly used in investigating inflammation in biomaterials [15]. In one study, J774A.1 macrophage cell lines were incubated in poly (lactic-co-glycolic acid (PLGA) microparticles. As a result, increased interleukin (IL)-1β and tumor necrosis factor (TNF)-α protein levels were observed. THP-1, human leukemia monocytic cell line, is also used in detecting monocyte/macrophage activity.

A recent study was performed using subcutaneously injected poly-L-lactide (PLLA) on humans. CD68+ macrophages (regarded as M1 macrophage markers) were surrounding the PLLA microparticles immediately. Collagen synthesis was not increased immediately but within 6–8 weeks of the injection. Collagen III was surrounding the PLLA particles while collagen I was only found in the periphery of the granulomatous reactions. Similarly, other studies used subcutaneously implanted PLLA in the back of rats and in WAG rats, revealing encapsulation, consisting of macrophages as part of the foreign body granulomatous reaction. Inflammatory response intensity was decreased throughout the 6 months while an increase was noticed 12 months later due to the PLLA fragmentation. Formation of such response was suggested to be induced by sharp-edged materials that cause inflammatory response in the surrounding tissue.

A study on implanted polymer (copolymer of poly-L-lactide-co-ecaprolactone in a 70:30 ratio) devices for rotator cuff tear repair was performed on Sprague-Dawley rats. Although the test showed favorable tolerability results, one of the animal models developed fibrosarcoma at the implantation site. It was concluded that the induction of the fibrosarcoma was related to the rodent-predilection response [8, 10, 13].

6 The Role of the Toxicologic Pathologist in Medical Device Evaluation

When there is a change in the physiology, morphology, growth, development, reproduction, or in the life expectancy of a material or system influencing negatively on its performance or its capacity to react to stress, it is defined as adverse. Thus, the identification of an adverse effect has a major role in assessing biocompatibility. Certain criteria should be followed in order for an effect to be determined as less adverse: (1) no changes in the device function or its surrounding tissue, (2) the response is adaptive, (3) the effect is transient, (4) the severity is limited, (5) changes are limited or independent and do not affect other parameters, (6) the effect is not a precursor or part of a progressive change, (7) it is a secondary effect of another adverse effect, (8) a consequence of the experimental model [3].

In short-term studies, in which the performance of the device is being evaluated, histopathology may not be necessary. In studies that involve similar and repetitive device assessment with predictable findings and complications, there are ready protocols and therefore well-trained staff that may interpret the observations during a necropsy without the need of a pathologist [5]. However, in long-term studies performed to determine the safety, efficacy, biological effect or when an implantable device is expected to result in alteration of tissue, an experienced toxicologic pathologist is preferred to evaluate the histopathology of the changes. The goal of the pathologist is to interpret the toxicological findings as "adverse" or "nonadverse".

Upon histopathological evaluation, the Board Certified pathologist is expected to include a conclusive statement concerning the potential adverse or non-adverse effects of each treatment-related lesion. The adversity judgment should be based on the criteria described in position papers published by the Society of Toxicologic Pathology (STP) and the European Society of Toxicologic Pathology (ESTP) [3, 7].

This statement is expected to refer to the animal species used and the experimental conditions specific for a study and will help to determine the No Observed Adverse Effect Level (NOAEL) (and Pass/Fail, in case of need). Parameters which may be taken into consideration for the determination of adversity include the presence of ulceration, necrosis, mineralization and thrombosis, and potential recovery, if this phase is included in the study design [1]. In particular, the severity grade and extension of such potential adverse lesions will be considered [14]. Lesions that are focal and of minor grade (up to grade 2 of 4) will potentially be considered as not adverse. However, extensive lesions, and of higher grade than 2, may be considered as adverse. In any case, the determination of adversity should always be case-by-case, and the rationale for the suggested adversity should be justified with appropriate references. Examples of histopathological determination of adversity are shown by [4, 9, 11].

The adverse pathology findings should not be considered as an adverse effect independently but should also consider the clinical adverse effects and outcomes [7]. Additional imaging techniques such as micro-computed tomography (micro CT; 3D) or orthogonal (2D) should be considered for more accurate interpretation and also to support the results documented by the pathologist [6].

7 New Developments and the Future of Biocompatibility

The field of biomaterials continues to expand with increased developments of promising materials that might help improve performance of an implanted device. Precise in vivo and in vitro studies should be performed in order to evaluate the biomaterial biocompatibility. One of the major concerns is the healing of some implants where a fibrous capsule may be formed thus hindering the clinical application of the device. Insufficient vascularity near the implant-tissue interface is usually associated with poor outcomes. Therefore, vascularized tissue which is similar to normal tissue structure may be used. The use of extracellular matrix (ECM)-based materials (prepared by the decellularization of mammalian tissue) has been realized to be effective in repair and reconstruction of tissues. Decellularized ECM derived from small intestinal submucosa (SIS), for example, has been found to heal with minimal fibrosis and excellent vascularity. This is believed to be due to the degradation of the ECM into bioactive peptides by the macrophages. Neonatal cell culture-derived soluble ECMs have also been suggested. Biomaterials with porous features (having interconnecting open spaces) have shown to also heal with increased vascularization and reduced fibrotic tissue. A method was developed to make the size of the pores in the range of 30–40 microns and angiogenic healing with mild fibrosis was observed. It has been demonstrated that implanted spheres, 1.5 mm and above, heal with significantly higher FBR when compared to smaller spheres. Triazole-containing molecules have also been suggested to inhibit FBR. Another study suggested nonfouling zwitterionic hydrogels being resistant to FBR [7].

8 Conclusion

The protein and biomaterials' surface interaction is very important when designing a biocompatible material since the composition of the chemical plays a critical role in the interaction. A desirable biologic response with minimal adverse effects remains the main target in producing a biocompatible material. Therefore, the use of standardized tests helps us better understand the material behavior and its safety in relation to the tissues.

Quiz

1. The biomaterial is considered biocompatible when:

    (a) The response is aggressive.
    (b) There are multinucleated foreign-body giant cells present.
    (c) The response is mild and resolving (i.e., recovery) spontaneously.
    (d) There is formation of thick fibrous encapsulation surrounding the implant.
    Answer: c. The biomaterial is considered biocompatible when a response is described as a mild inflammatory reaction (between 3 and 6 weeks), resolving (i.e., recovery) spontaneously and leading to thin fibrous encapsulation that isolates the implant from the tissue.

2. The multi-step approach of evaluating the biocompatibility of medical devices and implantable drug delivery systems includes the following steps:

    (a) Initial safety evaluation, material screening, product analysis, and finally product testing.
    (b) Initial material screening, preclinical safety evaluation, product testing, and finally product analysis.
    (c) Initial material screening, product testing, product analysis, and finally safety evaluation.
    (d) Initial product testing, material screening, safety evaluation, and finally product analysis.

    Answer: b. The multi-step approach of evaluating the biocompatibility of medical devices and implantable drug delivery systems begins with initial material screening, safety evaluation, product testing, and finally product analysis.

3. Good Laboratory Practice (GLP) regulations apply to:

    (a) Studies regarding choosing technicians suitable to work in the laboratory.
    (b) Studies regarding laboratory products.
    (c) Studies regarding preclinical safety evaluation.
    (d) Studies regarding laboratory techniques.

    Answer: c. All biocompatibility testings should comply with Good Laboratory Practice (GLP) regulations which apply to the preclinical studies regarding safety. The GLP assures that the studies are conducted in accordance with regulatory submissions.

Biocompatibility of Polymers

4. Which of the following is *not* among the important factors that should be investigated when evaluating the effect of the drug on the device and its interaction within the biological environment:

   (a) Hemocompatibility
   (b) Hepatocompatibility
   (c) Cytotoxicity
   (d) Irritation

   Answer: b. The effect of the drug on the device and its interaction within the biological environment should be investigated for cytotoxicity, irritation, and hemocompatibility.

5. Cell counting (viable/non-viable cells) technique is a method used to measure:

   (a) Carcinogenicity screening.
   (b) Cytotoxicity screening.
   (c) Hemocompatibility.
   (d) Cell death.

   Answer: b. Cell counting (viable/non-viable cells) technique using a cytometer, or an automatic cell counter, is a basic initial step in cytotoxicity screening.

6. Which of the following statements about cell death is false?

   (a) Caspase activation, nuclear fragmentation, and apoptotic body formation are characteristics of necrotic cells.
   (b) Necrosis is due to cellular insult secondary to an injury or blood supply insufficiency.
   (c) Apoptosis is a programmed process that occurs as a homeostatic mechanism to maintain the cell population.
   (d) Apoptosis occurs as a defensive process.

   Answer: a. Characteristics of an apoptotic cell include caspase activation, nuclear fragmentation, and apoptotic body formation.

7. Which of the following factors do not provide a biocompatible environment:

   (a) Minimal platelet adhesion.
   (b) Increased thrombogenic potential.
   (c) Reduced inflammatory response.
   (d) Minimal platelet activation.

   Answer: b. A surface with minimal platelet adhesion and activation with reduced thrombogenic potential as well as inflammation is needed to provide an ideal biocompatible environment with a thrombosis-resistant material device.

8. Which macrophage cell lines are mostly used in the assessment of biomaterial inflammation?

   (a) Human macrophage cell line.
   (b) Mouse-derived J774A.1 and RAW264.7 macrophage cell line.
   (c) Fish macrophage cell line.
   (d) C57BL/6 Mouse Bone Marrow Macrophage cell lines.

Answer: b. The mouse-derived J774A.1 and RAW264.7 macrophage cell lines are mostly used in investigating inflammation in biomaterials.

9. When evaluating adverse effects, all the following criteria are considered to be less adverse except?

   (a) No changes in the device function or its surrounding tissue.
   (b) The response is adaptive.
   (c) The changes are unlimited and might affect other parameters.
   (d) The effect is not a precursor or part of a progressive change.

   Answer: c. Certain criteria should be followed in order for an effect to be determined as less adverse: (1) no changes in the device function or its surrounding tissue, (2) the response is adaptive, (3) the effect is transient, (4) the severity is limited, (5) changes are limited or independent and do not affect other parameters, (6) the effect is not a precursor or part of a progressive change, (7) it is a secondary effect of another adverse effect, (8) a consequence of the experimental model.

10. Which of the following is not considered a parameter when determining adversity upon histopathological evaluation of preclinical studies?

    (a) Necrosis
    (b) Ulceration
    (c) Thrombosis
    (d) Fibrosis

    Answer: d. Parameters which may be taken into consideration for the determination of adversity include the presence of ulceration, necrosis, mineralization and thrombosis, and potential recovery.

References

1. Baldrick P, Cosenza ME, Alapatt T, Bolon B, Rhodes M, Waterson I (2020) Toxicology paradise: sorting out adverse and non-adverse findings in animal toxicity studies. Int J Toxicol 39(5):365–378. https://doi.org/10.1177/1091581820935089
2. Bernard M, Jubeli E, Pungente MD, Yagoubi N (2018) Biocompatibility of polymer-based biomaterials and medical devices-regulations: in vitro screening and risk-management. Biomater Sci 6(8):2025–2053. https://doi.org/10.1039/c8bm00518d
3. Kerlin R, Bolon B, Burkhardt J, Francke S, Greaves P, Meador V, Popp J (2016) Scientific and Regulatory Policy Committee: recommended ("best") practices for determining, communicating, and using adverse effect data from nonclinical studies. Toxicol Pathol 44(2):147–162. https://doi.org/10.1177/0192623315623265
4. Litvin G, Klein I, Litvin Y, Klaiman G, Nyska A (2021) CorNeat KPro: ocular implantation study in rabbits. Cornea 40(9):1165–1174. https://doi.org/10.1097/ICO.0000000000002798
5. Nikula KJ, Funk K (2016) Regulatory forum opinion piece: an experienced pathologist should be present at necropsy for novel medical device studies. Toxicol Pathol 44(1):9–11. https://doi.org/10.1177/0192623315617035

6. O'Brien MT, Schuh JACL, Wancket LM, Cramer SD, Funk KA, Jackson ND, Kannan K, Keane K, Nyska A, Rousselle SD, Schucker A, Thomas VS, Tunev S (2022) Scientific and Regulatory Policy Committee points to consider for medical device implant site evaluation in nonclinical studies. Toxicol Pathol 50(4):512–530. https://doi.org/10.1177/01926233221103202
7. Palazzi X, Burkhardt JE, Caplain H, Dellarco V, Fant P, Foster JR, Francke S, Germann P, Gröters S, Harada T, Harleman J, Inui K, Kaufmann W, Lenz B, Nagai H, Pohlmeyer-Esch G, Schulte A, Skydsgaard M, Tomlinson L et al (2016) Characterizing "adversity" of pathology findings in nonclinical toxicity studies: results from the 4th ESTP international expert workshop. Toxicol Pathol 44(6):810–824. https://doi.org/10.1177/0192623316642527
8. Ramot Y, Haim-Zada M, Domb AJ, Nyska A (2016) Biocompatibility and safety of PLA and its copolymers. Adv Drug Deliv Rev 107:153–162. https://doi.org/10.1016/j.addr.2016.03.012
9. Ramot Y, Nedvetzki S, Rosenfeld S, Rousselle SD, Nyska A, Emanuel N (2020) D-PLEX100 in an abdominal surgery incision model in miniature swine: safety study. Toxicol Pathol 48(5):677–685. https://doi.org/10.1177/0192623320928902
10. Ramot Y, Nyska A, Markovitz E, Dekel A, Klaiman G, Zada MH, Domb AJ, Maronpot RR (2015) Long-term local and systemic safety of poly(l-lactide-co-epsilon-caprolactone) after subcutaneous and intra-articular implantation in rats. Toxicol Pathol 43(8):1127–1140. https://doi.org/10.1177/0192623315600275
11. Ramot Y, Steiner M, Lavie Y, Ezov N, Laub O, Cohen E, Schwartz Y, Nyska A (2021) Safety and efficacy of sFilm-FS, a novel biodegradable fibrin sealant göttingen minipigs. J Toxicol Pathol 34(4):319–330. https://doi.org/10.1293/tox.2021-0030
12. Ratner BD, Schoen FJ (2020) The concept and assessment of biocompatibility. In: Biomaterials science, 4th edn. Elsevier, New York. https://doi.org/10.1016/b978-0-12-816137-1.00056-8
13. Rousselle SD, Ramot Y, Nyska A, Jackson ND (2019) Pathology of bioabsorbable implants in preclinical studies. Toxicol Pathol 47(3):358–378. https://doi.org/10.1177/0192623318816681
14. Schafer KA, Eighmy J, Fikes JD, Halpern WG, Hukkanen RR, Long GG, Meseck EK, Patrick DJ, Thibodeau MS, Wood CE, Francke S (2018) Use of severity grades to characterize histopathologic changes. Toxicol Pathol 46(3):256–265. https://doi.org/10.1177/0192623318761348
15. Schmalz G, Galler KM (2017) Biocompatibility of biomaterials – lessons learned and considerations for the design of novel materials. Dent Mater 33(4):382–393. https://doi.org/10.1016/j.dental.2017.01.011
16. Williams DF (2017) A paradigm for the evaluation of tissue-engineering biomaterials and templates. Tissue Eng C Methods 23(12):926–937. https://doi.org/10.1089/ten.tec.2017.0181

Sterilization Techniques of Biomaterials (Implants and Medical Devices)

Chau Chun Beh

Abstract Biomaterials comprise natural and synthetic components including polymers, tissues, living cells, metals, ceramics, etc. They are used to replace or repair the living tissues and organs that are malfunctioning. Hence, it is essential to ensure the biomaterials are safe and sterilized before being used in the body. Particularly, we learned that sterilization is crucial from the most recent pandemic caused by the COVID-19 virus. The most commonly used sterilization methods include steam-autoclaving, dry-heat, radiation processes (gamma, X-rays, electron beam, and ultraviolet), gas plasma, and ethylene oxide treatment. Supercritical carbon dioxide has been investigated extensively as an alternative to conventional sterilization methods in recent years. The sterilization techniques with their advantages and disadvantages will be detailed in this chapter.Graphical Abstract

---

C. C. Beh (✉)
Western Australian School of Mines: Minerals, Energy & Chemical Engineering, Curtin University, Perth, WA, Australia
e-mail: jane.beh@curtin.edu.au

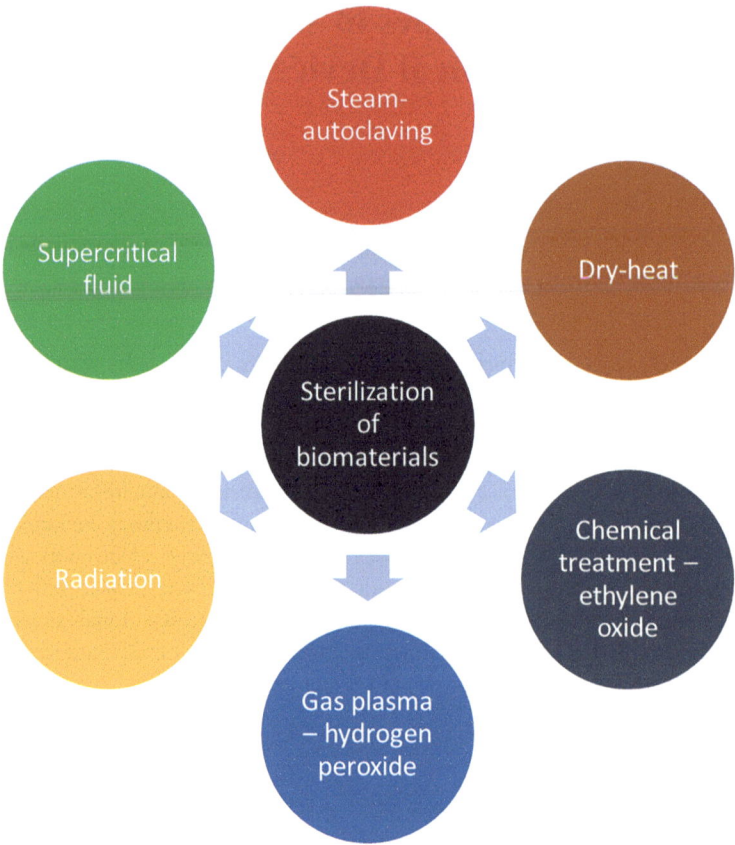

Keywords Sterilization · Steam-autoclaving · Radiation · Supercritical fluid · Ethylene oxide

1 Introduction

Biomaterials find useful applications in implants and medical devices. Biomaterials are composed of natural or synthetic materials such as polymers, metals, ceramics, living cells, tissues, and many more. They are generally safe in the body and biodegradable, and some may gradually be eliminated from the body after releasing the targeted drugs or achieving their functionality. For instance, medical implants, regenerated tissues, nanoparticles, biosensors, and drug delivery systems are typical biomaterials. Biomaterials are used to repair, assist, or replace living tissues and organs that are not functioning at an acceptable level.

Sterilization of biomaterials is crucial as these materials have to be safe before being used or integrated into the body. The risk of infection caused by the use of unsterilized biomaterials can be as severe as leading to complications and death. It is essential to understand the pros and cons of various types of sterilization as there is no one sterilization process that is suitable for all materials. In fact, some processes may damage the functionality of the biomaterials. Assurance of sterility is significant in good manufacturing practices and in patient safety. The acceptable sterility assurance level (SAL) is one out of a million or 10^{-6}. Generally, most sterilization processes have been designed to achieve a minimum SAL of 10^{-6}.

Sterility assurance level (SAL) is a unit of measurement to estimate the sterility level of the substance. Sterility assurance level is a probability of a non-sterile unit surviving sterilization. The acceptable SAL is 10^{-6} or 1 in 1,000,000 likelihood of an organism surviving the sterilization process. Sometimes, log reduction is used as a measurement to express the amount or percentage of living microorganisms eliminated after sterilization. For instance, a 2-log reduction indicates that the number of microorganisms has been reduced by 10^{-2} or 100-fold. The sterility assurance level can be calculated by doubling the log reduction number, which indicates the amount of microorganisms eliminated. For example, a 4-log reduction is translated to a SAL of 10^{-2} and a 12-log reduction is translated to a SAL of 10^{-6}. Hence, SAL is not equal to the log reduction.

2 Sterilization Techniques

All types of biomaterials that are used within the body or contacted with corporeal fluids are sterilized to avoid the risk of introducing harmful microorganisms into the body. Sterilization refers to the inactivation or elimination of all living microorganisms including bacteria, spores, fungi, and viruses. A sterilization process alters the structure of the macromolecules in the pathogenic microorganisms that result in their death and halting their reproduction process. For instance, ethylene oxide is useful in replacing hydrogen atoms on molecules that need to maintain life and halting life-supporting functionality. Nevertheless, there is no single sterilization process that is able to sterilize all types of medical devices or biomaterials. Various sterilization processes may lead to various adverse effects on the biomaterials including changes in physical, mechanical, and biological properties.

Several sterilization methods such as heat, steam-autoclaving, irradiation (gamma-rays, X-rays, electron beam, and ultraviolet), gas plasma, and chemical (ethylene oxide) treatment are commonly used for sterilizing biomaterials. Recently, sterilization by supercritical fluid technology (carbon dioxide) has been studied extensively. The efficiency of sterilization can be measured by sterility assurance level. The benefits and limitations of various sterilization methods have been summarized in Table 1.

Table 1 The benefits and limitations of various sterilization methods

Sterilization method	Benefits	Limitations
Steam-autoclaving	Cost-effective Short processing duration No toxic residues Safe for the environmental	Not for electric device components, degradable polymers, oil-based materials, and biomaterials that are sensitive to heat and moisture
Dry-heat sterilization	Simple No toxic residues High penetration Suitable for oil-based materials	Cause oxidation of polymers like polyamide and nylon Cause degradation of polymers, compromised thermomechanical properties, and change in drug release profile
Ethylene oxide treatment	Low temperature Effective	Toxic residues in products Carcinogenic Flammable Explosive Affect structural properties of biopolymers
Gas plasma	Low temperature Fast Less toxic than ethylene oxide Eliminate highly resistant spores	Reactive species left in biomaterials Affect chemical and mechanical properties of biopolymers
Gamma irradiation	High penetration Low temperature Effective	May cause degradation or cross-link of polymers (change in structural properties)
X-rays	Higher penetration than gamma rays and electron beam Clean process No toxic residues No thermal damage to biopolymers Low temperature	Extra shielding required Not for continuous bulk sterilization
Electron beam	Tuneable penetration depth (reduce risk of harming biological components) Short treatment time	Low penetration Limited to low density and small substances Dose is less uniform than gamma irradiation
Ultraviolet irradiation	Effective for vegetative bacteria Low cost Low temperature	Not effective for bacterial spores, prions, and numerous viruses Affect molecular weight and tensile strength of biopolymers
Supercritical fluid (CO_2)	No toxic residues No chemical reagents Low temperature	Mild to high operating pressure May affect the morphology and porosity of biopolymers

Information extracted from Ref. [1]

2.1 Steam-Autoclaving

Steam-autoclaving or moist-heat sterilization is a simple, easy-to-apply, fast, safe, non-toxic, and inexpensive method. It is the most commonly used technique in hospitals and healthcare facilities for sterilizing heat-resistant surgical equipment and intravenous fluids. The essential metabolic and structural components of microorganisms are destroyed in the heated steam environment. The process involves relatively less expensive equipment using saturated steam. In this method, biomaterials are placed in an autoclave and exposed to saturated steam under high temperatures of 121–134 °C for a few minutes to 20 min, depending on the heating profile and applied pressure [1]. This process causes the destruction of microorganisms by irreversible denaturation of both enzyme and structural protein components. Steam-autoclaving has been used for biomedical polymers such as polyvinyl chloride, polyurethanes, and polytetrafluoroethylene for several applications including hematic circuits, catheters, and vascular prostheses.

On the contrary, steam-autoclaving is not suitable for biomaterials that are made of degradable biomedical polymers and oil-based materials. Steam-autoclaving is also not to be used when electric device components are involved. Steam-autoclaving is restricted by the dimensional stability, glass transition temperature of polymers, and hydrolytic resistance when multiple sterilization cycles are needed. Hence, steam-autoclaving is suitable for polymers such as polyvinyl chloride, polyacetals, low-density polyethylene, and polyamides. Increased moisture and pressure can cause a significant change in the mechanical properties of biopolymers due to hydrolysis and degradation. Degraded polymers from autoclaving can cause low drug loading or a higher drug release rate.

2.2 Dry-Heat Sterilization

Dry-heat sterilization is one of the earliest sterilization methods due to its simplicity and feasibility. The sterilization process uses an oven with a high temperature ranging from 150 to 170 °C for a duration of 60 to 150 min [2]. The differences between dry-heat sterilization and steam-autoclaving are the temperature range and the duration of the process. Dry-heat sterilization uses higher temperatures and longer exposure times to kill microorganisms and bacterial spores, as compared to steam-autoclaving.

During the sterilization, the heat is first absorbed into the exterior surface of the material, and then passed inward to the next layer until the sterilization temperature is reached. Dry-heat is suitable for oil-based materials, biomaterials with complex designs, and biomedical devices with closed cavities. The sterilization eliminates microorganisms by deep dehydration and protein denaturation. Dry-heat method is used mainly for thermo-resistance materials such as glass and steel, but not for biopolymers. Dry heat sterilization is suitable for temperature-resistant polymers such as polytetrafluoroethylene (PTFE) and silicone rubber. However, polymers like polyamide and nylon could undergo oxidation regardless of the drying temperature being less than their melting temperatures. Polymers that melt or soften during the procedure will degrade and lead to compromised thermomechanical properties and changes in the drug release profile of the materials.

2.3 Chemical Treatment – Ethylene Oxide

Ethylene oxide (EtO) is a colorless, flammable, and commonly used gas in the chemical sterilization process. The chemical sterilization by EtO is operated at a low temperature, hence, it is favorable to thermally labile, radiation- and moisture-sensitive biomaterials. The EtO technique involves diffusion and permeation of EtO gas. The procedure comprises gas removal by creating a vacuum in the sterilization vessel, humidification, EtO exposure, and air washes. The inactivation of microorganisms by EtO is achieved by the chemical alkylation following interaction with EtO gas. The alkylation of carboxylic, hydroxylic, and sulfhydrilic components in nucleic acids results in the change of morphology of proteins and subsequently, cell death. The chemical sterilization by EtO is conducted in two stages: (i) vacuum is applied to the autoclave, and (ii) EtO gas is introduced at a concentration of 600–1200 mg/L. The process parameters such as humidity, operating temperature, and sterilization cycle duration are controlled within 40–90%, 30–50 °C, and 2–8 h, respectively [1, 2].

The chemical sterilization process using EtO is an effective, flexible, and robust method for eliminating microorganisms since EtO is compatible with a wide range of biomaterials, especially medical devices that contain electronic components. However, biopolymers may not be suitable for EtO sterilization as EtO may affect the molecular weight, mechanical properties, degradation rate, and surface chemistry. The toxicity of EtO may cause deterioration of polymers regardless of low operating temperature. The polymer properties can be significantly affected in terms of molecular weight and glass transition temperature, in particular polymers that contain COOH functional group and polyethylene (PEG). Significant degradation was observed on PEG-based bioresorbable biopolymers where EtO esterified the carboxylic acid groups in desaminotyrosol-tyrosine. Therefore, chemical composition is an important factor when selecting a sterilization method. In addition, EtO-sterilized polymer-based biomaterials with larger pores or finer structures are less stable than non-porous samples.

The EtO technique requires comprehensive process control and monitoring to remove residuals on the sterilized substances without any leaks of flammable, explosive, and carcinogenic gas during the process. The toxic residues from EtO may be left on the surface or within the biomaterials, which will result in extreme skin irritation and other medical complications if incorporated into the body. Furthermore, the residual EtO contents (EtO and its secondary products including ethylene chlorohydrin and ethylene glycol) left in the polymers have mutagenic, carcinogenic, and allergenic effects.

In order to decrease the toxic effects and flammability of EtO, inert gases such as fluorinated hydrocarbons and carbon dioxide are mixed with EtO during the sterilization process. It is crucial to allow the residual gas to dissipate from the biomaterials to an acceptable level after the sterilization process. Hence, a long vacuum aeration time or repeated air washings is necessary to remove all EtO gas toxic residues.

2.4 Gas Plasma – Hydrogen Peroxide

Gas plasma sterilization was established as an alternative to the EtO sterilization method because the gas plasma method uses less toxic materials than EtO processing and can be more cost-effective than irradiation sterilization. The gas plasma sterilization process uses a combination of hydrogen peroxide (H_2O_2) vapor and low-temperature gas plasma ranging between 25 and 70 °C [3]. Gas plasma sterilization is useful for thermally labile biomaterials. Generally, reactive gas mixtures with high oxygen contents can be used for gas plasma sterilization to increase sterilization efficiency. Reactive gas plasma is effective to eliminate highly resistant microorganisms such as spores.

In gas plasma sterilization, biomaterials are placed in a chamber. The moisture content within the chamber is removed under a vacuum. The chamber is then sealed with a set pressure and 60% of H_2O_2 is vaporized into the chamber [1]. The sterilization of microorganisms occurs as the vapor diffuses throughout the chamber. Plasma is produced as H_2O_2 vapor is converted into reactive and biocidal free radicals by an electric field or a magnetic field. Gas plasma sterilization has been demonstrated to inactivate microorganisms such as vegetative bacteria, mycobacteria, spores, bacterial endospores, fungi, yeasts, and viruses. However, gas plasma sterilization is not suitable for materials with long lumens since the penetration of gas is difficult, as well as porous materials as the materials may absorb H_2O_2 before the gas is converted into active plasma form.

2.5 Radiation Process

Inactivation of microorganisms such as bacteria, fungi, and viruses can be achieved by ionizing radiation such as gamma (Υ), X-ray and electron beam, as well as non-ionizing radiation like ultraviolet radiation (UV). Sterilization by ionizing radiation uses high-intensity and short-wavelength radiation to generate harmful effects for inhibiting cell division of microorganisms and genetic damage. On the contrary, sterilization by non-ionizing radiation uses long-wavelength and low-energy radiation. Hence, non-ionizing radiation is suitable for sterilizing the surfaces of materials as it does not penetrate substances. The sterilization process by radiation can be completed by direct ionization or indirect reactions of the free radicals generated during the procedure.

The radiation sterilization process is beneficial for thermally labile biomaterials with a porous structure. The penetration and intensity of the high-energy radiation can be controlled with precise and uniform dosage distribution. Particularly, the radiation process is useful for the synthesis and modification of polymers for biomedical applications without involving toxic additives. Several synthesis processes of polymers can make use of radiation treatment when the generated free radicals are chemically active and used during the modification procedures including crosslinking of polymers and polymerization.

Gamma (Υ) irradiation is the most popular sterilization technique among the radiation processes. The energy sources used for Υ-irradiation are cobalt 60 (Co-60) or cesium 137 (Cs-137) [4]. Co-60 is a non-flammable, insoluble, non-dispersible, and non-fissionable metal, hence, it is the commonly used energy source. Co-60 has a half-life of 5.3 years with a decay process of emitting electrons and Υ-rays to convert into a non-radioactive nickel-60 (Ni-60). The energy of Υ-rays is relatively low and does not cause radioactive effects on the sterilized substances. The dose of radiation is determined by the density and size of the substances, temperature, and water content. A single standard dose of Υ-irradiation for sterilization is 25,000 Gray (25 kGy). Gamma rays have a relatively high penetrating power of up to 50 cm. High doses of Υ-rays are required for substances involving stubborn microorganisms including viruses, parasites, and helminths. Gamma irradiation is commonly used to sterilize surgical sutures, metallic bone implants, knee and hip prostheses, and other biomaterials. However, Υ-irradiation is not suitable for certain polymers due to their sensitivities and it may cause degradation or cross-link of the polymers.

X-ray radiation is a form of electromagnetic energy produced by several sources such as X-ray tubes, fast-protons, and electron beam accelerators [5]. X-rays are generated when the electrons are accelerated by a high energy source and interact with high atomic number nuclei of metal atoms such as tungsten, molybdenum, copper, gallium, indium, and silver. The accelerated electrons slow down as they interact with the metal atoms, which causes a release of energy in the form of X-rays. Metals with high atomic numbers produce high conversion efficiency of X-rays, while elements with low atomic numbers produce low X-rays conversion. The energy of X-rays is determined by the energy of the electrons. The wavelengths of X-rays are shorter than those of UV-rays, but longer than Υ-rays. X-rays have higher penetrating ability than Υ-rays and electron beams. Sterilization by X-rays is a clean process without toxic residues, and they do not cause thermal damage to biopolymers as the procedures are operated at a relatively low temperature.

Electron beam (E-beam) radiation is a sterilization process with tuneable penetration depth to reduce the risk of harming biological components. E-beam sterilization is performed under an inert atmosphere. E-beam is generated by electron beam accelerators or generators (from 10^{-13} J to 20×10^{-13} J). E-beams have lower penetration (approximately 5 cm of the materials) than Υ-rays, hence, E-beam is limited to sterilizing low-density and small substances. The dose distribution of E-beam is less uniform than Y-irradiation, due to the penetration depth. Nevertheless, sterilization by E-beam has a shorter treatment time with higher doses, higher throughout with lesser damaging impacts on the substances, and lower cost than Y-irradiation. The scan pattern and direction of an E-beam can be controlled by strong magnets due to electrons being negatively charged [1].

UV irradiation is commonly used to sterilize surfaces of materials and biopolymer implants. UV radiation is electromagnetic with wavelengths shorter than visible light. The electromagnetic spectrum comprises energies with both electrical and magnetic properties that can be categorized based on wavelength and photonic interaction with substances. The UV wavelengths range between the high-energy X-rays (<100 nm) and the lower-energy visual spectrum (>400 nm). Ionization with a change in the atomic charge of matter results from the interactions between the energy and the substances under a wavelength of less than 100 nm. Increasing the wavelengths could increase electron excitation and decrease ionization from the interaction between the energy and the substances. UV wavelengths are categorized into four groups: (i) "Vacuum UV", the most energetic wavelengths (<200 nm) that interact with oxygen atoms and organic molecules at low doses, (ii) "UV C", the "germicidal" spectrum (200–280 nm) that has biocidal effects on bacteria, (iii) "UV B", the synthesis of Vitamin D and the "sun burning" effect (280–315 nm), and (iv) "UV A", the light generated by black light fixtures (315–400 nm) [6].

UV irradiation causes the excitation of electrons and the accumulation of photoproducts. UV irradiation is effective to inactivate vegetative bacteria. On the contrary, UV irradiation is not effective to sterilize bacterial spores, prions, and numerous viruses. The commonly used UV wavelength for sterilization is within "UV C" range, which is from 200 to 280 nm, particularly, 260 nm is reported to be the most effective. Sterilization by UV can be optimized based on the wavelength and the duration of exposure, especially for biopolymers since UV can affect the molecular weight and tensile strength of biopolymers.

There are several limitations of radiation sterilization. The radiation process involves specialized equipment. Hence, the sterilization process is more expensive than steam-autoclaving and dry-heat treatment.

2.6 Supercritical Fluid

Supercritical fluid (SCF) is a pure substance at a pressure and a temperature that are above its critical point on a three-phase diagram, as illustrated in Fig. 1. The unique characteristics of SCF include gas-like viscosities, liquid-like densities, and diffusivities intermediate to liquids and gases. Among all the gases, carbon dioxide (CO_2) is the most commonly used gas in SCF technology because CO_2 is non-toxic, inexpensive, non-flammable, environmentally friendly, and widely available.

Backdated in the 1980s, SCF was first used as a sterilization tool for the food industries. Since then, high-pressure CO_2 has become an attractive alternative

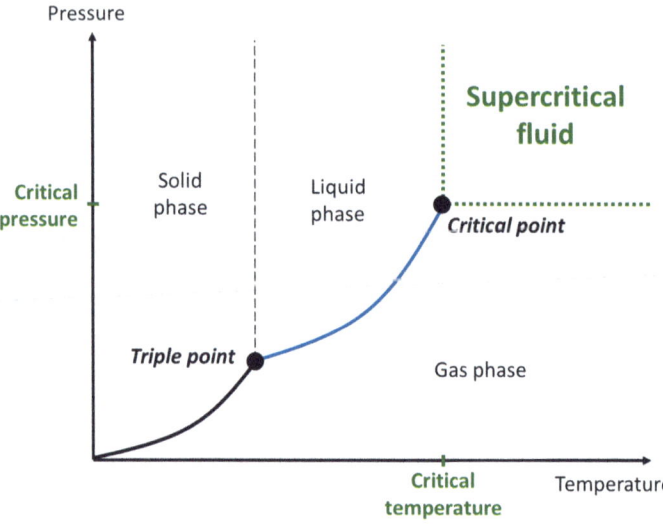

Fig. 1 A three-phase diagram of a pure substance

sterilization tool for biopolymers and the inactivation of bacteria and viruses in biomedical applications. Compared to the EtO sterilization, CO_2 is a more appealing sterilization method as CO_2 does not leave any toxic residue and is non-reactive to polymers, hence, there will not be chain scission or cross-linking reaction occur. CO_2 can be an alternative sterilization route for thermolabile materials compared to autoclaving or high-heat treatment.

After several patents for utilizing SCF as a sterilization method were introduced, SCF has recently gained attention as a useful sterilization tool, especially for biomedical applications. This is because supercritical CO_2 (scCO_2) does not affect the biochemical and biomechanical properties of bone fragments, tendons, and acellular dermal matrices. Supercritical CO_2 sterilization has been demonstrated to inactivate various types of microorganisms such as bacteria, fungi, and yeasts.

The sterilizing effect of scCO_2 is generated by decreasing the cytoplasmic pH by forming carbonic acid, damaging cell membranes, extracting or inactivating the key enzymes, and inducing shear forces upon depressurization. In addition, the increment of temperature contributes to the increased diffusivity of CO_2 and permeability of cell membranes, which causes inactivation of cells. The high diffusivity of scCO_2 helps to reduce the duration required to perform sterilization. Supercritical CO_2 is capable to achieve a sterility assurance level of 6 logs or 10^{-6} at relatively low temperatures and process times.

A schematic diagram of a standard scCO_2 setup for sterilization is shown in Fig. 2. In scCO_2 sterilization, a biomaterial is placed in a high-pressure vessel. The working rig usually involves a water bath or a temperature-controlled environment to set to a desired operating sterilization temperature. The system is pressurized to the operating pressure after the system is heated. The polymer matrix is washed

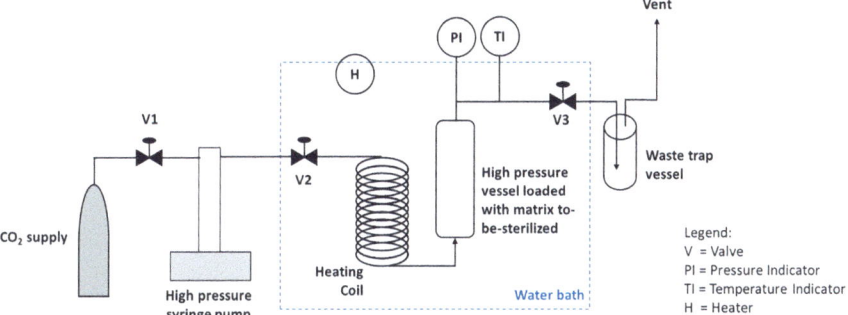

Fig. 2 A schematic diagram of a standard supercritical CO_2 setup for sterilization

with CO_2 via a vent system under a consistent flow rate at the set pressure. After a desired reaction time duration, the sterilized matrix is collected from the high-pressure vessel upon depressurization.

As $scCO_2$ is non-polar and inert, additives with low molecular weight can be useful to enhance the solvent strength and chemical action of $scCO_2$ during the sterilization process. A small amount of additives such as hydrogen peroxide (H_2O_2), tert-butyl hydroperoxide, and peracetic acid are combined with $scCO_2$ treatment for better sterilization. The presence of additives ensures the inactivation of microorganisms including bacterial endospores of various bacterial species. Furthermore, by having a small amount of additives, milder operating conditions and shorter exposure time are adequate to achieve sterilization of biomaterials. However, the chemical additives may cause oxidation and depolymerization of polymers, which results in changes in the chemical and physical properties of the biomaterials. Hence, it is essential to examine the amount of additives used and appropriate characterization procedures after sterilization.

Dillow and co-workers have demonstrated the use of $scCO_2$ as a tool to sterilize Gram-positive (*Staphylococcus aureus*, *Bacillus cereus*, and *Listeria innocua*) and Gram-negative (*Salmonella salford*, *Psoriasis vulgaris*, *Legionella dunnifii*, *Pseudomonas aeruginosa*, and *Escherichia coli*) microorganisms [7]. The operating pressure has been set at 205 bar while the temperature was varied from 34 to 60 °C. During the sterilization process, the system was applied with a continuous cycle of depressurization and re-pressurization (approximately five cycles per hour) to enhance the driving force for the mass transport of CO_2 where the differential pressure of the cycle is more than 100 bar. The effect of sterilization was expressed in a log ratio of the number of active microorganisms after sterilization to the number of microorganisms before sterilization. Generally, a sterility assurance level of 10^{-6} and above can be achieved by $scCO_2$ at lower temperatures and process duration, as compared with the dry heat treatment and steam-autoclaving. The results have been plotted in the chart shown in Fig. 3.

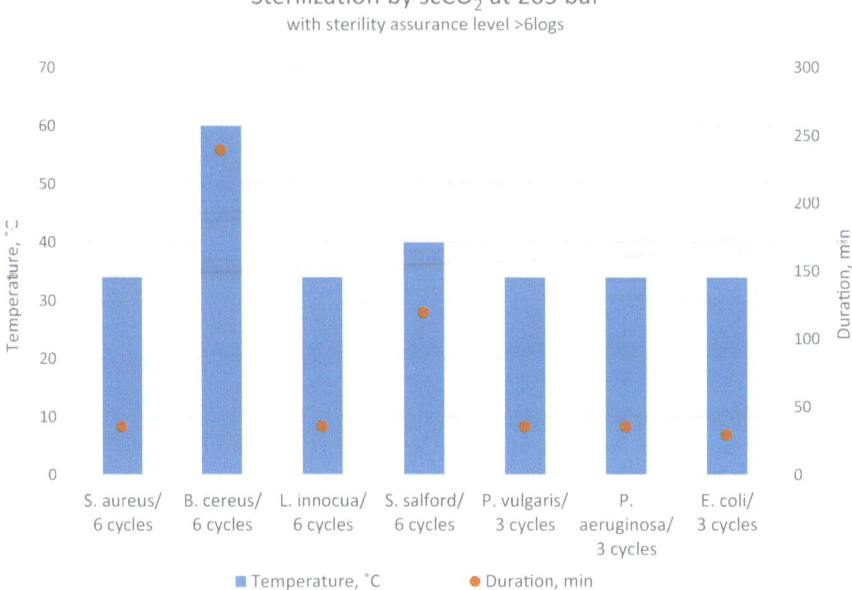

Fig. 3 Inactivation of Gram-positive and Gram-negative microorganisms by supercritical CO_2 at 205 bar. (Chart plotted using the data in Dillow et al. [7])

3 Conclusion

Biomaterials play important roles in tissue engineering, implants, medical devices, and drug delivery systems. It is crucial to ensure the biomaterials will not lead to complications or severe infections in patients. Hence, the sterilization of biomaterials is a significant step to achieve a minimum sterility assurance level of 10^{-6}. However, there is no single sterilization process that can be used for all types of biomaterials. Therefore, the benefits and drawbacks of each sterilization technique have to be taken into consideration for each biomaterial.

Quiz

1. Which of the following sterilization methods is suitable for all types of materials?

    (A) Steam-autoclaving
    (B) Ethylene oxide treatment
    (C) Radiation process
    (D) None of the above

There is no single sterilization process that can fit all purposes. All processes lead to various adverse effects on biomaterials including changes in physical, mechanical, and biological properties.

2. What is the acceptable sterility assurance level?

   (A) 10^{-4}
   (B) 10^{-5}
   (C) 10^{-6}
   (D) None of the above

   A minimum sterility assurance level of 10^{-6} is important in good manufacturing practices and patient safety.

3. What is the operating temperature range of steam-autoclaving sterilization?

   (A) 90–100 °C
   (B) 100–110 °C
   (C) 121–134 °C
   (D) 151–164 °C

   The optimum temperature range has been investigated, i.e., 121–134 °C. An operating temperature lower than the above range will not achieve a sterility assurance level of 10^{-6} within the same time frame.

4. Which of the following is not suitable for steam-autoclaving sterilization?

   (A) Catheters
   (B) Vascular prostheses
   (C) Hematic circuits
   (D) Electric device components

   The moisture content from steam-autoclaving will damage the electric device components.

5. Gas plasma sterilization was introduced as an alternative to which of the following sterilization method?

   (A) Steam-autoclaving
   (B) Dry-heat
   (C) Ethylene oxide treatment
   (D) Supercritical fluid

   This is due to gas plasma using less toxic materials than ethylene oxide.

6. Chemical sterilization using ethylene oxide (EtO) may not be suitable for biopolymers due to EtO may affect their…?

   (A) Degradation rate
   (B) Molecular weight
   (C) Mechanical properties
   (D) All the above

The biopolymer properties can be significantly affected, especially the polymers that contain COOH functional group and PEG. Significant degradation was observed on PEG-based bioresorbable biopolymers where EtO esterified the carboxylic acid groups in desaminotyrosol-tyrosine.

7. Which of the following procedure is applicable for reducing the toxic effects and flammability of EtO?

   (A) Mix EtO with water during the sterilization process
   (B) Mix EtO with inert gases such as fluorinated hydrocarbons and CO_2 during the sterilization process
   (C) Heat the sterilization vessel up to 200 °C
   (D) None of the above

It is important to remove all EtO toxic residues as they may lead to extreme skin irritation and medical complications if the biomaterials are incorporated into the body. In addition, the residual EtO contents including its secondary products such as ethylene chlorohydrin and ethylene glycol have mutagenic, carcinogenic, and allergenic effects. The procedure above (mixing with inert gases) will help in reducing the toxic effects of EtO.

8. Which of the following is non-ionizing radiation?

   (A) X-ray
   (B) Electron beam
   (C) Gamma radiation
   (D) Ultraviolet radiation

UV is a non-ionizing radiation. It uses long-wavelength and low-energy radiation while the ionizing radiation (gamma, X-ray, and electron beam) uses high-intensity and short-wavelength radiation.

9. What is the standard dose of gamma irradiation for sterilization?

   (A) 5 kGy
   (B) 15 kGy
   (C) 25 kGy
   (D) 50 kGy

A single standard dose of Υ-irradiation for sterilization is 25,000 Gray (25 kGy). Gamma rays have a relatively high penetrating power of up to 50 cm. High doses of Υ-rays are required for substances involving stubborn microorganisms including viruses, parasites, and helminths.

10. Which of the following sterilization method may contain toxic residues?

    (A) Ethylene oxide treatment
    (B) Steam-autoclaving
    (C) Dry-heat
    (D) Supercritical CO_2

The other three options do not involve chemicals. Hence, they will not have toxic residues.

References

1. Beh CC, Farah S, Langer R, Jaklenec A (2019) Chapter 12. Methods for sterilization of biopolymers for biomedical applications. In: Antimicrobial materials for biomedical applications. The Royal Society of Chemistry, London, pp 325–347
2. Tessarolo F, Nollo G (2008) Sterilization of biomedical materials. In: Encyclopedia of biomaterials and biomedical engineering. Informa Healthcare, New York, pp 2501–2510
3. Dai Z, Ronholm J, Tian Y, Sethi B, Cao X (2016) Sterilization techniques for biodegradable scaffolds in tissue engineering applications. J Tissue Eng 7:2041731416648810
4. Singh R, Singh D, Singh A (2016) Radiation sterilization of tissue allografts: a review. World J Radiol 8(4):355
5. Fairand BP (2001) Radiation sterilization for health care products: x-ray, gamma, and electron beam. CRC Press, Boca Raton
6. Kowalski W (2009) UVGI disinfection theory. In: Ultraviolet germicidal irradiation handbook. Springer, Berlin, pp 17–50
7. Dillow AK, Dehghani F, Hrkach JS, Foster NR, Langer R (1999) Bacterial inactivation by using near- and supercritical carbon dioxide. Proc Natl Acad Sci U S A 96(18):10344–10348

Index

B
Biocompatibility, 4–7, 11, 13–15, 18, 20, 21, 23–26, 28, 30, 32, 35, 39, 41–44, 46, 52, 58, 63, 65, 67, 68, 73, 81, 88–91, 103, 104, 148, 153–155, 161, 164–166, 168, 170–174, 176–181, 187, 225–227, 236–242, 247–250
Biomaterials, 2–32, 35, 47, 62, 65–67, 81, 82, 85, 89, 91, 97–116, 120–131, 133, 137, 138, 148, 149, 153, 160–181, 184–196, 210, 211, 229, 233, 236, 237, 242, 243, 245, 247, 249–252, 256–268
Biomaterials history, 4, 162
Biomedical applications, 6, 7, 9, 11, 13, 15, 20–22, 24–26, 41, 42, 51, 59, 60, 62, 65–67, 72–89, 106, 125, 150, 190, 210, 211, 262, 264

C
Cardiac implants, 168, 169
Classification, 3, 9–26, 38, 58, 73–77, 88, 120–122, 137, 138, 154, 156, 201–203, 229

D
Degradation mechanisms, 37, 39
Diagnosis, 28–29, 136
Drug crystals, 148, 152, 153, 155
Drug delivery, 5, 35, 65, 76, 122, 171, 199, 239
Drug delivery systems (DDSs), 7, 10, 14, 22, 25, 28, 35, 41, 46, 50, 60–64, 66, 67, 86, 104, 136, 137, 142–149, 152–156, 171, 238, 239, 250, 256, 266

Drug release kinetics, 137, 140, 151

E
Elastic modulus, 20, 67, 78, 80, 85, 121, 124, 125, 127, 128, 131, 132
Ethylene oxide (EtO), 8, 26, 185, 241, 243, 257, 258, 260–261, 266–268

F
Foreign-body reaction, 238

H
Hydrogels, 14, 42, 60, 72, 121, 148, 168, 185, 198, 249

I
Implants, 2, 45, 58, 82, 121, 194, 237, 256

L
Lipids, 28, 34, 50, 98, 100–103, 106–108, 110, 112–116, 148, 149, 160, 227
Living materials, 184–190, 192–196

M
Manufacturing, 13, 34, 98, 103, 104, 106–109, 113, 116, 122, 130, 136, 153, 161, 174, 179–181, 201, 215, 230, 240, 241, 257, 267
Mechanical design, 122, 130

Mechanical tests, 79, 88, 125–130, 133
Medical implants, 20, 31, 161, 256

N
Natural and semi-natural polymers, 55–67
Natural & synthetic biomaterials, 9, 160, 167, 188
Natural versus synthetic polymers, 21, 121, 148, 168, 170, 171
Neural tissue implants, 171, 173
Nucleic acid delivery, 97–100, 102, 104–106, 108–111, 114, 115

O
Ocular implants, 161, 174, 175

P
Pharmacokinetics (PK), 137–140, 146–148, 154, 204–207, 209, 210, 225–229
Polymer characterization, 198–233
Polymer conjugates, 199, 202, 204, 207, 209–212, 215, 221, 222, 226, 229, 230
Polymers, 10, 34, 56, 72, 98, 120, 143, 186, 198, 236, 256
Polysaccharides, 9, 10, 12–16, 34, 35, 50, 58–63, 68, 69, 104, 160, 166, 185, 193, 200–210
Protein, 9, 34, 58, 73, 97, 140, 160, 185, 199, 237, 259
Protein conjugation, 214–215, 221, 230, 233

R
Radiation, 261–263, 266, 268

RNA therapeutics, 97, 106–108, 112, 113

S
Safety, 98, 99, 103, 106, 111, 116, 177, 204, 225, 230, 238, 239, 241, 242, 245, 248, 250, 257, 267
Sensing, 187, 190, 193, 194
Skeletal implants, 161–165
Skin implants, 165, 166, 180
Smart hydrogels, 72–89
Steam-autoclaving, 257–259, 263, 265–268
Sterilization, 4, 7, 8, 106, 237, 241, 256–268
Stimuli responsive, 77, 84–86, 88, 153, 203, 220
Supercritical fluid (SCF), 257, 258, 263–267

T
Therapeutic polymers, 199, 204, 211–214, 219–221, 224, 225, 230
Tissue engineering, 5, 10, 11, 13–15, 22–29, 35, 39, 41, 43, 45, 46, 51, 53, 54, 60, 62–66, 69, 73, 78, 79, 82, 84, 85, 87–89, 122, 165–168, 170–173, 266
Tissue engineering & regeneration, 29
Toxicology, 226, 239

V
Viscoelasticity, 79, 121–123

W
Wound healing, 11–13, 15, 18, 25, 29, 47, 60, 62, 63, 66, 73, 84–89, 92, 122, 165, 166, 189, 194, 201, 245, 247

SPRINGER NATURE

GPSR Compliance

The European Union's (EU) General Product Safety Regulation (GPSR) is a set of rules that requires consumer products to be safe and our obligations to ensure this.

If you have any concerns about our products, you can contact us on ProductSafety@springernature.com

In case Publisher is established outside the EU, the EU authorized representative is:

Springer Nature Customer Service Center GmbH
Europaplatz 3
69115 Heidelberg, Germany

The manufacturer's authorised representative in the EU is Springer Nature Customer Service Centre GmbH, Europaplatz 3, 69115 Heidelberg, Germany. If you have any concerns regarding our products, please contact ProductSafety@springernature.com

Printed and bound by CPI Group (UK) Ltd, Croydon, CR0 4YY
26/03/2026
02078969-0003